Advanced Functional Materials

Advanced Functional Materials

A Perspective from Theory and Experiment

Biplab Sanyal
Department of Physics and Astronomy
Uppsala University
Sweden

Olle Eriksson
Department of Physics and Astronomy
Uppsala University
Sweden

ELSEVIER
AMSTERDAM • BOSTON • HEIDELBERG • LONDON
NEW YORK • OXFORD • PARIS • SAN DIEGO
SAN FRANCISCO • SINGAPORE • SYDNEY • TOKYO

Elsevier
Radarweg 29, PO Box 211, 1000 AE Amsterdam, The Netherlands
Linacre House, Jordan Hill, Oxford OX2 8DP, UK

Library of Congress Cataloging-in-Publication Data
A catalog record for this book is available from the Library of Congress

British Library Cataloguing-in-Publication Data
A catalogue record for this book is available from the British Library

ISBN: 978-0-444-53681-5
ISSN: 1875-4023

For information on all Elsevier publications
visit our Web site at store.elsevier.com

Printed and bound in Great Britain
10 11 12 10 9 8 7 6 5 4 3 2 1

Contents

Foreword

The present book is a part of the Elsevier book series, "Science and Technology of Atomic, Molecular, Condensed Matter and Biological Systems." This series deals with subjects of current interest which have led to advances in the basic understanding of these systems, or are potentially capable of use for applications in basic science or technology, or are already being used for such applications. The series is aimed at a broad audience of physicists, chemists, and biologists involved in these disciplines and their interfaces, as well as related applications in fields such as materials science and technology, bioscience and biotechnology.

Other features of the series include some emphasis on the electronic structure aspects of the understanding of the systems as well as their applications, although not exclusively so. One of the reasons for this emphasis is that both the static and dynamic aspects of many current systems of interest involve the binding or transport of electrons. Also the main process of characterization of condensed matter and biological systems often involves the understanding of their electronic properties. Another feature of the book series is the emphasis on both theoretical and experimental aspects of the subjects presented. This is because the current quantitative understanding of the electronic properties of many properties of atomic, molecular, condensed matter, and biological systems has become possible through both theory and experiment. This close interplay of theory and experiment will be a common theme of most of the books of the series. Thus, in a particular subject of current interest, there will be an attempt to deal with their theoretical and experimental aspects, either in a single book, or if necessary, in separate dedicated books. It is hoped that with all these planned features of the books in this series, each of the books in any specific subject or system will be useful to a broad range of the scientific community with both basic and applied interests.

The first book of this series, "Nanoclusters: A Bridge Across Disciplines", a multi-author book with two Volume Editors from USA, Professor Purusottam Jena, Department of Physics, Virginia Commonwealth University, and Professor A. Welford Castleman Jr., Department of Chemistry, Pennsylvania State University, was published in the last quarter of 2010.

The present multi-author book, "Advanced Functional Materials–A Perspective from Theory and Experiment", is the second in the series and is edited by Professor Biplab Sanyal and Professor Olle Eriksson, both from the Department of Physics and Astronomy, Uppsala University, Sweden.

This book deals with functional materials that are currently at the frontiers of materials science and technology research and development efforts. It is concerned with three classes of functional materials, magnetic semiconductors, multiferroics combining ferroelectricity and magnetism and the novel material graphene, which belongs to the burgeoning field of carbon-based nanomaterials, ranging from fullerenes and nanotubes to the two-dimensional systems related to graphite. Because of the wide popularity of these functional materials for technological applications, there was a need for a book of the present type aimed at the sizable scientific and technological communities interested in both the basic and applied aspects of these functional materials and the understanding of their very interesting properties as discussed in the Introductory chapter. The two Volume Editors have very effectively responded to this need. They are both very accomplished and productive theoretical research scientists. Their current combined research interests span the structures, at both the atomic as well as electronic levels, and the magnetic, spectroscopic, and transport properties of the three advanced functional materials covered here. With their continued strong research interests and productivities in the structures and properties of these functional materials, they have a thorough knowledge of the research activities in these materials in Europe and other parts of the world. They have therefore been able to assemble a strong and active group of authors who have deep knowledge of the nature and properties of the three functional materials. This excellent group of authors, with the Volume Editors also acting as authors themselves, have contributed a comprehensive groups of chapters with emphasis on the theory as well as experimental aspects of the structures, properties, and preparation of each of the three functional materials covered in the book.

In conclusion, in view of the comprehensive nature of this book, in my opinion, it has the strong potential to be very valuable, for both novice as well as experienced scientists interested in getting a good background in both the theoretical and experimental aspects of the three advanced functional materials as well as the current and likely additional expected applied achievements. It can also be used very effectively in graduate courses in materials science and applied condensed matter physics and chemical engineering. It fulfills all the general aims of this series described earlier and I expect it to be a valuable asset furthering important basic and technological developments in the field of functional materials.

Tara Prasad Das
State University of New York at Albany, USA

Preface

The human race has been exposed to tremendous technological innovations throughout the centuries. The very heart of any advancement of technology relies on the developments of new and functionalized materials. We can summarize a plethora of novel materials properties, which are used in devices and tools, that makes our everyday life better. Therefore, the fundamental understanding of the properties of materials serves as the first and foremost step in developing advanced functionalities and new technologies. The last century has experienced many outstanding discoveries in the materials world through novel experiments and theory, e.g., transistor technology based on semiconductors, superconductivity, and high-temperature superconductors; various flavors of the quantum Hall effect; fullerenes and carbon nanotubes; shape memory materials; catalytic surfaces; new permanent magnets; batteries based on Li-absorbing materials; photonic crystals; layer-by-layer growth of tailored materials; hard coating surfaces; novel characterization tools; and new elements like Pu. Since the list of significant discoveries in the world of materials science and solid-state physics is very long, this book deals with just a few recent breakthroughs in materials research with a perspective from both experimental and theoretical science. Diluted magnetic semiconductors, grapheme, and multiferroics are topics discussed in this book.

"Spintronics," or spin-based electronics, created an enormous sensation in the 1990s after the realization that not only the charge, but also the spin of an electron, can be utilized in electronics to achieve better functionalities. Injection, transport, and detection of spins became an important focus in materials science research. In 2007, Albert Fert and Peter Grunberg were awarded the Nobel Prize in physics for discovering "giant magnetoresistance," a phenomenon that boosted the magnetic recording industry enormously, and opened many doors for spintronics investigations in general. The new idea of having spintronics in semiconductors engaged the practitioners to import magnetism into conventional semiconductors, e.g., Si, GaAs, ZnO, etc., by doping a small amount of transition metal elements and giving rise to a new class of materials called diluted magnetic semiconductors (DMS). The challenge, however, was highly nontrivial concerning the experimental synthesis, reliable characterization, and a thorough understanding of the emerging phenomena by state-of-the-art theoretical methods. Today, the challenge is to develop suitable DMS materials for room-temperature device applications.

Multiferroic materials possess several order parameters, e.g., ferro(antiferro) magnetism, ferroelectricity, and ferroelasticity, coexisting in a single material.

The goal is to control magnetism by electric field or vice versa in the so-called magnetoelectric multiferroics. These materials usually belong to the class of complex oxides where exotic structural, magnetic, and electronic properties pose a great challenge to experiments and theory. Noncollinear magnetism, electron correlation, relativistic spin-orbit coupling, etc., are often encountered in these complex oxides, causing centrosymmetry to break and, hence, the appearance of electric polarization.

One of the most important scientific discoveries in the last decade is the experimental realization of graphene, which is a single atomic layer of carbon with sp^2 character demonstrating the first observation of a truly 2D crystal. Graphene is an extremely interesting material due to its unique electronic spectrum resembling massless Dirac Fermions; the presence of extremely high mobility (200,000 cm^2/Vs), which is attractive for transistor applications; the observation of minimum quantum conductivity; Klein tunneling; a half-integer quantum Hall effect; and many other unique properties. Hence, it is not surprising to observe the recent explosion of activities in this field, with an enormous amount of publications of fundamental and applied investigations of this exotic material. The potential of graphene was recognized by awarding the 2010 Nobel Prize in physics to Andrei Geim and Konstantin Novoselov for synthesizing graphene for the first time. A tremendous research effort has been undertaken worldwide to try to understand and manipulate the electronic properties of graphene along with research on its magnetic, optical, mechanical, and transport properties by means of both theory and experiments. Functionalized graphene has potential applications in sensors, catalysts, fuel cells, hydrogen storage, microelectronics components, etc. Also, graphene nanoribbons have been the focus due to their edge effects and interaction with different chemical entities.

In this book, we have covered these three important fields of research both from theoretical and experimental aspects, and its chapters contain the latest developments in these fields. The diluted magnetic semiconductors are described with emphasis on crystal growth and understanding the magnetic properties from an electronic structure point of view, where various mechanisms for the exchange interaction naturally appear. The possibility that the electronic structure of these materials could be correlated is also discussed. Various aspects of crystal chemistry and its coupling to the functionality of the multiferroic materials are also presented, together with the current understanding of the electronic structure of this complex group of materials. Finally, various aspects of chemical functionalization of graphene are described, both from an experimental and theoretical point of view.

It is our sincere hope that this book will be useful both to the newcomers as well as to the practitioners in these fields of research.

Biplab Sanyal
Olle Eriksson
(Uppsala University, Sweden)

Contributors

Elsevier Publishing Company Series: Science and Technology of Atomic, Molecular, Condensed Matter and Biological Systems

Advanced Functional Materials-a perspective from theory and experiment

Biplab Sanyal and Olle Eriksson (Volume Editors)
Dept. of Physics & Astronomy, Uppsala University, Sweden

Spintronics or spin-based electronics has been one of the forefront topics of research for the last decade due to its enormous potential applications in technological devices directly connected to the daily usage of modern society. Apart from a major application in magnetic storage devices, semiconductor spintronics has been proposed and partially verified as a better alternative to conventional charge based electronics. An enormous effort has been put forward worldwide both experimentally and theoretically to understand the basic properties of magnetic semiconductors to be used as a potential candidate in this new technology. In this book, these issues will be discussed from both theoretical and experimental points of view, especially on the electron correlations, electronic structure characterization by spectroscopic methods and growth of these materials. Apart from magnetic semiconductors, multiferroic materials will also be studied as an example of multifunctional materials where one can combine ferroelectricity and magnetism to exhibit novel properties. Lastly, we will review the exotic properties of the most sensational material of today, i.e., graphene, the two dimensional form of graphite with a focus on magnetism, which is a hot and debatable topic regarding sp magnetism in carbon materials . In summary, this book will serve as an excellent basis of the understanding of advanced multifunctional materials from both theory and experiment.

The expected readers of this book will be graduate students and already established researchers in this field. The book will mostly cover basic science with a touch of applied aspects. The book will cover aspects of Physics, Chemistry, Condensed Matter Science and Materials Science.

There are no other books published on the issue of advanced functional materials. There are a few books on spintronics, but no single book covering the physics of magnetic semiconductors, multiferroics, and graphene. The present book is unique in this respect. The other strong point is the inclusion of complementary contributions on theoretical and experimental issues.

Chapters

(1) Theory of Diluted Magnetic Semiconductors

Assoc. Prof. Biplab Sanyal (biplab.sanyal@physics.uu.se)
Prof. Olle Eriksson (olle.eriksson@physics.uu.se)
Dr. Igor di Marco (igor.dimarco@physics.uu.se)
Dr. Patrik Thunström (patrik.thunstrom@physics.uu.se)

Division of Materials Theory, Dept. of Physics and Astronomy,
Uppsala University, Sweden

Dr. Johan Hellsvik (johan.hellsvik@spin.cnr.it)
Dr. Diana M. Iusan (diana.iusan@spin.cnr.it)

Consiglio Nazionale delle Ricerche – Istituto Nazionale per la Fisica della Materia (CNR-INFM)
Laboratorio Regionale CASTI, L'Aquila (Italy)

Dr. Lars Bergqvist (lbergqv@kth.se)
Tillämpad Materialfysik, KTH, Stockholm, Sweden

Prof. Peter Dederichs (p.h.dederichs@fz-juelich.de)
IFF, Forschungszentrum Juelich, Germany

(2) Molecular Beam Epitaxial Growth and Magnetotransport Properties of GaMnAs

Dr. Janusz Sadowski (Janusz.Sadowski@maxlab.lu.se)
MAX-Lab, Lund University, Sweden and
Institute of Physics, Polish Academy of Sciences, Poland
Dr. Jaroslaw Domagala
Institute of Physics, Polish Academy of Sciences, Poland

(3) Spectroscopic Investigations of Electronic Structure

Dr. Ronny Knut (ronny.knut@physics.uu.se)
Prof. Olof Karis (olof.karis@physics.uu.se)
Dept. of Physics and Astronomy, Uppsala University, Sweden

(4) Theory of Magnetism in Graphene

Prof. Oleg V. Yazyev (oleg.yazyev@epfl.ch)
Institute of Theoretical Physics, EPFL, Switzerland

Prof. Mikhail Katsnelson (m.katsnelson@science.ru.nl)
Theoretical Physics, Radboud University Nijmegen, The Netherlands

(5) Synthesis of Graphene and Derivatives

Prof. Helena Grennberg (helena.grennberg@kemi.uu.se)
Department of Organic Chemistry, Uppsala University, Sweden

Prof. Ulf Jansson (ulf.jansson@kemi.uu.se)
Department of Materials Chemistry, Uppsala University, Sweden

(6) Multiferroics : Theory, Mechanisms and Materials

Dr. Paolo Barone (paolo.barone@aquila.infn.it)
Dr. Silvia Picozzi (silvia.picozzi@aquila.infn.it)
Consiglio Nazionale delle Ricerche – Istituto Nazionale per la Fisica della Materia (CNR-INFM)
Laboratorio Regionale CASTI, L'Aquila (Italy)

Assoc. Prof. Biplab Sanyal (biplab.sanyal@physics.uu.se)
Division of Materials Theory, Dept. of Physics and Astronomy,
Uppsala University, Sweden

(7) Magnetoelectric Complex Metal Oxides: Main Features of Preparation, Structure and Properties

Prof. Sergey A. Ivanov (ivan@cc.nifhi.ac.ru)
X-ray laboratory, Department of Inorganic Materials, Karpov Institute of Physical Chemistry
Moscow K-64 Russia

Chapter 1

Theory of Diluted Magnetic Semiconductors

Olle Eriksson*, Diana M. Iusan†, Patrik Thunstrom*, Johan Hellsvik†, Igor di Marco*, Lars Bergqvist‡, Peter Dederichs†† and Biplab Sanyal*

**Division of Materials Theory, Department of Physics and Astronomy, Uppsala University, Box 516, 75120 Uppsala, Sweden*

†Consiglio Nazionale delle Ricerche Istituto Nazionale per la Fisica della Materia (CNR-INFM), Laboratorio Regionale CASTI, LAquila, Italy

‡Tillampad Materialfysik, KTH, Stockholm, Sweden

††IFF, Forschungszentrum Juelich, Germany

Chapter Outline Head

I. INTRODUCTION

A very promising way to obtain novel devices is to connect different physical properties of materials, thus potentially enriching the functionality of these devices. This line of thought is pursued in the field of spintronics, where one goal is to find materials whose magnetic properties can be tuned by electrical fields and vice versa. Among the most promising candidates are a group of materials called diluted magnetic semiconductors (DMS), materials with a low concentra-

Advanced Functional Materials. http://dx.doi.org/10.1016/B978-0-44-453681-5.00001-7

tion (typically 5–10%) of magnetic elements embedded into a semiconducting host. Their potential in spintronic applications has led to very intensive investigations, both experimentally [1], as well as theoretically [2-5]. These materials show complexity regarding the crystal chemistry, the magnetic properties, as well as transport behavior (for a short summary of these materials see Ref. [6]).

II. EXPERIMENTAL BACKGROUND OF SELECTED MATERIALS

Concise reports on the magnetic properties of DMS materials were made as early as in the 1970s. Cochrane et al. evidenced ferromagnetic properties of the $(GeTe)_{1-x}(MnTe)_x$ alloys, attributed to an RKKY-like interaction for low values of x [7]. Later on, Story et al. were the first to demonstrate the dependence of the magnetic properties of PbSnMnTe semiconductors on the carrier concentration [8]. Also, Spałek et al. investigated the magnetic properties of a large number of Mn-doped II–VI semiconductors, and evaluated the exchange integral for the nearest-neighbor and next-nearest-neighbor Mn atoms [9]. An important moment in the field of diluted magnetic semiconductors was the discovery of carrier-mediated ferromagnetism of Mn-doped InAs and Mn-doped GaAs, the latter with a T_C as high as 110 K [1] (which in subsequent works has been pushed up to 170 K for a Mn concentration of 7–8%[10]). This was soon followed by the theoretical predictions by Dietl et al. [2] of high transition temperatures of a series of Mn-doped semiconductors. The theory was built on Zener's model of p–d exchange for ferromagnetism [11], where the exchange splitting of localized d-electrons of magnetic impurity atoms (e.g., Mn) was argued to induce a hole in the valence band (e.g., of GaAs), with the related valence band moment coupling anti-ferromagnetically to the impurity d-moment. The exchange coupling between the localized moments was assumed to be mediated by the conduction carriers of the host, and a long ranged character of the exchange interaction was proposed. This implies that even for dilute doping concentrations, with Mn atoms distributed far apart, a significant ferromagnetic coupling between transition metal moments might be obtained. Based on a mean field analysis, ordering temperatures above room temperature were predicted for several semiconducting hosts with only a few percent of transition metals. The picture of long-ranged interactions has been challenged from calculations based on first principles theory, that suggest a much more short ranged interaction in general, with magnetic percolation as a key ingredient [4,5].

In practical applications, an ordering temperature above room temperature is necessary, and two materials were predicted [2] to be very promising in this regard, namely Mn-doped GaN and Mn-doped ZnO. The consequence of this prediction was the exponential increase in the interest in the field, and numerous groups have been involved in the growth and characterization of these materials by a variety of techniques. Unfortunately, it has been very difficult experimentally to unambiguously reach sufficiently high values of the ordering temperature. As a matter of fact no unanimous ordering temperature well above room temperature has been established.

As Mn-doped GaN and Mn-doped ZnO have been predicted to be room-temperature ferromagnets [2], much of the attention has been focusing on these systems. Some early studies of Mn-doped GaN reported ferromagnetic properties [12,13]. More recently, Yoshii et al. reported ferromagnetism above room temperature, which was suppressed below 10 K [14]. In contrast, ferromagnetism with a low T_C of 25 K was found for Mn-implanted p-type GaN [15], and paramagnetism was observed in other recent experiments [16,17]. Studies by Cui et al. showed that GaN with a Mn content of ~4%, grown by MOCVD, was ferromagnetic only due to the appearance of a secondary Mn_xN_y phase [17].

The magnetic properties of Mn-doped ZnO also continue to be lively debated, and the scientific community is also here divided into two groups, either claiming intrinsic room temperature ferromagnetism or the lack of it. Ferromagnetism at room temperature or above was for instance reported in Refs. [18–20]. In a different series of experiments it was found that the ferromagnetism is dependent on the growth conditions. For instance, Xu et al. observed ferromagnetic properties for Mn:ZnO films prepared under N_2 atmosphere, while the films prepared under O_2 atmosphere were paramagnetic [21]. Similarly, Hou et al. showed that the ferromagnetism of Mn:ZnO thin films is destroyed by annealing in oxygen [22]. Sharma and Varma found that hydrogenation induces week ferromagnetism in otherwise paramagnetic Mn:ZnO samples [23]. In these cases the observed ferromagnetism was attributed to the presence of defects. Also, in a study by Yan et al. the ferromagnetism of highly resistive samples was attributed to zinc vacancies [24]. Ferromagnetic properties were also observed for Mn:ZnO co-doped with N [25], or P [26]. Yang et al. showed a clear dependence of the ferromagnetic properties on the electron concentration in n-type Mn:ZnO co-doped with Ga, for which the magnetic moment decreases with decreasing electron concentration [27]. Paramagnetism of Mn:ZnO has also been reported by several groups [28–33]. Studies by Blasco et al. of 4% TM-doped ZnO (TM = Mn, Fe, Co) point out that the ferromagnetism at room temperature is not intrinsic, but due to the presence of secondary phases, often below the detection limits of XRD.

Co-doped ZnO, the counterpart of Mn-doped ZnO, has also received a lot of attention. The reports are controversial for this system as well, and claims of both high T_C ferromagnetism [34,35] and paramagnetism [36] were made. In addition, antiferromagnetic Co–Co interactions for n-type Co:ZnO were reported [36]. Recently, Zhou et al. have shown that ferromagnetism of Co:ZnO is due to the formation of magnetic precipitates and they emphasized the difficulty of using XRD for their detection [37].

From the description above it seems that the experimental situation in DMS materials seems unclear, but it may be argued on general grounds that DMS materials with a homogeneous distribution of magnetic impurity atoms (e.g., Mn or Co) should have substantial magnetic moments of these impurities (of order μ_B/atom). This is indeed what first principles theory results in for the presently considered materials (see, e.g., Ref. [50-52]). Hence one may suspect that reports of large ordering temperatures for materials with a tiny magnetic moment for the magnetic impurity atoms are caused by antiferromagnetic clusters with uncompensated spins at the surface of the cluster.

There are several unsolved issues of DMS materials, e.g., the nature and strength of the interatomic exchange interaction, the dynamical properties of the magnetization, as well as the possible influence of electron correlations on the electronic structure and magnetic properties. Several reviews with various focus have been published for diluted magnetic semiconductors [38–41]. Here we summarize some of the more recent aspects of diluted magnetic semiconductors, with emphasis of electron correlations, magnetization dynamics and possible maximum ordering temperatures. We start however with a short description of the general features of magnetism and exchange in these materials, based on density functional theory in the local spin density approximation or parametrized Hartree–Fock like theory as given by the LDA + U method.

III. FIRST PRINCIPLES THEORY

First principles theory of DMS materials relies on the theorems of Ref. [42], which turns the many-body problem of the electronic structure into a one-electron problem with an effective potential. The latter is often approximated by the so-called local density approximation (LDA) [43]. By now several numerical methods for finding a solution to this effective one-electron problem have been suggested and found to give reliable results, both as concerns the magnetic moments and inter-atomic exchange interaction between magnetic atoms of a solid. Hence first principles calculations based on, e.g., linear muffin-tin orbitals [44–46] are found to agree with results of the KKR method [47]. For close packed solids it is also by now well known that the atomic sphere approximation gives results in agreement with full-potential calculations [48]. Since these calculations can be made in a spin-polarized fashion, it is possible to calculate magnetic moments of a material, and if electron correlation is not a too dominant feature, experiment and theory agree with an error often not exceeding a few percent.

First principles theory also offers the possibility to evaluate the pair exchange interaction between atomic spins, and hence it can be concluded that the parameters of the Heisenberg Hamiltonian are possible to calculate, without adjustable parameters. The most common approach is the use of Liechtenstein's theory [49], where the exchange interaction between two spins is calculated using a classical Heisenberg Hamiltonian and making use of the magnetic force theorem. The expression for the pair-exchange parameter reads

$$J_{ij} = \frac{1}{4\pi} \int^{E_F} \mathrm{d}E\, Im\, Tr_L(\Delta_i T_{\uparrow}^{ij} \Delta_j T_{\downarrow}^{ji}), \tag{1}$$

where $\Delta_i = t_{i\uparrow}^{-1} - t_{i\downarrow}^{-1}$, t being the on-site scattering matrix. T is the scattering path operator which is related to the off-diagonal element of the Green's function. Tr_L is the trace over the orbital indices of the scattering matrices. The first principles calculations presented in this chapter made use of a Green's function Korringa–Kohn–Rostoker (KKR) method within the atomic sphere approximation (ASA) [46]. The local density approximation was used for the exchange–correlation functional, together with an *spdf* basis. In order to treat the chemical and magnetic disorder due to the impurities in the system we have employed the coherent potential approximation (CPA). The calculations were performed with a Mn concentration of 5%. Vacancies in different concentrations at the cation site were incorporated into the system in order to obtain a hole concentration of 3.0×10^{20} cm^{-3}.

The values of calculated exchange parameters among the Mn spins embedded in all semiconducting hosts considered in this work are given in Table 1, up to the 16th neighboring shell. Note that a positive (negative) J_{ij} signifies a ferromagnetic (antiferromagnetic) interaction. We have for computational simplicity considered only the cubic zinc-blende structure of these materials, which is the observed structure for all materials listed in the table, except for GaN and ZnO, where the hexagonal wurtzite structure is the stable form. The two structures are however similar enough that one can expect that first principles calculations of magnetic moments and exchange interaction are similar for both structures. The data in Table 1 lists the calculated moments projected onto the Mn atom, when it is substituted (with a concentration of ~5%) for the cation of the compounds listed in the table. It is clear that for all studied systems one finds a large magnetic moment of the Mn atom, in the range ~4 μ_B/atom. This seems to be a value most first principles calculations reach for the majority of the DMS materials studied so far. Table 1 also lists the exchange interaction between Mn atoms, as a function of distance, or rather as a function of coordination shell. It is clear from the table that the interaction is dominated by the nearest neighbor contribution, which is ferromagnetic for all studied materials. The exchange interaction between atoms further away is quickly decaying; after four shells it is in general less than one percent of the nearest neighbor interaction. One may note that for some distances the interaction is actually negative and hence of an antiferromagnetic nature. However, the overall ordering of all materials listed in Table 1 is ferromagnetic. The short ranged nature of the inter-atomic exchange of DMS systems has led to the conclusion that magnetism in these materials is heavily influenced by percolation [4,5].

TABLE 1 Calculated Mn projected magnetic moments and exchange parameters for various diluted magnetic semiconductors incorporating a Mn concentration of 5%.

	AlP	AlAs	GaN	GaP	GaAs	GaSb	InP	InAs	ZnO	ZnSe	ZnTe
μ (μ_B)	4.00	3.94	4.00	4.00	3.96	3.94	3.84	3.94	4.60	3.94	3.90
J_{01}	3.045	2.733	3.686	2.048	1.811	1.062	1.645	1.271	0.762	2.003	0.975
J_{02}	0.013	0.015	−0.054	0.016	0.019	−0.022	0.054	0.001	−0.060	0.122	0.134
J_{03}	0.181	0.235	0.057	0.206	0.232	0.157	0.250	0.188	0.037	0.316	0.260
J_{04}	0.482	0.505	0.349	0.408	0.403	0.222	0.342	0.263	0.132	0.406	0.250
J_{05}	0.007	0.005	0.002	0.002	−0.009	−0.039	−0.034	−0.049	−0.000	0.003	−0.036
J_{06}	0.025	0.043	0.010	0.054	0.070	0.057	0.070	0.057	0.002	0.090	0.097
J_{07}	0.037	0.054	0.021	0.056	0.069	0.045	0.054	0.048	0.008	0.070	0.058
J_{08}	0.000	0.000	0.001	−0.001	−0.004	0.001	−0.011	0.000	−0.002	−0.002	−0.007
J_{09}	0.001	0.001	0.001	0.000	−0.002	−0.003	−0.010	−0.003	−0.001	−0.000	−0.008
J_{10}	0.106	0.127	0.060	0.116	0.135	0.099	0.111	0.109	0.017	0.105	0.094
J_{11}	0.002	0.002	0.002	0.001	0.001	0.001	−0.004	0.003	−0.000	0.004	0.000
J_{12}	0.001	0.004	0.002	0.006	0.010	−0.008	0.002	−0.005	0.000	0.018	0.007
J_{13}	0.005	0.009	0.004	0.010	0.015	0.004	0.004	0.004	−0.000	0.012	0.003
J_{14}	0.000	0.000	0.000	0.000	0.000	0.006	0.003	0.010	−0.001	0.000	0.007
J_{15}	0.007	0.012	0.004	0.014	0.023	0.019	0.016	0.020	0.000	0.018	0.018
J_{16}	0.000	0.001	0.000	0.001	0.001	0.006	0.001	0.007	−0.001	0.001	0.003

IV. MAGNETIZATION DYNAMICS

Recently the dynamical properties of spin-systems have attracted intense focus. This is spurred both by technical developments, like pump–probe measurements using intense lasers, as well as a technological drive, where the speed of reading and writing information in a magnetic medium is of technological importance. However, the dynamics of magnetism offers also a way to interpret the nature of the exchange interactions of a material, e.g., if it is magnetically ordered or a spin-glass. The latter aspect is of particular importance for DMS materials, since for some cases spin-glass behavior has been demonstrated experimentally, whereas for some materials speculations about a spin-glass material have not been confirmed. In this sub-section we outline one approach with which to undertake theoretical studies of magnetization dynamics of DMS materials, namely the use of atomistic spin dynamics (ASD) simulations.

Observing that the dominant microscopic magnetic moments are large and stable in many of these materials, a Langevin atomistic spin dynamics (ASD) approach can be used for simulations of the magnetization dynamics [50–52]. This technique relies on the adiabatic approximation and coarse graining to atomic magnetic moments. The central entity of the technique is the stochastic Landau–Lifshitz–Gilbert equation

$$\frac{\mathrm{d}\mathbf{m}_i}{\mathrm{d}t} = -\gamma \mathbf{m}_i \times [\mathbf{B}_i + \mathbf{b}_i(t)] - \gamma \frac{\alpha}{m} \mathbf{m}_i \times \frac{\mathrm{d}\mathbf{m}_i}{\mathrm{d}t}, \tag{2}$$

for the atomic moments, $\mathbf{m}_i$, in an effective field, $\mathbf{B}_i$. The first term on the right hand side describes gyromagnetic precession, the second term is a dissipative damping term on the Gilbert form. γ is the gyromagnetic ratio and $\mathbf{b}_i(t)$ is a stochastic magnetic field with a Gaussian distribution. The magnitude of that field is related to the phenomenological damping parameter α through the fluctuation dissipation theorem. The local effective field $\mathbf{B}_i$ is calculated from a parametrized magnetic Hamiltonian that can contain contributions from isotropic and anisotropic exchange, single-ion magnetocrystalline anisotropy (MAE), and Zeeman energy. When focusing on shorter length scales, the influence of magnetostatic interactions can typically be neglected. For a comprehensive description of the ASD approach, see, e.g., Refs. [50–52].

As described in the previous section, the Heisenberg exchange interactions in a DMS can be calculated using DFT methods. Recently Hals et al. used scattering matrix theory to calculate the Gilbert damping tensor for GaMnAs [53]. The formalism takes into account the sample shape and treats chemical disorder in a non-perturbative way. It is well known from experiments that the sample quality can strongly influence the damping, with dirtier samples typically exhibiting a larger damping than cleaner and well-annealed samples. This is captured by the scattering matrix technique. Furthermore it was proposed that for smaller samples it is the sample shape that is the main cause for anisotropic damping, rather than the anisotropies of the crystal structure.

Insights into the microscopic interactions can alternatively be inferred from various experimental techniques. Of importance are here not least magnetic resonance (MR) techniques which can give insight into exchange stiffness, the effective damping as well as MAE and shape anisotropy. To disentangle the contributions to the MR signal can be a challenging task, which can be further complicated by surface effects and sample impurities. Bihler et al. have recently analyzed the magnetic resonances in GaMnAs taking into account the sample geometry, and also effects from surface spin pinning [54].

The theory of magnetic resonance has also been relevant in optical experiments on DMS. Rozkotova et al. investigated the transient polar Kerr rotation achieved by ultrashort 1.54 eV laser pulses with a fluence of 10 $\mu J/cm^2$ [55]. It was established that the oscillatory part of the Kerr signal, probed with a linear polarized probe pulse is connected to the ferromagnetic order and shows only a weak dependence on the polarization of the pump pulse. The pump pulse is exciting a comparably small number of carriers of the GaMnAs sample and consequently the ferromagnetism is only weakly perturbed [56]. It was proposed that the phonons emitted by photo-injected carriers cause a change in the MAE, thereby initiating magnetic precession which reveals itself in the oscillating Kerr signal. Using more intense pump pulses, with a fluence in the range of 1–10 mJ/cm^2, Zahn et al. investigated the carrier and magnetization dynamics in GaMnAs [56]. The laser pulse this time achieves a strong, but not complete demagnetization of the sample. The carrier dynamics was studied through the reflectivity signal and it was observed that the recombination of carriers proceeds over several nanoseconds. As demonstrated by Scherbakov et al., coherent magnetic precession in GaMnAs can be initiated by picosecond acoustic pulses [57]. The MAE in GaMnAs is highly sensitive to strain. In contrast to the optical pump methods, the change of the anisotropy is here achieved without affecting the concentration of charge carriers.

In simulations based on Eq. (2) one has the freedom to choose freely the initial spin configuration and to include or exclude the various terms in the Hamiltonian. Using exchange parameters calculated for different As antisite concentrations, Langevin spin dynamics simulations have been performed on GaMnAs, focusing on how the magnetization responds to changes in the temperature [58]. The pair correlation function for equal times is defined as

$$\begin{aligned} G_{ij}(t) &= \langle \mathbf{m}_i(t) \cdot \mathbf{m}_j(t) \rangle \\ &= \langle \mathbf{m}(\mathbf{r}_i, t) \cdot \mathbf{m}(\mathbf{r}_j, t) \rangle, \end{aligned} \tag{3}$$

where $\langle \cdots \rangle$ represents an average over all atoms in the simulation box. Starting the simulations from either a completely disordered (high temperature limit) or ferromagnetic (low temperature limit) spin configuration, the evolution in time of the order parameter and $G_{ij}(t)$ were investigated. The pair correlation function, $G_{ij}(t)$, depends not only on the distance between Mn atoms but also, as an immediate consequence of the directional dependence of the exchange interaction, on the direction in the lattice in which $G_{ij}(t)$ is calculated. In Figures 1 and 2 the pair correlation function is plotted at different times as a function of distance between atoms for two different starting configurations of the atomic moments; a random distribution (denoted rDLM in Figs. 1 and 2) and a ferromagnetic distribution (denoted FM in Figs 2.a and 2). The simulations are performed for As antisite concentrations of 0.75% and 1.25% at a temperature of 100 K. For the lower antisite concentration, this temperature is below the Curie temperature, whereas for the higher antisite concentration it is above. In these figures the pair correlation function approaches its equilibrated value after a sufficiently long time of the simulations. For the FM configuration in Fig.1 this is seen to take place after approximately 5 ps. It is also clear from Fig.1 that the FM starting configuration reaches equilibrium faster than a rDLM configuration. It can be observed in Figure 2 that substantial short range correlation occurs also above the Curie temperature. The rate of magnetic relaxation was in Ref. [58] also studied by following the spin autocorrelation function and comparisons were performed with other materials, e.g., typical spin-glass materials

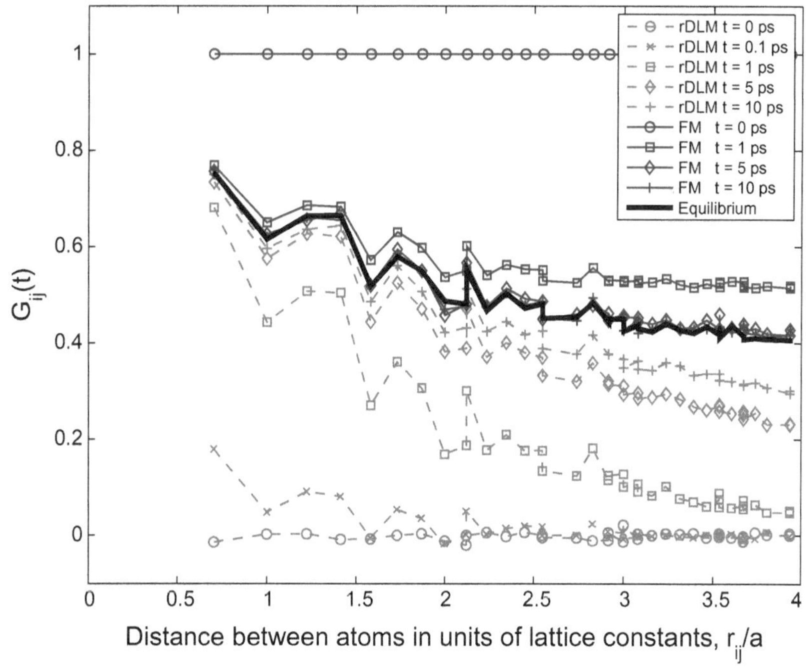

FIGURE 1 Time evolution of the pair correlation function $G_{ij}(t)$ starting from ferromagnetic (blue solid line) respective random (red dashed line) configurations. Values are obtained for an As antisite concentration of 0.75% and at $T = 100$ K. The equilibrium pair correlation is shown by the black thick line. (For interpretation of the references to color in this figure legend, the reader is referred to the web version of this book.)

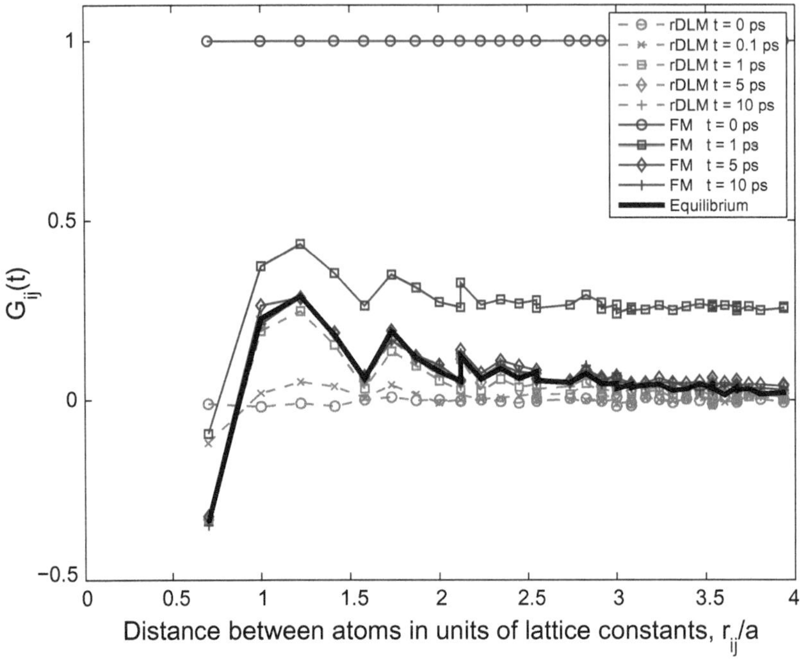

FIGURE 2 Same as Figure 1 but for an As antisite concentration of 1.25%.

like Mn-doped Cu. It was concluded [58] that in the dilute limit Mn-doped GaAs is not a spin-glass, a situation which most likely applies also to other DMS materials. It was also found [58] that the relaxation, as a consequence of the comparably weak exchange field associated with the dilution, is a slower process in a DMS material than in ordinary ferromagnetic Fe.

V. DYNAMICAL MEAN FIELD THEORY

So far we have discussed theoretical results based on the local spin density approximation (LSDA), in which correlation effects are included to some extent in the exchange and correlation part of the function. However, it is clear that this way to treat correlations has limitations, in particular for transition metals forming compounds with pnictides and chalcogenides (oxides are a typical example). Hence one may suspect that also for DMS materials, e.g., Mn-doped GaAs, a more direct method to include electron correlation is needed. So far, the LDA + U method has been tried and one can conclude that, e.g., for Mn-doped GaAs the magnetic properties are not dramatically changed, when compared to calculations based on LSDA. This holds both for the magnetic moment as well as the ordering temperature. However, neither LSDA or LDA + U seems to reproduce the measured valence band spectrum of, e.g., Mn-doped GaAs, suggesting a complex form of correlations in DMS materials. A recent experimental finding of Mn-doped GaAs shows [59] that the Fermi level is situated essentially in the band gap, in a fashion similar to what LSDA or LDA + U calculations give [5]. The assumptions of the model of Ref. [2], result in a situation where the Fermi level cuts through the top of the valence band, and with the Mn d-states situated at significantly lower energy, seemingly in contrast to the experiment of Ref. [59].

The discussion above point to that a thorough analysis of the electronic structure of Mn-doped GaAs is highly needed, in which the effects of electron–electron are included at a level beyond a static Hartree–Fock level, as provided by the LDA + U method. We present here the electronic structure and magnetic properties using dynamical mean field theory and LDA + U, in combination with a full-potential electronic structure method. We have for this calculation used a full-potential linear-muffin tin orbital method [60–63], in which electron correlations are included using a combination density-functional theory in local density approximation and dynamical mean-field theory [65] (LDA + DMFT scheme). The supercell Ga_7MnAs_8 was considered, corresponding to a Mn concentration of 12.5%, with the experimental lattice constant $a = 10.681$ a.u. Our calculations made use of the following basis functions: 4s, 4p, and 3d orbitals for Mn, Ga, and As, with two tail energies. Convergence of the total energy up to the meV leads to a **k**-mesh of 343 **k**-points in the full Brillouin Zone. Integration was carried on both with the Gaussian smearing method. Finally the exchange–correlation functional has been described using the LDA of Perdew and Wang.

The correlation effects were included by means of the LDA + DMFT scheme within the implementation described in Refs. [62,63], recently extended with the inclusion of the full self-consistency over the electron density [66]. The effective impurity problem has been solved using an exact diagonalization (ED) procedure, developed in Ref. [68]. We refer to this work for further details concerning the calculations. The values of the electron–electron Coulomb repulsion, as given by the Hubbard U, was 6 eV, which represents an interaction strength in a commonly accepted range. The value of the exchange energy has been fixed to $J = 0.7$ eV. It should be noted here that the quality of the ED calculations depends on how many ligand functions are allowed to hybridize with the correlated orbitals (Mn d states in the present case), and the result presented here is not obtained from an optimized set of such hybridizing functions, due to the computational cost of a full set of 'hybridizing functions'. Hence, our current results using ED at the impurity step of DMFT, represent an indication of the possibility in the capability for theory to reproduce observed spectral properties, awaiting a more complete calculation [69].

In Fig. 3 we compare the calculated DOS from LSDA, LDA+U and DMFT theory. It is clear that the three different approximations result in different shapes and structures of the DOS. Theory based on DMFT results in the overall widest

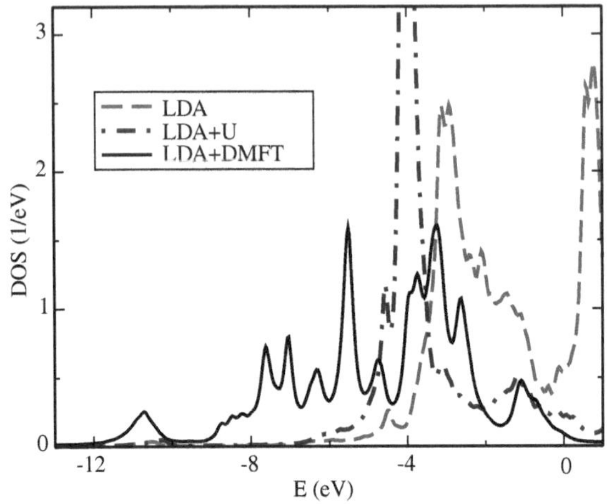

FIGURE 3 Calculated DOS with LDA (dashed line), LDA+U (dashed-dotted line) and DMFT theory (full line). The Fermi level is at zero energy.

spectrum, which range from the Fermi level down to approximately 10 eV binding energy. In this calculation there are three conspicuous features; a smaller peak at 1 eV binding energy, a large peak centered at 3 eV binding energy with a width of roughly 2 eV, and finally a broad feature which ranges from 4 to 8 eV binding energy in which one narrow feature at 5.5 eV binding energy has highest intensity. The LDA+U calculation essentially has only one main feature, the large peak at 4 eV binding energy. There is in addition a smaller intensity extending all the way up to the Fermi level. Finally, the calculation based on LSDA has a broader feature which extends from the Fermi level down to 4 eV binding energy. Both the LDA+U and the LSDA calculations have little or no intensity at higher binding energies than 4 eV. An experimental verification on which of the three approximation is the best the describe the electronic structure of Mn doped GaAs would hence be to compare the data in Fig.3 with the an experimental photo-electron spectrum.

VI. HOW HIGH CAN THE ORDERING TEMPERATURES OF DILUTED MAGNETIC SEMICONDUCTORS BECOME?

In the sections above we have addressed some of the pertinent issues of electron correlations, the magnetization dynamics and the exchange interactions, in general, of DMS systems. We end this chapter by a comparison between two of the most commonly discussed theories, i.e., the more long ranged Zener pd-exchange in mean-field theory [2], and the more short ranged exchange, as given by LSDA and LDA + U, and Monte Carlo simulations [4]. We will present arguments that first principles theory in combination with MC simulations offers a different picture compared to the mean-field theory, both regarding the mechanism for the exchange coupling, the range of the interaction and the expected ordering temperatures, with a considerably better overall agreement with experimental observations. In order to illustrate this we make a detailed analysis based on this theory, for all materials studied in Ref. [2]. In order to make a detailed comparison between the two theories, it is very important that exactly the same materials parameters are considered, and we have for this reason chosen for our calculations the same materials parameters as those considered in Ref. [2]. Hence we consider a 5% concentration of Mn atoms as well as the same concentration of charge carriers as in Ref. [2].

The calculated exchange parameters listed in Table 1 were used to evaluate the Curie temperatures of the diluted magnetic semiconductors by means of Monte Carlo (MC) simulations and by the mean field approximation (MFA). The Monte Carlo simulations were done by employing the single flip Metropolis algorithm. The critical temperatures were obtained from a fourth-order cumulant crossing method [70]. The fourth-order cumulant of the order parameter (magnetization) reads

$$U_L = 1 - \frac{\langle M^4 \rangle}{3\langle M^2 \rangle^2}. \tag{4}$$

The critical temperature is obtained from the intersection point of the curves of U_L for different system sizes L.

In the MFA of the Heisenberg model the Curie temperature is given by

$$T_C^{\mathrm{MFA}} = \frac{2x}{3k_B} \sum_{j \neq 0} J_{0j}, \tag{5}$$

where x is the concentration of Mn and k_B is the Boltzmann constant.

Without any experimental parameter as input we calculated in this way the Curie temperature of all materials listed in Table 1. Our calculated values are compared to those predicted by Ref. [2] in Figure 4. We note that in general the different theories give very different values of the ordering temperature. In some cases the difference is orders of magnitude (e.g., Mn-doped ZnO and GaN), and in some cases it is of the order of 100–400% (for Mn-doped AlP and AlAs). Only for a few cases there is some similarity between, e.g., the data of Ref. [1] and the Monte Carlo data of Figure 1 (in Mn doped InAs and GaSb), and judging by the general trend of Figure 1 any agreement is only fortuitous. Overall the results based on first principles theory in combination with Monte Carlo simulations are in better agreement with experimental observations. For instance, none of the high ordering temperatures predicted in Ref. [2] have been unambiguously observed in experiments, especially for wide band-gap DMS. In several of the experimental reports it is concluded that ordering temperatures above room temperature are caused by secondary phases or precipitates, and that

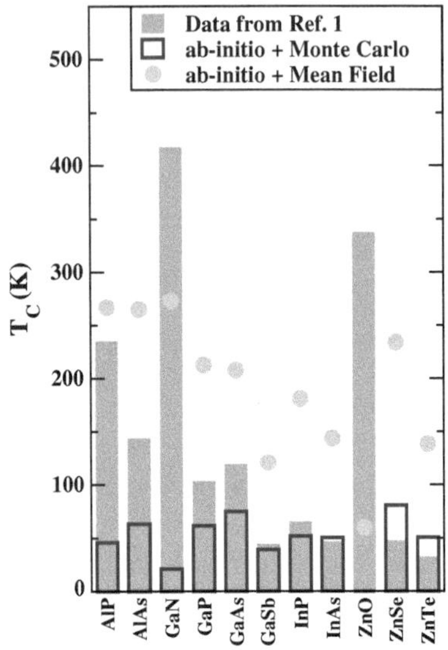

FIGURE 4 Calculated ordering temperatures of Mn doped in selected semiconducting hosts. The bars represents data of Ref. [1], whereas the black bars represent the current theory, and the filled circles represent mean field calculations using the exchange parameters of Table 1.

a random distribution of magnetic impurities does not exhibit room temperature ferromagnetism. The first principles data in Figure 4 are consistent with these observations.

The reason behind the large differences between the two theoretical models which are compared in Figure 4 is clear from the analysis of Ref. [4,5]. For any diluted magnetic semiconductor, first principles calculations result in interatomic exchange interactions which are damped exponentially with distance. This is rigorously true for the wide band-gap semiconductors where impurity band exists in the band gap resulting in a sharp decrease of the exchange interaction parameters, due to the presence of disorder and an energy gap in the spectrum. In these cases, the difference between the two theories becomes absolutely transparent. This fact follows the analysis of Haldane and Anderson [71] and is missing in the theory of Ref. [2]. A short ranged interaction leads to that mean-field theories are less reliable and that magnetic percolation becomes important for a homogeneously disordered dilute spin system. The theory presented here, which involves Monte Carlo simulations that sample disordered configurations offers a way to obtain qualitative values of ordering temperatures in this case, whereas a mean field-based theory would result in ordering temperatures which are much too high [4].

In order to illustrate this point we also evaluated, as mentioned above, mean field values of the ordering temperature using the exchange parameters in Table 1, i.e., the same exchange parameters as used in the Monte Carlo simulations. These mean field values are also shown in Figure 4, and it may be seen that this results in much too large values of the ordering temperature. Since the analysis of Ref. [2] is based on a mean field treatment, the large disagreement between the yellow bars and filled circles in Figure 4 must signal that the exchange interactions calculated here are very different than the ones estimated in Ref. [2]. For this reason it seems that any coincidence between the ordering temperature of the approach used in Ref. [2] and the data presented here, e.g. for Mn-doped GaSb, is only fortuitous. The difference in the exchange parameters of Table 1, and those considered in Ref. [2] is due to that the former are given by a delicate balance between super-, double-, and p–d exchange, whereas the latter only consider p–d exchange. First principles theory show a balance of the different mechanisms behind the inter atomic exchange, which is different for different materials, and it is this capability to obtain materials specific information which makes this theory a powerful tool for studies of magnetism in general.

The Monte Carlo data in Figure 4 shows that a low concentration (5%) of Mn atoms is insufficient for obtaining room temperature ferromagnetism in III–V and II–VI semiconductors, in contrast to a mean field analysis. The reason for this is the short ranged nature of the exchange interaction, and the associated magnetic percolation phenomenon, which is the main obstacle to address when one tries to design diluted magnetic semiconductors with ordering temperatures above room temperature. These facts suggest that chemical ordering between magnetic dopants is a very important material property to be able to control and manipulate experimentally. Spinodal decomposition, clustering, nanolaminates, and delta-doping are examples of routes for controlling it, and should be considered as key pathways for achieving room temperature ordering temperatures of diluted magnetic semiconductors, in particular for wide-gap semiconductors. As a matter of fact, it has been shown from materials simulations that a careful control of the chemical modulation in diluted magnetic semiconductors can increase the ordering temperature of these materials with over one order of magnitude [72]. It has also been suggested that super-paramagnetic blocking phenomena play an important role for magnetic response of decomposed DMS[73].

VII. CONCLUSION

In this chapter we have reviewed some of the more recent developments in the science of diluted magnetic semiconductors, including magnetization dynamics, the effects of electron correlations, and finally we have presented a realistic estimate on the possible ordering temperature one can expect in these materials. From atomistic spin dynamics it would seem that the dynamics of Mn-doped GaAs does not exhibit a spin-glass behavior [58], at least for Mn concentrations below ~10%, in contrast to some claims [67]. Instead it has been pointed out [58] that the dynamics of this system is that of a slow magnet, which is caused by the weakened local Weiss-field, due to the dilute concentration of magnetic atoms of the material.

The effects of electron correlations here have been investigated using dynamical mean-field theory in combination with an impurity solver based on DMFT and exact diagonalization. Inclusion of electron correlations in DMFT results in Mn states which have a main feature at $\sim$3 eV binding energy, and that extend all the way up to the Fermi level. This fact should be possible to very experimentally.

Realistic estimates of the ordering temperature of several III–IV and II–V DMS materials, with 5% Mn and a homogenous distribution of the Mn atoms, show that a Curie temperature above 100 K is unlikely. Although the theoretical method used to calculate the Curie temperature is expected to be associated with some error, it has been shown in the past to reproduce ordering temperatures of, e.g., bcc Fe with an error not exceeding 20%. The data in Figure 4 is also expected to not be associated with an error bigger than $\sim$20%, and hence not to exceed 100 K. This conclusion is in sharp contrast to the theoretical estimates of Ref. [2], also shown in Figure 4, where considerably higher ordering temperatures are reported. The main reason behind the different values of the ordering temperature presented here, and the one of Ref. [2] lies in the range of the exchange interaction. In Ref. [2] it is assumed to be rather long-ranged, motivating a mean-field analysis, whereas the first principles calculations of inter-atomic exchange result in a much more short ranged behavior (see Table 1), that point to magnetic percolation being important.

In Figure 5 we contrast the picture of exchange interaction in DMS materials as resulting from the analysis of Ref. [2] (left part of the figure) with a picture which is emerging from first principles theory [5] (right part). The left part shows an exchange interaction which is long ranged and has no angular component to it, whereas the exchange interaction in the right part, suggested by first principles theory is much more short ranged and also reflects the orbital character of the atoms building up the lattice. In the insets of Figure 5 a schematic picture of the directional character of the exchange interaction is shown, where in particular the strong directional character of the exchange as given by first principles theory stands out. It is in particular the $\langle 1\,1\,0 \rangle$ direction that has a particularly strong interatomic-exchange [5], a theoretical result which is in agreement with experimental observations.

Although the analysis presented here seems to exclude an immediate technological use of DMS materials at room temperature, at least in the dilute limit of 5–10% doping, it seems that these materials possess several intriguing features, involving complexity in the magnetization dynamics and electron correlation effects which lie beyond a static approximation (like LDA + U). Hence the field of DMS materials is certainly worth-while to pursue, primarily from a fundamental scientific point of view.

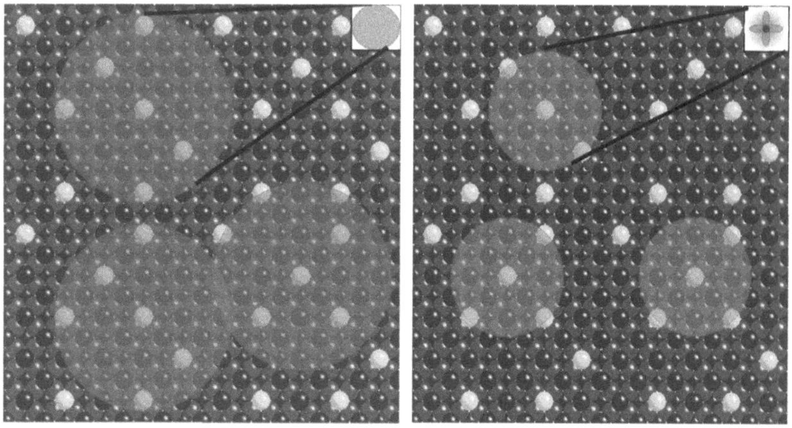

FIGURE 5 Schematic picture of interatomic exchange interaction in DMS materials. magnetic atoms (e.g., Mn) are shown in bright colour whereas the atomic species of the semiconducting material (e.g., Ga and As) are shown in darker color. The range of the interaction is specified by the extend of the shaded/grey region around some of the Mn atoms. To the left we show the more common picture of an exchange interaction which is long ranged and has only little directional dependence. In the right figure we show the main results from an analysis of first principles results, i.e., an exchange interaction which is short ranged and has a large angular dependence. In the top right corner of each figure the range and strength of the exchange interaction around a given magnetic atom is shown in gray scale, where a deeper gray color indicates a stronger interaction.

REFERENCES

[1] Ohno H. Science 1998;281:951–6.
[2] Dietl T et al. Zener model description of ferromagnetism in zinc-blende magnetic semiconductors. Science 2000;287:1019–22.
[3] Litvinov VI, Dugaev VK. Phys Rev Lett 2001;86:5593.
[4] Bergqvist L et al. Magnetic percolation in diluted magnetic semiconductors. Phys Rev Lett 2004;93:137202.
[5] Sato K et al. Low-temperature ferromagnetism in (Ga, Mn)N: ab initio calculations. Phys Rev B 2004;70:201202.
[6] Pulizzi F. Dilute for impact. Nature Milestones. doi:10.1038/nphys872.
[7] Cochrane RW, Plischke M, Ström-Olsen JO. Magnetization studies of $(GeTe)_{1-x}(MnTe)_x$ pseudobinary alloys. Phys Rev B 1974;9:3013–21.
[8] Story T et al. Carrier-concentration-induced ferromagnetism in PbSnMnTe. Phys Rev Lett 1986;56:777–9.
[9] Spałek J et al. Magnetic susceptibility of semimagnetic semiconductors: the high-temperature regime and the role of superexchange. Phys Rev B 1986;33:3407–18.
[10] Edmonds KW et al. Hall effect and hole densities in $Ga_{1-x}Mn_xAs$. Appl Phys Lett 2002;81:3010–12.
[11] Zener C. Interaction between the d shells in the transition metals. Phys Rev 1950;81:440–4.
[12] Reed ML et al. Room temperature magnetic (Ga, Mn)N: a new material for spin electronic devices. Mater Lett 2001;51:500–3.
[13] Sasaki T et al. Magnetic and transport characteristics on high Curie temperature ferromagnet of Mn-doped GaN. J Appl Phys 2002;91:7911–3.
[14] Yoshii S et al. Evidence for carrier-induced high-T_C ferromagnetism in Mn-doped GaN film. EPL 2007;78:37006.

[15] Shi Y et al. The influence of implantation temperature on the magnetism and structure of Mn^+ implanted p-GaN films. Physica B 2007;388:82.

[16] Kaminski M et al. New chemical method of obtaining thick $Ga_{1-x}Mn_xN$ layers: prospective spintronic material. Chem Mater 2007;19:3139–43.

[17] Cui XG et al. Structural and magnetic properties in Mn-doped GaN grown by metal organic chemical vapor deposition. Appl Phys Lett 2008;92:152116.

[18] Jayakumar OD, Gopalakrishnan IK, Kulshrestha SK. On the room temperature ferromagnetism of Mn doped ZnO. Physica B 2006;381:194–8.

[19] Thakur P et al. X-ray absorption and magnetic circular dichroism characterizations of Mn doped ZnO. Appl Phys Lett 2007;91:162503.

[20] Jayakumar OD et al. Significant enhancement of room temperature ferromagnetism in surfactant coated polycrystalline Mn doped ZnO particles. J Alloys Comp 2007;438:258–62.

[21] Xu Q et al. Room temperature ferromagnetism in Mn-doped ZnO films mediated by acceptor defects. Appl Phys Lett 2007;91:092503.

[22] Hou D-L et al. Magnetic properties of Mn-doped ZnO powder and thin films. Mater Sci Eng B 2007;138:184–8.

[23] Sharma VK, Varma GD. Oxygen vacancies induced room temperature ferromagnetism in hydrogenated Mn-doped ZnO. J Appl Phys 2007;102:056105.

[24] Yan W et al. Zn vacancy induced room-temperature ferromagnetism in Mn-doped ZnO. Appl Phys Lett 2007;91:062113.

[25] Yan W et al. Structures and magnetic properties of (Mn, N)-codoped ZnO thin films. Appl Phys Lett 2007;90:242509.

[26] Ivill M et al. Magnetization dependence on carrier doping in epitaxial ZnO thin films co-doped with Mn and P. J Appl Phys 2007;101:123909.

[27] Yang Z et al. Electron concentration dependent magnetization and magnetic anisotropy in ZnO:Mn thin films. Appl Phys Lett 2008;92:042111.

[28] Wójcik A et al. Low temperature growth of ZnMnO: a way to avoid inclusions of foreign phases and spinodal decomposition. Appl Phys Lett 2007;90:082502.

[29] Masuko K et al. Preparation and the magnetic property of ZnMnO thin films on $(000\bar{1})$ ZnO single crystal substrate. J Magn Magn Mater 2007;310:e711–3.

[30] Savchuk AI et al. Optical and magneto-optical properties of ZnMnO and ZnMnFeO single crystals and thin films. Phys Stat Sol (a) 2007;204:106–11.

[31] Yang JH et al. Structure and magnetic properties of Mn-doped ZnO semiconductors. Solid State Commun 2007;143:566–9.

[32] Blasco J et al. Magnetic properties of doped ZnO prepared by different synthetic routes. J Magn Magn Mater 2007;316:e177–80.

[33] Kane MH et al. Magnetic and optical properties of single crystals of transition metal doped ZnO. Phys Stat Sol (b) 2007;244:1462–7.

[34] Liu GL et al. High T_C ferromagnetism of $Zn_{(1-x)}Co_xO$ diluted magnetic semiconductors grown by oxygen plasma-assisted molecular beam epitaxy. Appl Phys Lett 2007;90:052504.

[35] Song C et al. The magnetic properties of Co-doped ZnO diluted magnetic insulator films prepared by direct current reactive magnetron co-sputtering. J Magn Magn Mater 2007;309:25–30.

[36] Sati P et al. Antiferromagnetic interactions in single crystalline $Zn_{1-x}Co_xO$ thin films. Phys Rev Lett 2007;98:137204.

[37] Zhou S et al. Using X-ray diffraction to identify precipitates in transition metal doped semiconductors. J Appl Phys 2008;103:07D530.

[38] Zunger A. In: Zeits F, Ehrenreich H, Turnbull D, editors. Solid state physics, vol. 39. New York: Academic; 1986. p. 275.

[39] Jungwirth T, Sinova J, Masek J, et al. Rev Mod Phys 2006;78:809–64.

[40] Sato K, Bergqvist L, Kudrnovsky J, et al. Rev Mod Phys 2010;82:1633–90.

[41] Dietl T. Nat Mater 2010;9:965–74.

[42] Hohenberg P, Kohn W. Phys Rev 1964;136:864; Kohn W, Sham LJ. Phys Rev. 1965;140:1133.

[43] von Barth U, Hedin L. A local exchange–correlation potential for the spin polarized case. I. J Phys C 1972;5:1629–42.

[44] Andersen OK. Phys Rev B 1975;12:3060.

[45] Skriver HL. The LMTO method: muffin tin orbitals and electronic structure. Berlin: Springer-Verlag; 1984.
[46] Abrikosov IA, Simak SI, Johansson B. Locally self-consistent Greens function approach to the electronic structure problem. Phys Rev B 1997;56:9319.
[47] Korringa J. Physica 1947;13:392; Kohn W, Rostoker N. Phys Rev 1954;94:1111.
[48] Wills JM, Eriksson O, Alouani M, Price D. In: Dreyssé H, editor. Electronic structure and physical properties of solids. Lecture notes in physics. Mont St. Odile: Springer; 1998.
[49] Liechtenstein AI, Katsnelson MI, Antropov VP, Gubanov VA. Local spin density functional approach to the theory of exchange interactions in ferromagnetic metals and alloys. J Magn Magn Mater 1987;67:65.
[50] Skubic B, Hellsvik J, Nordström L, Eriksson O. A method for atomistic spin dynamics, development and examples. J Phys: Condens Matter 2008;20:315203.
[51] Skubic Björn. Spin dynamics and magnetic multilayers. Thesis, Uppsala University, 2007.
[52] Hellsvik Johan. Atomistic spin dynamics, theory and applications. Thesis, Uppsala University, 2010.
[53] Hals Kjetil M, Brataas Arne. Phys Rev B 2011;84:104404.
[54] Bihler C, Schoch W, Limmer W, Goennenwein S, Brandt M. Phys Rev B 2009;79:045205.
[55] Rozkotová E, Němec P, Horodyská P, Sprinzl D, Trojánek F, Malý P, et al. Appl Phys Lett 2008;92:122507.
[56] Zahn JP, Gamouras A, March S, Liu X, Furdyna JK, Hall KC. J Appl Phys 2010;107:033908.
[57] Scherbakov A, Salasyuk A, Akimov A, Liu X, Bombeck M, Brüggemann C, et al. Phys Rev Lett 2010;105:117204.
[58] Hellsvik J, Skubic B, Nordström L, Sanyal B, Nordblad P, Svedlindh P, et al. Phys Rev B 2008;78:144419.
[59] Ohya S et al. Phys Rev Lett 2010;104:167204.
[60] Wills JM, Cooper BR. Phys Rev B 1987;36:3809.
[61] Wills JM, Eriksson O, Alouani M, Price D. In: Dreyssé H, editor. Electronic structure and physical properties of solids. Lecture notes in physics. Mont St. Odile: Springer; 1998.
[62] Grechnev A, Di Marco I, Katsnelson MI, Lichtenstein AI, Wills John, Eriksson O. Phys Rev B 2007;76:35107; Thunström P, Di Marco I, Lebegue S, Katsnelson MI, Svane A, Eriksson O. Phys Rev B 2009;79:165104.
[63] Wills JM, Alouani M, Andersson P, Delin A, Eriksson O, Grechnev A. In: Full-potential electronic structure method, energy and force calculations with density functional and dynamical mean field theory. Springer series in solid-state sciences, vol. 167, 2010.
[64] Katsnelson MI, Irkhin VYu, Chioncel L, Lichtenstein AI, de Groot RA. Rev Mod Phys 2008;80:315.
[65] A review can be found in: Georges A, Kotliar G, Krauth W, Rozenberg MJ. Rev Mod Phys 1996;68:13.
[66] Grånäs O et al. Computational Materials Science 55, 295 (2012).
[67] Wang WZ, Deng JJ, Lu J, Sun BQ, Zhao JH. Appl Phys Lett 2007;91:202503.
[68] Thunström P et al. (to be published).
[69] Di marco I et al. (to be published).
[70] Landau DP, Binder K. A guide to Monte Carlo simulations in statistical physics. Cambridge: Cambridge University Press; 2000.
[71] Haldane FDM, Anderson PW. Simple model of multiple charge states of transition-metal impurities in semiconductors. Phys Rev B 1976;13:2553–9.
[72] Fukushima T et al. Spinodal decomposition under layer by layer growth condition and high Curie temperature quasi-one-dimensional nano-structure in dilute magnetic semiconductors. Jpn J Appl Phys 2006;45:L416–8.
[73] Sato K et al. Super-paramagnetic blocking phenomena and room-temperature ferromagnetism in wide band-gap DMS (Ga, Mn)N. Jpn J Appl Phys 2007;46:L682–4.

Chapter 2

Molecular Beam Epitaxial Growth and Magnetotransport Properties of GaMnAs

Janusz Sadowski[*,†] **and Jaroslaw Z. Domagala**[†]
[*] *MAX-IV laboratory, Lund University, P.O. Box 118, 221 00 Lund, Sweden*
[†] *Institute of Physics, Polish Academy of Sciences, al. Lotnikow 32/46, 02-668 Warszawa, Poland*

Chapter Outline Head

I. INTRODUCTION

GaMnAs ternary alloy with Mn ions occupying Ga sites in the GaAs host lattice became, during last decade, a canonical ferromagnetic semiconductor. Materials combining ferromagnetic and semiconducting properties have been known for a long time; for example, some rare earth chalcogenides like EuS, and EuO exhibit ferromagnetic (FM) phase transition at the temperatures in the range of 30–60 K [1]; however, GaMnAs belongs to the smaller class of materials in which FM phase transition is due to the exchange interaction between spins of transition

Advanced Functional Materials. http://dx.doi.org/10.1016/B978-0-44-453681-5.00002-9

metal ions (Mn located at Ga site in GaAs carriers magnetic moment of 5 μ_B, due to the spin orientation of a half filled 3d shell) and free carriers present in the semiconducting host material. This type of interaction between free carriers and magnetic ions of a transition metal (TM) located in a semiconductor matrix was first reported by Story et al. in 1986 for PbSnMnTe a – narrow gap IV–VI semiconductor [2] then the RKKY model was successfully applied to describe the interactions between fee carriers and TM ions yielding, ferromagnetic, paramagnetic, or spin-glass behavior, depending on the concentration of TM ions and free carriers. A decade later, similar properties were discovered in heavily p-type doped II-Mn-VI ternary alloys CdMnTe [3] and ZnMnTe [4], but in 1996 the first paper reporting FM phase transition in GaMnAs ternary alloy was reported by Ohno et al. [5]. The efforts to crystallize ternary alloy of III–V semiconductor (InAs) and Mn had been initiated already in 1989, when the successful realization of monocrystalline InMnAs layers by low temperature molecular beam epitaxy (MBE) growth was reported [6]. However, this time no FM phase transition in InMnAs was observed, some experimental features pointing for the FM phase in InMnAs were reported later on, in 1992 [7], but they were interpreted in the bound magnetic polaron scheme, so no FM coupling between diluted Mn ions due to the interactions with free carriers was postulated. Such a type of diluted III-Mn-V ferromagnet was reported for the first time by Ohno et al. in 1996 [5] and already that time quite high paramagnetic-to-ferromagnetic temperature (T_C) value of 110 K was attained in thin GaMnAs MBE grown layer, with concentration of Mn equal to 5.3%, and concentration of valence band holes in the range of 10^{20} cm^{-3} [5]. This report followed by the theoretical description applying Zener model for ferromagnetism [8] developed in the 1950s for transition metals, which was employed by Dietl et al. and published in the seminal paper [9] generated a vast interest in different types of TM doped semiconductors with tetrahedral coordination of transition metal cations. Dietl's model not only described the magnetic properties of GaMnAs (such as T_C and magnetic anisotropy), but also predicted (quite high in some cases) T_C values for other III–V and also IV–VI semiconductors such as GaN, and ZnO, respectively, alloyed with Mn and sufficiently highly doped to p-type.

These predictions generated a huge interest in the scientific community (the Dietl's paper published in Science has been cited over 5000 times (as for January 2012)) and large number of groups started then to investigate different semiconductors doped not only with Mn but also with other TM elements, such as Co, Cr, Fe, V, Ni [10–14]. Shortly afterwards a lot of reports claiming intrinsic ferromagnetic properties of different semiconductors doped with TM elements appeared, the most famous examples being Mn doped GaN [15], and Cr doped ZnTe [11]. However, later on all these claims have been negatively verified [16–18]. It turned out that the nano-precipitates of concentrated TM phases, either pure or alloyed with some elements of the semiconductor matrix are responsible for the measured FM properties. In some semiconductor materials like GaN, or ZnO it is also suggested that the FM phase transition without TM metal elements may occur, due to the magnetic moments that might be associated with the point defects (mostly vacancies) in the semiconductor lattices. By now there are only few examples of diluted ferromagnetic semiconductors, e.g., PbSnMnTe [2], GeMnTe [19] and GaMnAs [20] (also InMnAs [20], GaMnSb [21], InMnSb [22], GaMnP [23] and heavily p-doped II–Mn–VI semiconductors [3,4]) with well established mechanism of carrier-induced FM phase transition. The T_C temperatures are the highest for GeMnTe and GaMnAs (close to 200 K for both [19,24]), however GaMnAs is the most extensively studied and best known due to the well developed preparation methods and technological importance of GaAs-based compounds.

II. MOLECULAR BEAM EPITAXY GROWTH OF GAMNAS

Molecular beam epitaxy (MBE) is a versatile method of deposition of monocrystalline layered structures of reach variety of materials such as semiconductors, metals and even organic materials. It has been initiated by Arthur and Cho in late 60s and early 70s [25,26] and then further developed in 70s by research groups at Bell Labs (Cho and Arthur) and IBM (Chang and Esaki) [27]. The essential features of MBE technology like formation of heterostructures, remote doping, etc. were demonstrated; later on, in early 80s the MBE setups started to be commercially available. GaAs based structures are the "flagship" materials developed with use of the MBE technique resulting in important achievements both in basic physics, like for example high carrier mobility structures as well as in new application devices like laser structures, radiation detectors and others. p-Type doping of GaAs with Mn during MBE growth was tried already in mid 70s [28] but Mn has not been widely used as a p-type dopant in GaAs. Later on the great success of Mn doped II–VI semiconductors associated with their dilute magnetic semiconductor properties [29] motivated the search for possibilities of exploiting similar functionalities in III–V semiconductors. The key factor in successful formation of GaMnAs ternary alloy was to use very low substrate temperature during the epitaxial growth (below 300 °C) [30]. On the other hand crystallization of GaAs and its ternary alloys at that low temperatures leads to enhanced concentrations of point defects. In the case of low temperature grown GaAs (LT GaAs) which is usually crystallized with some excess As flux, the most abundant defects are As atoms located at Ga sites – As antisites (As_{Ga}) [31]. The defects typical for GaMnAs are described in the following paragraphs.

The simplified diagram describing the GaMnAs MBE growth is shown in Figure 1. As shown in Figure 1. GaMnAs has to be grown at temperatures below 300 °C, and the growth temperature has to be decreased with increasing Mn content. The maximum Mn content in uniform GaMnAs ternary alloy is close to 20%, and for this high Mn content material the substrate temperature during the MBE growth has to be well below 200 °C.

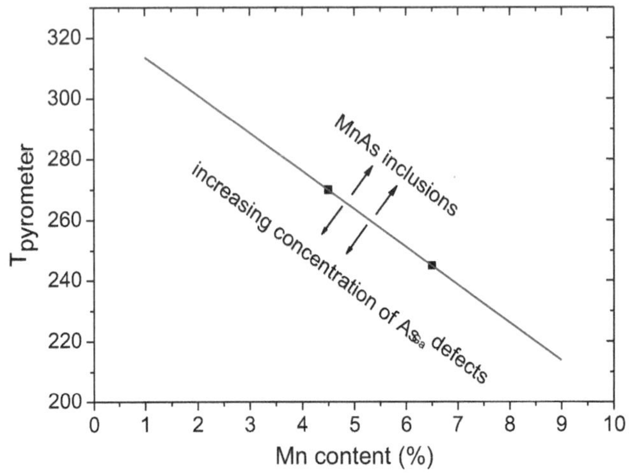

FIGURE 1 Schematic (simplified) diagram for GaMnAs growth by molecular beam epitaxy.

As for any other MBE grown material also the growth of GaMnAs can be controlled in situ by reflection high energy electron diffraction (RHEED) observed during the growth. The RHEED patterns give information about the crystalline perfection and smoothness of the grown layers; their dynamics allow controlling the growth rates and composition of multinary alloys.

Typical RHEED images registered during GaMnAs growth are shown in Figure 2. They were recorded with the RHEED beam parallel to the [1 1 0] and $[\bar{1}\,1\,0]$ azimuths of the GaMnAs(0 0 1) surface. The (1 × 2) reconstruction of GaMnAs(0 0 1) surface is clearly visible.

MBE growth of uniform GaMnAs can be effectively controlled by RHEED. The optimum quality material characterized by minimum density of defects

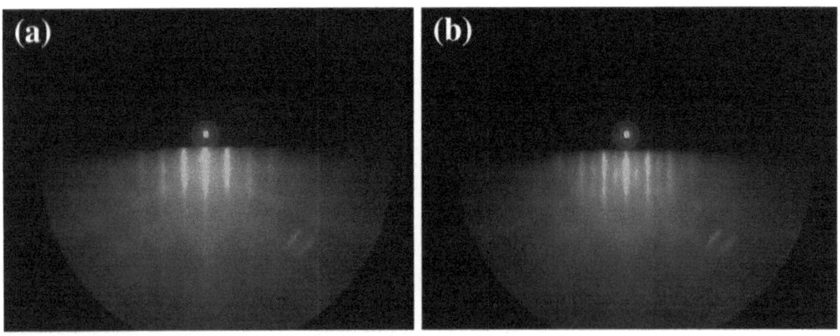

FIGURE 2 RHEED images of the (1 × 2) reconstructed GaMnAs(1 0 0) surface, recorded with 15 keV e-beam parallel to (a) [0 1 1], (b) $[0\,\bar{1}\,1]$ azimuth of GaMnAs surface.

has to be grown slightly below the "segregation line" shown in Figure 1. If the substrate temperature during GaMnAs deposition is too low then the concentration of As antisite defects increases; whereas at higher temperatures MnAs starts to segregate and 2-dimensional (2-D) layer-by-layer growth of uniform GaMnAs alloy is no longer possible. This is illustrated in Figure 3 showing RHEED images recorded during growth with MnAs phase segregation.

RHEED is a powerful tool in the MBE technology. It not only gives information on the surface structure and morphology, but can also be used for calibrations of composition of ternary (or multinary) alloys. The MBE growth is performed in excess flux of one of the components (as in the case of *III–V* arsenides, Ga in the case of GaN). Growing the ternary compound, i.e., adding the third element to the binary alloy by applying the molecular flux from effusion cell filled with the appropriate element, the growth rate of the layer crystallized from the molecular fluxes increases proportionally to the intensity of an additional flux. Knowing the sticking coefficient for the appropriate flux (usually equal to 1 for *III* elements in the MBE growth of *III–V* arsenides at standard conditions) one may calibrate the composition of the ternary compound in a simple way.

Having the growth rate of GaAs measured and equal to:

$$r_{\mathrm{GaAs}}$$

and measuring the growth rate of GaMnAs with the value

$$r_{\mathrm{GaMnAs}}.$$

The Mn contents (x) in $\mathrm{Ga}_{1-x}\mathrm{Mn}_x\mathrm{As}$ ternary alloy is equal to:

$$x = (r_{\mathrm{GaMnAs}} - r_{\mathrm{GaAs}})/r_{\mathrm{GaMnAs}}. \quad (1)$$

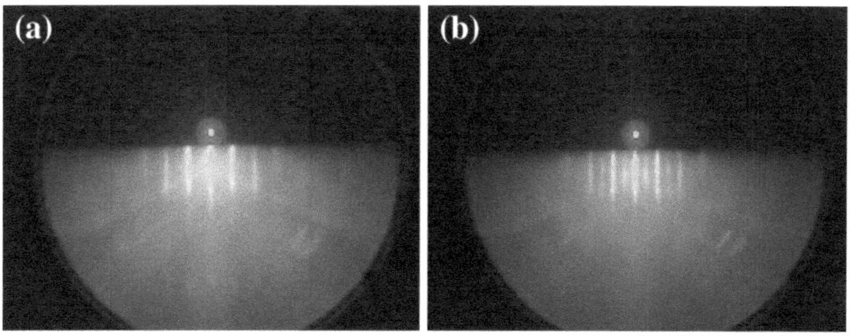

FIGURE 3 RHEED images recorded during MBE growth of GaMnAs at phase segregation conditions. The direction of RHEED beam is parallel to [0 1 1] azimuth of GaAs(1 0 0) substrate surface. (a) initial GaAs(1 0 0) surface with $c(4 \times 4)$ reconstruction, (b)–(d) images recorded after GaMnAs deposition time of 3, 6, and 11 min. The initial streaky RHEED patterns typical for a smooth surface evolve to spotty patterns proving the surface roughening due to appearance of 3-D islands of segregated MnAs.

Since most of the GaMnAs layers are grown with the Mn content in the range 1–10%. Evaluation of Mn contents in this low Mn composition regime is quite challenging and demands very precise measurements of the growth rates. Fortunately the growth conditions of GaMnAs are such that the damping of RHEED oscillations is very small – they can continue for more than 100 ML [32]. Thus the calibration of the growth rates can be performed with very high precision.

The reason of the small damping of RHEED oscillations is associated with GaMnAs growth conditions. In contrast to the standard growth conditions of GaAs-based III–V layers GaMnAs growth is only possible at very low substrate temperatures (below 300 °C), otherwise the MnAs phase separation occurs. In this low substrate temperature range the GaAs surface exposed to As flux is very As-rich [33]. During MBE growth at low temperature (LT) conditions excess. As at the surface is always present and it plays the role of a surfactant promoting the 2-D growth mode [34].

Figure 4 shows an example of RHEED oscillations for LT GaAs and GaMnAs, from which the Mn contents in GaMnAs ternary alloy was determined to be 3%.

A. Defects in GaMnAs

Since GaMnAs ternary alloy results from introducing Mn into LT GaAs it is natural to expect that all point defects typical to LT GaAs can occur also in

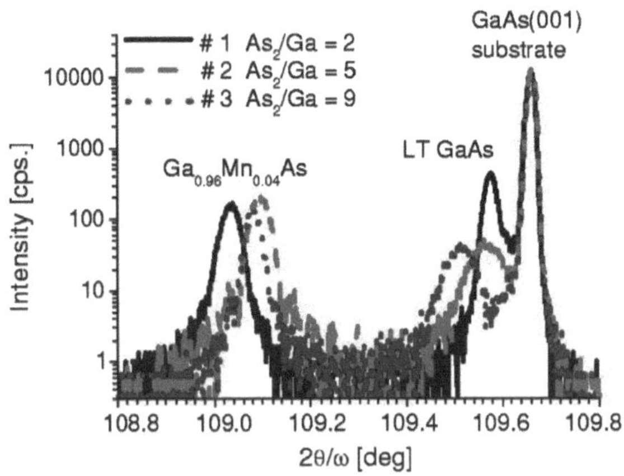

FIGURE 4 Intensity oscillations of specular RHEED beam recorded in $[0\bar{1}1]$ azimuth of GaMnAs(100) surface, during growth of LT GaAs (lower curve) and subsequent growth of GaMnAs (upper curve). The neighboring intensity maxima correspond to deposition of successive monolayers. The difference between the growth rates of GaMnAs and LT GaAs is proportional to the Mn content in GaMnAs ternary alloy (only Mn at Ga sites).

GaMnAs. The most abundant defects in LT GaAs are As antisites occurring in concentrations of up to about 0.5%.

In GaMnAs the same defects occur however it is difficult to directly measure their concentrations. Optical methods which have been used to measure $[As_{Ga}]$ in LT GaAs [31] cannot be applied to GaMnAs due to its poor optical properties. Similarly to LT GaAs, the concentrations of other defects like Ga vacancies are several orders of magnitude lower. The concentrations of Ga vacancies in GaMnAs have been measured by Tuomisto et al. [35] and their content was found to be below 10^{18} cm^{-3}, i.e., below 0.01 at.%.

The point defects which are specific to GaMnAs are Mn atoms located at interstitial sites (Mn_I), which have been first theoretically proposed by Masek and Maca [36] and then experimentally observed by Furdyna group University of Notre Dame, US [37]. The Mn interstitial defects can occur in two locations: As-coordinated and Ga coordinated [38]. Recent EXAFS measurement indicate that the first one (i.e., Ga-coordinated Mn interstitials) is preferential in GaMnAs [39].

Both As_{Ga} and Mn_I defects are double donors in GaAs host, i.e., they are effectively compensating the p-doping provided by Mn ions located at Ga sites (Mn_{Ga}). In the Zener mean field model [9], T_C depends on the density of Mn_{Ga} and the density of charge carriers according to the relation:

$$T_C \sim [Mn_{Ga}]\, p^{1/3}. \tag{2}$$

Due to the compensating effect of As_{Ga} and Mn_I defects, the net density of holes p is

$$p = [Mn_{Ga}] - 2([As_{Ga}] + [Mn_I]). \tag{3}$$

According to (1) and (2), maximizing T_C requires both maximum concentration of Mn_{Ga} and minimum concentration of As_{Ga} and Mn_I defects. These two demands are in fact contradictory, since increasing Mn content in GaMnAs requires a reduced growth temperature [20]. Low growth temperature, however, promotes the formation of As antisite defects [31]. Thus, there exists a Mn concentration that provides a compromise between these two tendencies. Experimental evidence suggests that this optimum Mn content is in the range of about 12 at.%. The maximum Mn content achieved so far in uniform GaMnAs is 20% [40], however growing GaMnAs with such extremely high Mn content is only possible for very thin layers (up to about 15 nm) for thicker layers the MnAs segregation occurs due to the accumulation of Mn at the growth surface. The concentrations of As_{Ga} compensating defects can attain about 0.5 at.% [41]. In case of Mn_I – up to about 20% of a total Mn content in GaMnAs can be located in the interstitial sites.

The direct evaluation of concentration of both As_{Ga} and Mn_I in GaMnAs is very difficult, but they both considerably increase the lattice parameter of GaMnAs. For As_{Ga} the increase of lattice parameter is known from earlier investigations of LT GaAs. For Mn_I and Mn_{Ga} the theoretical model developed by Masek et al. [42] determines the contribution of Mn in both locations to the lattice parameter of GaMnAs ternary alloy.

Within this model both defects increase the lattice parameter according to the formula:

$$a_{GaMnAs}(x, y, z) = a_0 + 0.02x + 0.69y + 1.05z, \quad (4)$$

where a_0 is the GaAs lattice constant, x, y and z are concentrations of Mn_{Ga}, As_{Ga} and Mn_I, respectively, normalized to Ga sites.

Applying formula (4) to the GaMnAs films grown with the same nominal concentration of Mn_{Ga} of 4%, as measured by RHEED oscillations and differing in concentration of As_{Ga} defects, which has been achieved by changing As/Ga flux ratio in subsequently grown GaMnAs films it is possible to estimate the concentration of As_{Ga} and Mn_I defects as presented in Table 1.

The X-ray diffraction measurements from which the values of lattice parameters used for calculating the concentration of defects presented in Table 1 are shown in Figure 5.

As can be seen in Table 1 the overall concentration of compensating defects in all three $Ga_{0.96}Mn_{0.04}As$ layers is close to 1%, which means 50% compensation of Mn_{Ga} acceptors (both As_{Ga} and Mn_I) are double donors so they compensate 2%, i.e., half of the Mn_{Ga} acceptors present in $Ga_{0.96}Mn_{0.04}As$ layers. This is consistent with T_C values for these samples (about 40 K) significantly lower than those described by the formula (2).

As shown in Table 1 the concentration of arsenic antisite defects [As_{Ga}] in GaMnAs can be controlled by As/Ga flux ratio during the MBE growth. It is also known that in LT GaAs the increase of the growth temperature leads to the decrease of [As_{Ga}], however in the case of GaMnAs the maximum growth temperature (T_g) is limited by the MnAs segregation threshold during the MBE growth. Thus at a given (maximum possible) T_g the As/Ga flux ratio should be minimized.

The postgrowth removal of compensating defects and corresponding increase of T_C and concentration of holes in GaMnAs is described in the next paragraphs.

TABLE 1 Concentrations of As_{Ga} and Mn_I defects in $Ga_{0.96}Mn_{0.04}As$ layers grown with different As_2/Ga flux ratios.

Sample	As_2/ Ga	X(Mn) [%]	a(LTGaAs) [Å]	a(GaMnAs) [Å]	AS_{Ga} [%]	Mn_I [%]
#1	2	4	5.6549	5.66408	0.20	0.82
#2	5	4	5.6551	5.66305	0.23	0.70
#3	9	4	5.6560	5.66325	0.35	0.64

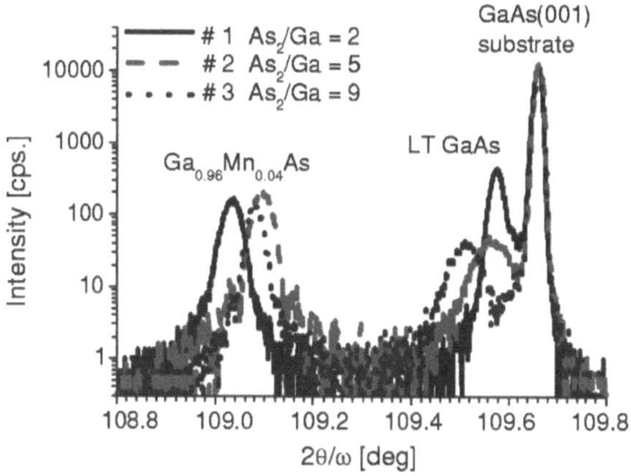

FIGURE 5 006 Bragg reflection measured in $2\theta/\omega$ mode with use of the Philips X'Pert diffractometer: three GaMnAs samples with the same concentration of Mn at Ga sites (4%) are grown at identical substrate temperature (230 °C) with different ratios of As_2/Ga fluxes. In each case GaMnAs layer is grown on thick LT GaAs buffer, whose lattice parameter was used to calculate the concentration of As antisites.

B. Removal of compensating defects by post-growth annealing

As_{Ga} defects in LT GaAs and in GaMnAs are stable up to the temperatures of about 400 °C. In contrast to As_{Ga} the Mn_I defects can be removed from the GaMnAs bulk by applying low temperature post-growth annealing. Due to the annealing the weakly bounded Mn interstitials diffuse to the GaMnAs surface and are passivated either by oxidation [43], or by binding with another reactive element, such as As deposited on GaMnAs surface as thick amorphous layer directly after the MBE growth [44]. The first method requires very long annealing times (up to 100 h) and gives considerable increase of T_C (from 60–100 K for as grown GaMnAs up to 170–190 K after annealing). The post-growth annealing method using amorphous As passivation layer allows to use much shorter annealing times, usually 1–3 h, and for sufficiently thin layers the surface of the annealed GaMnAs is well ordered and suitable for further epitaxial growth. The idea of this annealing method is schematically presented in Figure 6.

The high efficiency of the annealing process is confirmed by magnetization measurements (see Figure 7), which show a remarkable increase of T_C from about 30 to about 130 K after 3 h of annealing.

Since Mn interstitials are double donors partially compensating Mn_{Ga} acceptors, the annealing increases the concentration of holes 2–3 times. The saturation moment also increases significantly, since Mn interstitials tend to couple antiferromagnetically with Mn at Ga sites, making part of Mn_{Ga} atoms magnetically passive [38].

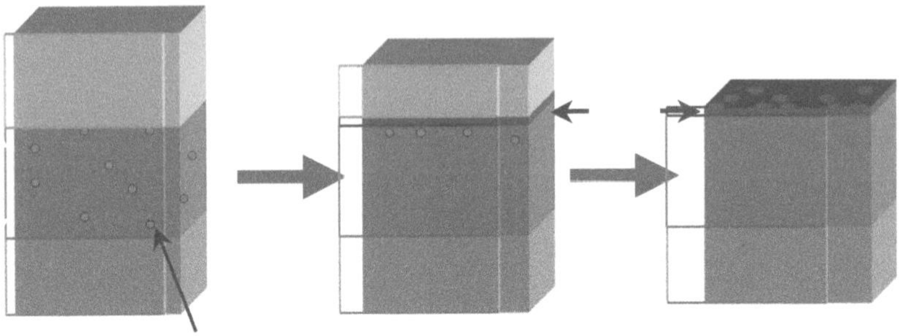

FIGURE 6 Procedure of removing Mn interstitials from GaMnAs by post-growth annealing with passivating amorphous arsenic layer.

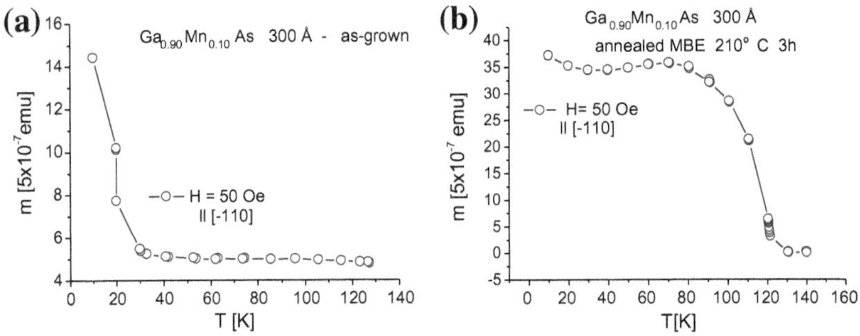

FIGURE 7 Temperature dependence of magnetic moment measured by SQUID for a 300 Å thick $Ga_{0.9}Mn_{0.1}As$ layer (a) before annealing and (b) after annealing with passivating As layer (right panel).

Besides magnetic and transport properties, post-growth annealing also affects the structural parameters of GaMnAs. Alloying with Mn causes a slight increase of the GaAs lattice constant. According to the formula (4) the main contribution to the lattice constant increase is due to Mn interstitials (however experiment shows that the influence of Mn at Ga sites on a_{GaMnAs} is larger than described by Masek model [45]). Thus upon LT post-growth annealing the GaMnAs lattice parameter should be reduced. This is indeed experimentally observed. Figure 8 shows results of X-ray diffraction measurements for a 700 Å thick $Ga_{0.94}Mn_{0.06}As$ layer. The lattice parameter reduction can be seen as shifts of diffraction peaks from GaMnAs layer to the higher diffraction angles. Already after 1 h annealing there is a remarkable lattice constant reduction. The effects saturates after annealing for 3 h. Annealing for longer time (30 h) does not lead to further lattice constant reduction. The broadening of the GaMnAs diffraction peaks is due to the small layer thickness.

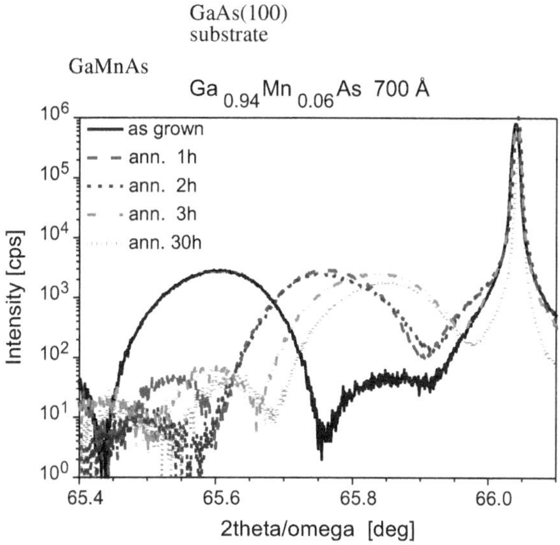

FIGURE 8 004 Bragg reflection from 700 Å thick $Ga_{0.94}Mn_{0.06}As$ layer grown on GaAs(1 0 0) substrate before annealing, and after post growth annealing under As capping at 190 °C for 1, 2, 3 and 30 h.

It was observed [43,44] that post-growth annealing of GaMnAs is only efficient for the layers which are not too thick. The thickness limit for the efficient annealing process is close to 1000 Å. It has been tentatively explained by formation of a surface layer blocking further out-diffusion of remaining Mn interstitial atoms. In the case of annealing in air this effect is not well understood since the structure of GaMnAs surface oxidized layer is not well known. Annealing in high vacuum, with the presence of amorphous As at the GaMnAs surface, leads to formation of thin (1–2 ML) MnAs layer in the zinc-blende phase [46]. For higher amount of Mn interstitials formation of MnAs dots at the surface occurs [47]. Figure 9 shows comparison of surface morphologies of 200 Å thick and 500 Å thick GaMnAs films after annealing with As capping layer. At the end of annealing process the As capping was completely desorbed, revealing the GaMnAs surface morphology.

Physical processes occurring during post growth annealing procedures used for improving magnetic and transport properties of GaMnAs are still not completely understood. It is well known that after reaching some optimum the extended annealing for too long time or at too high temperature degrades the magnetic properties of GaMnAs, i.e., causes decrease of T_C and increase of resistivity. The mechanism responsible for that is still not clarified, probably the second diffusion process, namely diffusion of Mn from Ga sites is also activated even at low temperatures. It is thus desirable to control the GaMnAs properties during post-growth annealing. One method used first by Nottingham [43] group

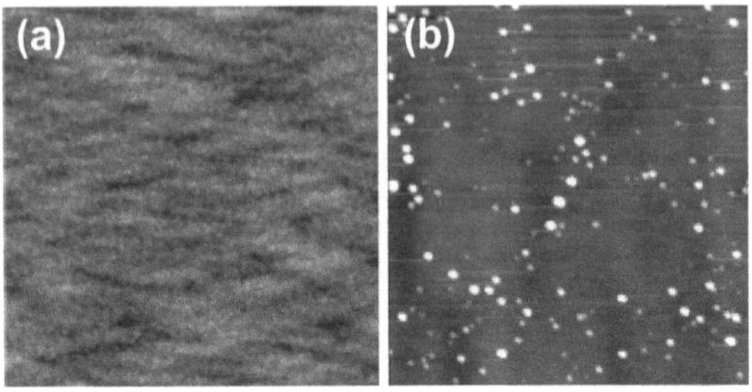

FIGURE 9 AFM images of (a) 200 Å, (b) 500 Å thick $Ga_{0.94}Mn_{0.06}As$ after annealing with As capping. At the end of annealing the As capping layer was completely desorbed. Mn from interstitial sites diffused to the surface reacts with As and makes MnAs dots visible as white spots in panel (b).

is to measure the GaMnAs resistivity during annealing. Since the important fraction of Mn interstitials is out-diffused to the substrate surface where they are passivated, the compensation of Mn_{Ga} acceptors is lowered and concentration of holes increases which is correlated with decreasing resistivity of GaMnAs layer. Another possibility is to measure the GaMnAs lattice parameter during post-growth annealing. Figure 10 shows the evolution of lattice parameter of three $Ga_{0.96}Mn_{0.06}$ As layers with different thicknesses of 200, 700 and 1000 Å. The GaMnAs layers were capped with amorphous As after the MBE growth. The post-growth annealing has been done in air after removing the samples from the MBE system.

As can be seen in Figure 10 the T_C increase in GaMnAs upon LT post growth annealing is correlated with the decrease of the lattice parameter. This confirms the theoretically predicted strong dependence of GaMnAs lattice parameter on concentration of Mn interstitials. Figure 10 also shows that annealing is less effective for thicker layers. The 1000 Å thick layer shows both lower T_C increase and lower decrease of the lattice parameter upon LT post growth annealing. This is probably due to the effect of surface completely saturated with MnAs or Mn oxides which are blocking the further surface out-diffusion of Mn interstitials from the bulk of GaMnAs layer.

The optimum LT post-growth annealing is a subtle process, since GaMnAs being a metastable compound, starts to decompose at temperatures only slightly higher than the growth temperatures. Extended annealings at temperatures above 300 °C lead to diffusion of Mn from Ga sites and formation of Mn-rich (MnAs) clusters in GaAs matrix of the former GaMnAs layer. This can be observed in the behavior of GaMnAs lattice parameter during the post-growth annealing process in the broader temperature range. Figure 11 shows the results of measurements of the lattice parameter of amorphous As capped 1500 Å thick $Ga_{0.93}Mn_{0.07}As$ layer during annealing in the temperatures between 50 and 550 °C. As a reference the lattice parameter of GaAs(100) substrate (a_{GaAs})

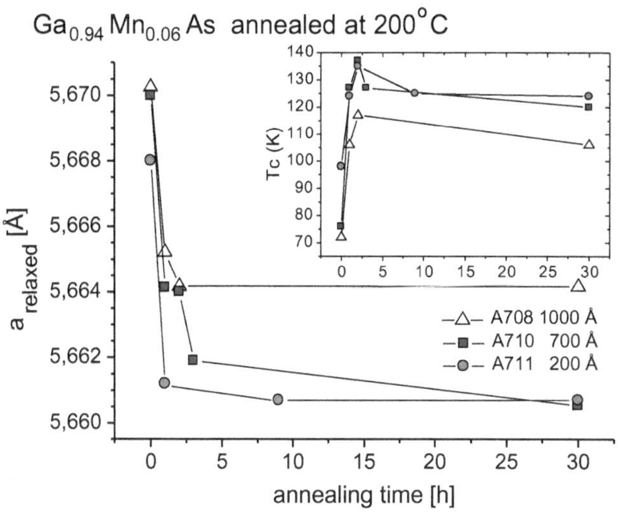

FIGURE 10 Evolution of lattice parameter of amorphous As capped $Ga_{0.94}Mn_{0.06}As$ layers differing in thickness, upon post-growth annealing at 200 °C. The inset shows changes in T_C measured by SQUID magnetometer.

was measured simultaneously, and shows typical monotonous increase with increasing annealing temperature. The lattice parameter of GaMnAs (a_{GaMnAs}) behaves differently. At low temperatures up to about 170 °C a_{GaMnAs} increases similarly to a_{GaAs} showing typical lattice parameter expansion with increasing temperature, there is no signature of diffusion processes in GaMnAs up to this temperature. In the temperature region of 170–230 °C a_{GaMnAs} increases with lower rate than a_{GaAs}. In fact subtracting the temperature induced lattice parameter increase leads to the conclusion that in this annealing temperature range the process inducing decrease of a_{GaMnAs} is activated. This can be the surface out-diffusion of Mn interstitials. The presence of amorphous As capping enhances the efficiency of this process (see also Figure 10). Finally at temperatures higher than 230 °C a_{GaMnAs} decreases with accelerating rate, which stabilizes above 280 °C. This can be due to the enhanced diffusion of Mn from Ga sites. At annealing temperatures higher than about 470 °C the lattice parameter of the layer increases again. In GaMnAs annealed at this and higher temperatures, the segregated MnAs nanocrystals have been detected by transmission electron microscopy [48–53]. At increasing annealing temperatures the size of the nanocrystals increase from 2–5 to 10–50 nm, and the local strain exerted by them on surrounding GaAs matrix is partially released by dislocations.

The results of GaMnAs annealing correlated with the measurements of the lattice parameter confirm that the optimum LT post-growth annealing should be performed in the narrow temperature range; annealing temperatures higher than

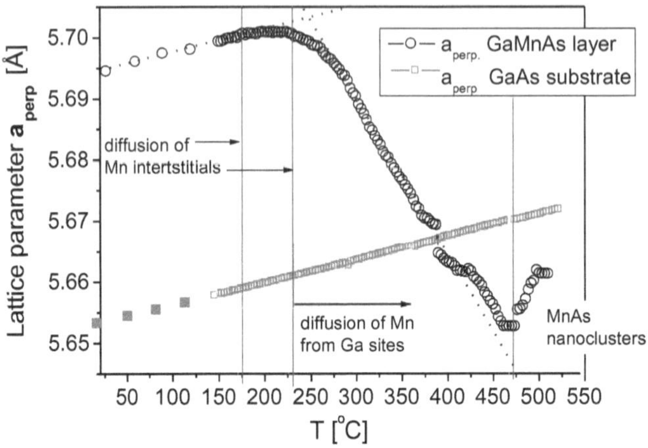

FIGURE 11 Changes of the lattice parameters of 1500 Å thick $Ga_{0.93}Mn_{0.07}As$ layer (circles) and GaAs(1 0 0) substrate (squares) during annealing at increasing temperatures, measured in situ with Philips X'Pert diffractometer.

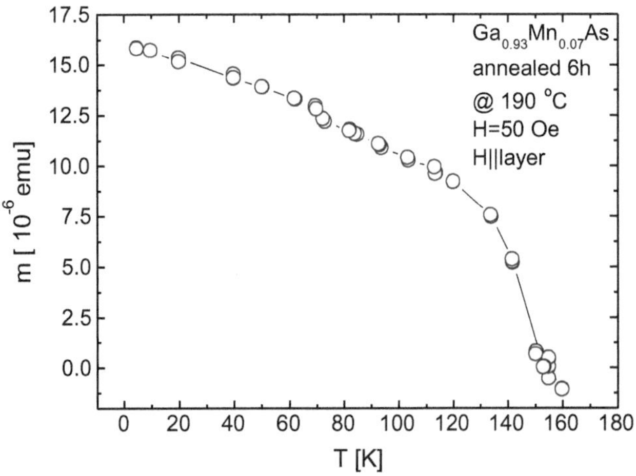

FIGURE 12 Temperature dependence of magnetic moment of 500 Å thick $Ga_{0.93}Mn_{0.07}As$ layer annealed on X-ray diffractometer at 190 °C. Ferromagnetic phase transition temperature after annealing is close to 160 K.

optimum cause diffusion of Mn from Ga sites which leads to the degradation of magnetic and transport properties of GaMnAs.

Figure 12 shows temperature dependence of magnetic moment of 500 Å thick $Ga_{0.93}Mn_{0.07}As$ layer annealed on X-ray diffractometer at a constant temperature of 190 °C with in situ control of the lattice constant variation. The annealing was stopped when the lattice constant variations saturated, which was observed after 6 h.

The optimized MBE growth and post-growth annealing procedures yield nowadays GaMnAs layers of T_C as high as about 200 K [24]. This is remarkably high as for diluted ferromagnetic semiconductor, but still too low in view of potential application in spintronic devices. However further progress in increasing T_C of GaMnAs cannot be excluded.

Even though the LT post-growth annealing is the well established method for T_C increase in GaMnAs, and is understood as Mn interstitals removal due to the out-diffusion towards the surface; there are also other interpretations of this effect. In some papers the authors postulate that Mn_I are redistributed within the bulk of GaMnAs and occupy random sites at which they do not have donor character [54]. However there is strong experimental evidence pointing on the diffusion of Mn interstitials and their passivation at the surface of GaMnAs layer [55–57].

III. MAGNETIC AND TRANSPORT PROPERTIES OF GAMNAS

A. Magnetic anisotropy of GaMnAs

GaMnAs layers are typically grown on GaAs(1 0 0) surface due to the relatively easy MBE growth and sufficiently good lattice matching between epilayer and the substrate. Slight increase of the GaMnAs lattice constant, with respect to GaAs causes the occurrence of compressive strain in the layers. The strain anisotropy between in-plane and perpendicular directions is reflected in the magnetic anisotropy of GaMnAs. For typical GaMnAs samples with compressive strain and sufficiently low degree of compensation of Mn_{Ga} acceptors the easy magnetic axis is in-plane. However the equivalent crystal directions at the (1 0 0) plane of zinc-blende structure are not magnetically equal in GaMnAs. There exists a distinct magnetic anisotropy between [0 1 1] and $[0\,\bar{1}\,1]$ directions, which so far has not been sufficiently explained. Figure 13 shows magnetic moment of 300 Å thick $Ga_{0.94}Mn_{0.06}As$ layer grown on GaAs(1 0 0) substrate measured with external magnetic field parallel to two orthogonal in-plane directions [0 1 1] and $[0\,\bar{1}\,1]$. The crystallographically equivalent directions are not magnetically equivalent – easy magnetization axis is parallel to $[0\,\bar{1}\,1]$.

As illustrated in Figure 13 GaMnAs(1 0 0) layer does not exhibit cubic magnetic anisotropy expected for zinc-blende structure. The uniaxial magnetic anisotropy component is always present in GaMnAs(1 0 0) layers. There are some suggestions that this is due to the preferential orientations of low density structural defects (stacking faults) occurring during GaMnAs MBE growth [57]; another possible explanation point out the possible nonuniform distribution of Mn in the bulk, due to the slight anisotropy of GaMnAs surface during the

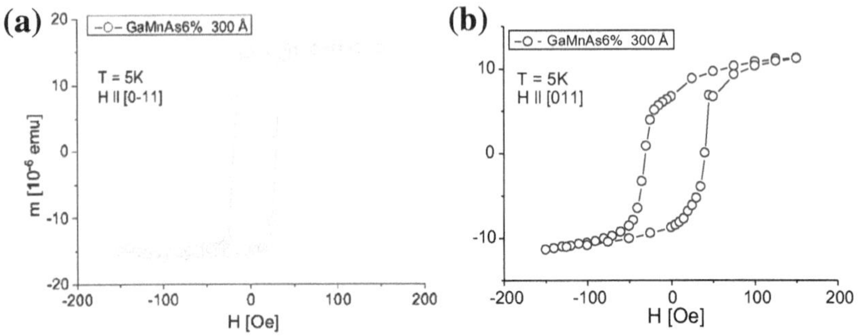

FIGURE 13 Hysteresis loops for 300 Å thick $Ga_{0.94}Mn_{0.06}As/GaAs(1\,0\,0)$ measured in two orthogonal in-plane directions. (a) $[0\,\bar{1}\,1]$ and (b) [0 1 1]. The easy magnetization axis is in the layer plane, parallel to the $[0\,\bar{1}\,1]$ azimuth of the GaMnAs(1 0 0) layer. Perpendicular in-plane direction which is crystallographically equivalent in the bulk is not an easy magnetization axis.

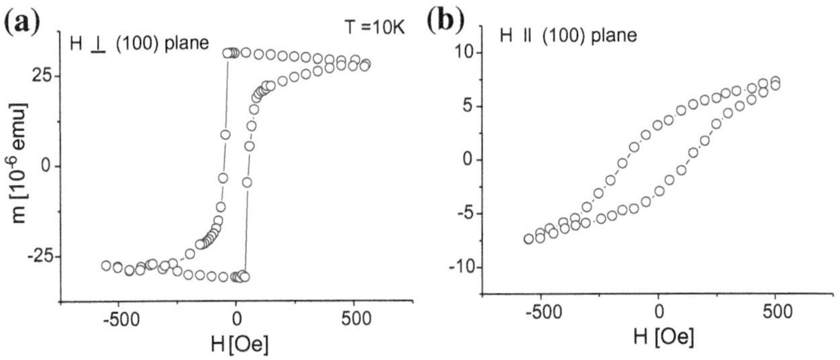

FIGURE 14 Hysteresis loops for 1000 Å thick $Ga_{0.94}Mn_{0.06}As$ grown on 1 μm thick, relaxed $In_{0.20}Ga_{0.80}As$ buffer deposited on GaAs(1 0 0). With external magnetic field oriented (a) perpendicular to the sample plane, and (b) parallel to the sample plane.

MBE growth caused by some preferential orientation of Mn reflected by (1 × 2) surface reconstruction [58], see also Figure 2.

If the strain in GaMnAs(1 0 0) layers is changed from compressive to tensile, then the easy magnetic axis of the typical layers with relatively high Mn content and high concentration of holes is perpendicular to the sample plane. This is shown in Figure 14 displaying hysteresis loops with magnetic field oriented: (a) perpendicular to the sample plane, and (b) in plane, parallel to the $[0\,\bar{1}\,1]$ direction. The measured sample is 1000 Å thick $Ga_{0.94}Mn_{0.06}As$ grown on 1 μm thick, relaxed $In_{0.20}Ga_{0.80}As$ buffer layer deposited on GaAs(1 0 0) substrate. The lattice parameter of $In_{0.20}Ga_{0.80}As$ is higher than that of $Ga_{0.94}Mn_{0.06}As$, thus the GaMnAs layer is in tensile strain state.

The nonstandard shape of a hysteresis loop measured at easy magnetic axis direction is due to the peculiar magnetic domain structure of GaMnAs grown on InGaAs buffer layers. The surface of relaxed InGaAs layers grown on GaAs(1 0 0) has cross-hatched undulation patterns reflecting the distribution of misfit dislocations in relaxed InGaAs buffer. These patterns are pinning the magnetic domain walls and influence the reorientation of magnetic domains upon external magnetic field reversal [59].

The strain state of GaMnAs layer with magnetic properties shown in Figure 14 has been verified by X-ray measurements. Figure 15 shows reciprocal space maps of asymmetrical ($\bar{2}\,\bar{2}\,4$) Bragg reflection, clearly displaying reciprocal space nodes of GaAs substrate, relaxed InGaAs buffer and GaMnAs layer completely strained to InGaAs.

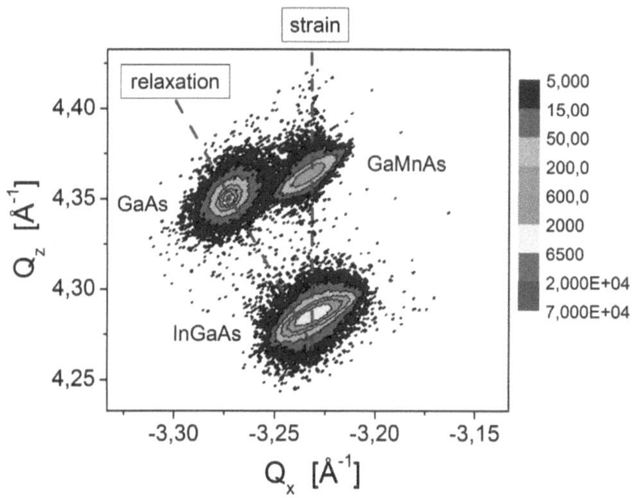

FIGURE 15 Reciprocal space map around $\bar{2}\bar{2}4$ Bragg refection of 1000 Å thick $Ga_{0.94}Mn_{0.06}As$ layer grown on 1 μm thick, relaxed $In_{0.20}Ga_{0.80}As$ buffer deposited on GaAs(1 0 0). GaMnAs layer is under tensile strain, fully strained to the InGaAs buffer.

In general magnetic anisotropy of GaMnAs(1 0 0) layers either under compressive or under tensile strain is more complex, and depends both on Mn content and concentration of holes. Figures 13 and 14 illustrate typical case of relatively high Mn content (above 5%) and high concentration of valence band holes. The variations of magnetic anisotropy induced by altering concentration of holes are quite well described by the Zener model in mean field approximation developed by Dietl et al.[60]. In particular when the concentration of holes is low, the orientation of easy magnetic axis is perpendicular to (1 0 0) plane for GaMnAs layers under compressive strain and in-plane for GaMnAs under tensile strain. This has been experimentally observed and opens interesting possibility of switching magnetization direction in GaMnAs between in-plane and perpendicular direction by changing concentration of holes for example with use of electrostatic gates. Such an affect has been already applied for in-plane rotation of easy magnetic axis [61].

B. Magnetotransport properties of GaMnAs

The transport of charge carriers in GaMnAs in the presence of external magnetic field can be phenomenologically described by anomalous Hall effect, according to the formula:

$$R_{\mathrm{Hall}} = R_0/d \cdot B + R_S/d \cdot M, \tag{5}$$

where R_0 is the normal Hall coefficient, *d* the sample thickness, *B* the magnetic field induction, R_S the anomalous Hall coefficient, and *M* the magnetization

According to the formula (5) at low magnetic fields the main contribution to Hall effect in GaMnAs is due to magnetization; at high magnetic fields, when magnetization is saturated the first, i.e., normal Hall effect is prevailing.

The coefficient R_S depends on the mechanism of diffusion of charge carriers and is a function of the longitudinal resistance (ρ_{xx}). The exact dependence of R_S on ρ_{xx} in GaMnAs is still being discussed. Until recently it was postulated, that depending on the concentration of holes in GaMnAs (associated with the Mn content and degree of compensation of Mn acceptors), R_S is proportional to $(\rho_{xx})^n$, where $n = 1$ or 2. The value of *n* exponent should be associated with the scattering mechanism of spin polarized charge carriers, namely "*side-jump scattering*", or "*skew-scattering*". Currently another model is considered, in which it is proposed that the dominant contribution to the anomalous Hall coefficient is due to the Berry phase of spin polarized holes and does not depend on their scattering mechanisms [62]. According to this model R_S is also proportional to $(\rho_{xx})^n$, where $n = 2$. Nowadays, it is believed that this model describes best the dependence of anomalous Hall coefficient on ρ_{xx} for GaMnAs with metallic transport properties (i.e., material with high hole concentration and low compensation of Mn acceptors), If these conditions are not met, which is the case for GaMnAs with low T_C values, the other effects are prevailing, such as scattering of charge carriers on defects; or change of the carrier transport mechanisms from diffusive to hopping conduction for samples with high degree of compensation. In these conditions the contribution from Berry phase to R_S is much weaker. So far there is no accurate theory describing transport in GaMnAs in this case (non-metallic samples). Usually the experimentally determined dependence of R_S on ρ_{xx} is used:

$$R_S \sim (\rho_{xx})^{\gamma}, \tag{6}$$

where $0.4 < \gamma < 0.6$ [63,64].

The value of γ equal to 0.4 is consistent with the anomalous Hall effect description proposed for dirty metals with diluted impurities [65].

Moreover, because the longitudinal resistance in GaMnAs depends on both temperature and magnetic induction B, the correct estimation of concentration of holes in GaMnAs from Hall effect measurements is not trivial. In small magnetic fields the second term of the formula (5) has dominant influence on the Hall resistance, so R_{Hall} is proportional to magnetization (steep increase at B slightly higher than 0). For the magnetic field, at which the magnetization of the material reaches saturation (about 0.1–0.2 T for GaMnAs) the normal Hall effect has dominant contribution to R_{Hall}, but the anomalous Hall term still cannot be neglected due to the magnetoresistance. In small fields, where anomalous Hall effect dominates, the longitudinal resistance (R_{xx}) slightly increases with increasing B (positive magnetoresistance), and then decreases, when B exceeds the value saturating the magnetization (negative magnetoresistance).

Figure 16 shows results of the Hall effect measurements of 300 Å thick $Ga_{0.94}Mn_{0.06}As$ with in-plane easy magnetic axis and T_C of about 140 K. The sample was post-growth annealed in the MBE system, with the presence of amorphous As capping for 3 h at 200 °C.

Below T_C Hall resistance shows typical behavior described above with dominating anomalous Hall effect at low fields. Measurements of longitudinal resistance show strongly temperature dependent magnetoresistance, reaching maximum at temperatures close to T_C. The resistance maximum close to T_C is a typical feature of GaMnAs samples with metallic conduction. It is believed to be due to critical scattering of charge carriers on inhomogenous sample containing ferromagnetic and paramagnetic regions. The sample used for Hall effect measurements shown in Figure 16 has in-plane easy magnetic axis. With Hall effect measurement geometry with magnetic field applied perpendicularly to the plane of GaMnAs layer, which is the hard magnetic axis, there is no hysteresis of R_{xy}. However the distinct effects due to the reorientation of magnetic domains and anisotropic magnetoresistance can be observed in planar Hall effect (PHE) geometry, i.e., with magnetic field oriented parallel to the plane of GaMnAs. The typical signature of planar Hall effect in GaMnAs is shown in Figure 17.

The appearance of Hall voltage with magnetic field applied parallel to the sample plane is due to anisotropic magnetoresistance in GaMnAs. The Hall voltage jumps are associated with the reorientation of magnetic domains upon reversal of in-plane magnetic field directions. The amplitude and detailed shapes of Hall voltage jumps are depending on orientation of easy magnetic axis, and external magnetic field directions with respect to the current flow direction (the Hall bar axis).

The detailed analysis of PHE measured at different angles of in-plane magnetic field with respect to the Hall bar axis enables to acquire accurate information on the magnetic anisotropy and its possible dependence on temperature and other factors.

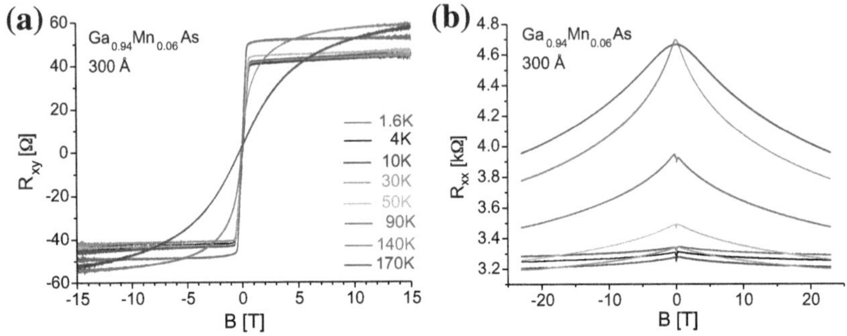

FIGURE 16 Hall resistance (left panel) and longitudinal resistance (right panel) of 300 Å thick $Ga_{0.94}Mn_{0.06}As$ layer measured at temperatures from 1.6 to 170 K and magnetic fields of up to 22 T. The ferromagnetic phase transition temperature of the sample is about 140 K. The easy magnetic axis is in-plane and parallel to the Hall bar axis.

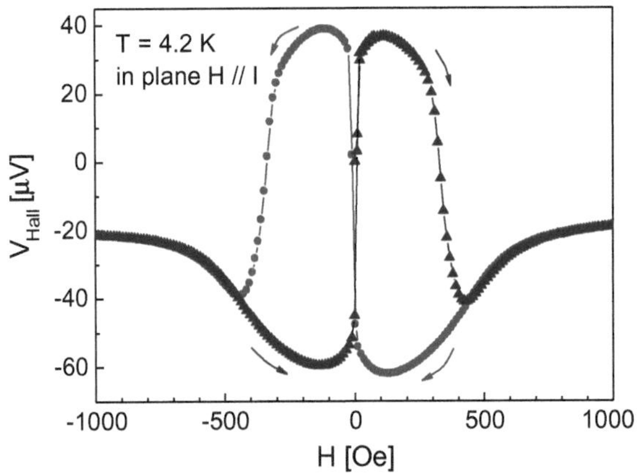

FIGURE 17 Planar Hall effect for 200 Å thick $Ga_{0.93}Mn_{0.07}As$ layer grown on GaAs(1 0 0). The arrows show the direction of magnetic field sweeps.

IV. SUMMARY

We have discussed the basic issues concerning molecular beam epitaxy growth of GaMnAs ferromagnetic semiconductor with moderate and high Mn content, in the range of 4–10%. The typical magnetic and transport properties of resulting GaMnAs samples with minimized concentrations of structural point-defects have been shown. The main defects, i.e., arsenic antisites and Mn interstitials can be controlled by suitable MBE growth conditions and appropriate post-growth annealing procedures. The optimized MBE growth and post-growth annealing procedures enable to reach remarkably high ferromagnetic phase transition temperatures in the range of 150–200 K in GaMnAs with high Mn content (above 7 at.%). Basic magnetic and transport properties of GaMnAs obtained in such optimized conditions have been investigated by SQUID magnetometry and Hall effect measurements.

Acknowledgements

The MBE system used for the growth of GaMnAs samples has been supported by the grants from Swedish Committee for Scientific Research (VR). The authors thank R. Mathieu from Uppsala University and P. Dziawa from IoP PASC Warszawa, for SQUID measurements. Hall effect measurements at high magnetic fields have been performed at Grenoble High Magnetic Fields laboratory with help of D. Maude and W. Desrat. Planar Hall effect measurements have been done at IoP PASC with help of Prof. T. Wosinski and his research group.

REFERENCES

[1] Passell L, Dietrich OW, Als-Nielsen J. Phys Rev B 1976;14:4897.
[2] Story T, Galazka RR, Frankel RB, Wolff PA. Phys Rev Lett 1986;56:777.
[3] Haury A, Wasiela A, Arnoult A, Cibert J, Tatarenko S, Dietl T, et al. Phys Rev Lett 1997;79:511.
[4] Ferrand D, Cibert J, Wasiela A, Bourgognon C, Tatarenko S, Fishman G, et al. Phys Rev B 2001;63:085201.
[5] Ohno H, Shen A, Matsukura F, Oiwa A, Endo A, Katsumoto S, et al. Appl Phys Lett 1996;69:363.
[6] Munekata H, Ohno H, von Molnar S, Segmüller Armin, Chang LL, Esaki L. Phys Rev Lett 1989;63:1849.
[7] Ohno H, Munekata H, Penney T, von Molnár S, Chang LL. Phys Rev Lett 1992;68:2664.
[8] Zener C. Phys Rev 1951;81:440.
[9] Dietl T, Ohno H, Matsukura F, Cibert J, Ferrand D. Science 2000;287:1019.
[10] Ando K, Saito H, Jin Zhengwu, Fukumura T, Kawasaki M, Matsumoto Y, et al. J Appl Phys 2001;89:7284.
[11] Saito H, Zayets V, Yamagata S, Ando K. Phys Rev Lett 2003;90:207202.
[12] Pearton SJ et al. J Vac Sci Technol A 2002;20:721.
[13] Hoa Hong Nguyen, Sakai Joe, Hassini Awatef. J Appl Phys 2005;97:10D312.
[14] Huang Rong-Tan, Hsu Chen-Feng, Kai Ji-Jung, Chen Fu-Rong, Chin Tsung-Shune. Appl Phys Lett 2005;87:202507.
[15] Sasaki Takahiko, Sonoda Saki, Yamamoto Yoshiyuki, Suga Ken-ichi, Shimizu Saburo, Kindo Kouichi, et al. J Appl Phys 2002;91:7911.
[16] Kuroda S, Nishizawa N, Takita K, Mitome M, Bando Y, Osuch K, et al. Nature Mater 2007;6:440.
[17] Zajac M, Gosk J, Grzanka E, Kaminska M, Twardowski A, Strojek B, et al. J Appl Phys 2003;93:4715.
[18] Dietl T. Nature Mater 2010;9:965.

[19] Fukuma Y, Asada H, Miyawaki S, Koyanagi T, Senba S, Goto K, et al. Appl Phys Lett 2008;93:252502.
[20] Ohno H. Nature Mater 2010;9:952.
[21] Nishitani Y, Endo M, Matsukura F, Ohno H. Physica E 2010;42:2681.
[22] Wojtowicz T et al. Appl Phys Lett 2003;82:4310.
[23] Scarpulla MA, Cardozo BL, Farshchi R, Hlaing Oo WM, McCluskey MD, Yu KM, et al. Phys Rev Lett 2005;95:207204.
[24] Chen L, Yang X, Yang F, Zhao J, Misuraca J, Xiong P, et al. Nano Lett. 2011;11:2584.
[25] Arthur Jr. JR. J Appl Phys 1968;39:4032.
[26] Cho AY. J Appl Phys 1970;41:782.
[27] Ludeke R, Chang LL, Esaki L. Appl Phys Lett 1973;23:201.
[28] Ilegems M, Dingle R, Rupp Jr LW. J Appl Phys 1975;46:3059.
[29] Furdyna JK. J Appl Phys 1988;64:R29.
[30] Ohno H. Science 1998;281:951.
[31] Liu X, Prasad A, Nishio J, Weber ER, Liliental-Weber Z, Walukiewicz W. Appl Phys Lett 1995;67:279.
[32] Sadowski J, Domagala JZ, Ba̧k-Misiuk J, Kolesnik S, Sawicki M, Swiatek K, et al. J Vac Sci Technol B 2000;18:1697.
[33] Suda A, Otsuka N. Surf Sci 2000;458:162.
[34] Shen A, Horikoshi Y, Ohno H, Guo SP. Appl Phys Lett 1997;71:1540.
[35] Tuomisto F, Pennanen K, Saarinen K, Sadowski J. Phys Rev Lett 2004;93:055505.
[36] Masek J, Maca F. Acta Phys Polonica A 2001;100:319.
[37] Yu KM, Walukiewicz W, Wojtowicz T, Kuryliszyn I, Liu X, Sasaki Y, et al. Phys Rev B 2002;65:201303.
[38] Blinowski J, Kacman P. Phys Rev B 2003;67:121204.
[39] Lawniczak-Jablonska K, Libera J, Wolska A, Klepka MT, Jakiela R, Sadowski J. Radiat Phys Chem 2009;78:S80.
[40] Chiba D, Nishitani Y, Matsukura F, Ohno H. Appl Phys Lett 2007;90:122503.
[41] Staab TEM, Nieminen RM, Gebauer J, Krause-Rehberg R, Luysberg M, Haugk M, et al. Phys Rev Lett 2001;87:045504.
[42] Masek J, Kudrnovsky J, Maca F. Phys Rev B 2003;67:153203.
[43] Edmonds KW, Boguslawski P, Wang KY, Campion RP, Novikov N, Farley NRS, et al. Phys Rev Lett 2004;92:037201.
[44] Adell M, Ilver L, Kanski J, Stanciu V, Svedlindh P, Sadowski J, et al. Appl Phys Lett 2005;86:112501.
[45] Zhao LX, Staddon CR, Wang KY, Edmonds KW, Campion RP, Gallagher BL, et al. Appl Phys Lett 2005;86:071902.
[46] Sadowski J. Acta Phys Polonica A 2008;114:1001.
[47] Sadowski J, Adell M, Kanski J, Ilver L, Janik E, Lusakowska E, et al. Appl Phys Lett 2005;87:263114.
[48] Moreno M, Jenichen B, Däweritz L, Ploog KH. J Vac Sci Technol B 2005;23:1700.
[49] Moreno M, Kaganer VM, Jenichen B, Trampert A, Däweritz L, Ploog KH. Phys Rev B 2005;72:115206.
[50] Kwiatkowski A, Wasik D, Kaminska M, Bozek R, Szczytko J, Twardowski A, et al. J Appl Phys 2007;101:113912.
[51] Rench DW, Schiffer P, Samarth N. Phys Rev B 2011;84:094434.
[52] Lawniczak-Jablonska K, Libera J, Wolska A, Klepka MT, Dluzewski P, Sadowski J, et al. Phys Status Solidi RRL 2011;5:62.
[53] Sadowski Janusz, Domagala Jaroslaw Z, Mathieu Roland, Kovács Andras, Kasama Takeshi, Dunin-Borkowski Rafal E, et al. Phys Rev B 2011;84:245306.
[54] Dobrowolska M, Tivakornsasithorn K, Liu X, Furdyna JK, Berciu M, Yu KM, et al. Nature Mater, 2012; 11:444.
[55] Adell M, Adell J, Kanski J, Sadowski J, Domagała JZ. Phys Rev B 2007;75:054415.
[56] Schmid B, Müller A, Sing M, Claessen R, Wenisch J, Gould C, et al. Phys Rev B 2008;78:075319.
[57] Kopecky M, Kub J, Maca F, Masek J, Pacherova O, Rushforth AW, et al. Phys Rev B 2011;83:235324.
[58] Birowska M, Sliwa C, Majewski JA, Dietl T. Phys Rev Lett 2012;108:237203.

[59] Thevenard L, Largeau L, Mauguin O, Patriarche G, Lemaître A, Vernier N, et al. Phys Rev B 2006;73:195331.
[60] Dietl T, Ohno H, Matsukura F. Phys Rev B 2001;63:195205.
[61] Chiba D, Sawicki M, Nishitani Y, Nakatani Y, Matsukura F, Ohno H. Nature 2008;455:515.
[62] Jungwirth T, Niu Qian, MacDonald AH. Phys Rev Lett 2002;88:207208.
[63] Shen S, Liu X, Ge Z, Furdyna JK, Dobrowolska M, Jaroszynski J. J Appl Phys 2008;103:07D134.
[64] Glunk M, Daeubler J, Schoch W, Sauer R, Limmer W. Phys Rev B 2009;80:125204.
[65] Onoda S, Sugimoto N, Nagaosa N. Phys Rev Lett 2006;97:126602.

Chapter 3

Spectroscopic Investigations of Electronic Structure

Ronny Knut and Olof Karis
Department of Physics and Astronomy, Uppsala University

Chapter Outline Head

X-ray-based spectroscopies are a collection of spectroscopical methods where the sample is subjected to X-rays and the response is measured as a function of energy, either of the incoming X-rays and/or from the electrons and photons emitted from the sample. We will discuss here several complementary techniques which together offers electronic structure information of both occupied and unoccupied electronic states with elemental specificity. A definitive advantage of X-ray-based spectroscopies is their sensitivity to low concentrations for instance impurities or dopants, which will become apparent in the following chapter where the methods are discussed in relation to transition metal doped oxides. The complexity of these materials offers an efficient way of describing the type of characterization obtained using spectroscopy.

Advanced Functional Materials. http://dx.doi.org/10.1016/B978-0-44-453681-5.00003-0

I. X-RAY ABSORPTION SPECTROSCOPY AND X-RAY MAGNETIC CIRCULAR DICHROISM

X-ray absorption spectroscopy (XAS) is a very powerful technique that is used in many different areas where understanding of electronic structure is relevant. For a more complete appreciation of the capabilities the technique offers, the reader is referred to the book by Stöhr [1]. X-ray spectroscopies in principle measure the transition probability going from an initial state (Ψ_i) to a final state (Ψ_f). For X-ray absorption we write

$$I_{abs} \propto |\langle \Psi_i | \epsilon \cdot \mathbf{r} | \Psi_f \rangle|^2 \delta_{E_f - E_i - \hbar\omega}, \quad (1)$$

where I_{abs} is the probability for absorption of incident excitation with energy $\hbar\omega$ and $\epsilon \cdot \mathbf{r}$ is the dipole operator representing this perturbation from incoming X-rays [1]. The delta function $\delta_{E_f - E_i - \hbar\omega}$ ensures that the energy is conserved in the process. X-ray absorption denotes a process where a core electron is transferred into an initially unoccupied state. Due to selection rules imposed by the dipole operator, the orbital quantum number is restricted to change by $\Delta l = \pm 1$. This implies, for instance, that core-level electrons occupying an s-orbit can only be transferred to a p-orbital. For transition metals the outer most d-orbitals are partly unoccupied and hence electrons can be transferred from a p- to d-orbital. The localized nature, sensitivity to crystal fields, and significance for magnetic properties of d-electrons make the p- to d-transition a highly studied absorption edge for transition metals. For 3d transition metals the 2p $\rightarrow$ 3d transition occurs between 400 and 1100 eV and is usually denoted as the L-edge. The dipole approximation is good when the X-ray wavelength is much larger than the diameter of the electron orbit, for photon energies above a few keV the quadrupole transition needs to be considered for accurate description.

If the energy of the X-rays is close to the onset of an absorption edge, the transferred electron obtains a state that is still bound to the atom which is often referred to as near edge X-ray absorption fine structure (NEXAFS) spectroscopy. Photon energies far above this onset results in core electrons going into the continuum, which is denoted as extended X-ray absorption fine structure (EXAFS) spectroscopy. These two cases yields very different information and will be discussed separately. We will begin by studying NEXAFS, which we from now on will denote as XAS, while EXAFS will be discussed in the end of this chapter in Section IV.

The absorption probability is proportional to the number of available states and XAS is thus a method for probing the unoccupied density of states in the presence of a core–hole, see Figure 1. The intensity of the light is measured both before and after the light has been transmitted through the sample. The intensity of the transmitted light I is related to the absorption cross-section

$$I = I_0 e^{-\mu x} \quad (2)$$

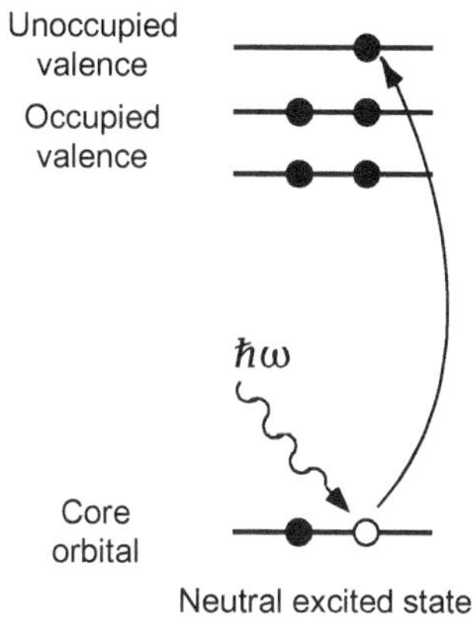

FIGURE 1 X-ray absorption spectroscopy. An atom absorbs a photon and a core electron is promoted to an unoccupied valence level, resulting in a neutral excited state.

where I_0 is the intensity of the impinging light, x is the sample thickness and μ is the absorption cross-section which depends on the X-ray energy. Since the intensity of the transmitted light decreases exponentially with the sample thickness and attenuation lengths of the order of tens of nm, this type of experiment can only be conducted for very thin samples. Instead it is common to use a more indirect way of measuring the absorption cross-section. After the absorption of light, the excited atom relaxes by filling the core hole with an electron from a higher occupied energy level. In this process the excess energy will be released by either photon or electron emission. For core levels accessible by soft X-rays the electron emission is much more probable to occur. Therefore a simple and common method for obtaining the absorption cross-section is to measure the number of secondary electrons leaving the sample. This is done either by measuring the current between ground and sample or by registering the emitted electrons using a multi-channel plate (MCP).

There is a high interest in integrating the well-established semiconductor technology with spin-based technology, which to date only has successful applications using metallic and insulating materials. It was shown that semiconductors doped with transition metal (TM) atoms can obtain a ferromagnetic state and consequently became interesting for semiconductor-based spintronic applications. One of the most studied materials is TM doped ZnO which was suggested, theoretically, to have promising magnetic properties above room temperature. However, even after a decade of studies there is still debate whether it is suitable as a magnetic semiconductor candidate. The reason for this uncertainty can mainly be attributed to very diverse experimental results, especially concerning magnetic properties. The following discussion concerning XAS will be presented as a case study of TM doped ZnO.

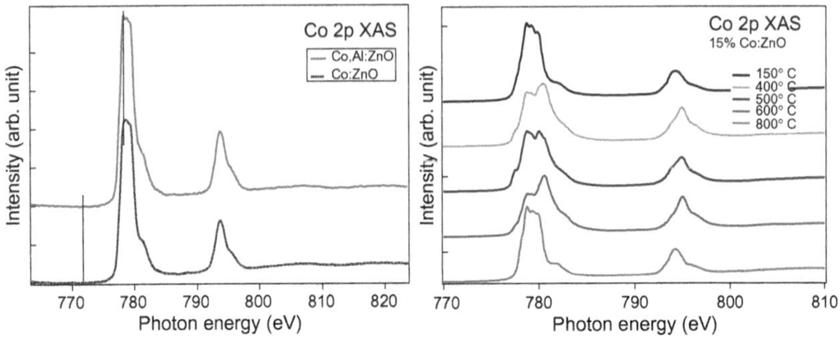

FIGURE 2 X-ray absorption spectroscopy of (left, from Ref. [2]) 5% Co-doped ZnO with the inclusion of 0.8% Al doping for one of the samples. The vertical lines indicate the energies used for on and off resonant photoelectron spectroscopy which will be discussed in the X-ray photoelectron spectroscopy chapter. (right, Knut et al., unpublished) 15% Co-doped ZnO annealed at various temperatures. At intermediate annealing temperatures a Co spinel is formed which changes the structure of the absorption.

In Figure 2(left) we show the Co $L_{2,3}$ absorption spectra for two 5% Co-doped ZnO samples, with and without 0.8% Al co-doping. Usually in XAS spectra, multiplet structures are observed in atoms with localized valence band electrons and appear due to the strong overlap of wave functions between the core hole and localized 3d electrons. This makes it possible to distinguish between metallic and ionic states for 3d transition metals since metallic states does not show multiplet features due to their delocalized nature. The multiplet structure of the peaks in Figure 2(left) is a clear indication that the Co is not in a metallic state. The spectra are similar to other results reported in the literature, which have been shown to correspond to Co^{2+} states in tetrahedral coordination, indicating Zn substitution [3–5]. These samples show a magnetic behavior which suggests that the Co ions order anti-ferromagnetically (AFM) in small clusters (~10 atoms). When the doping concentration is increased to 15% Co doping the structure of the absorption edges changes, as shown in Figure 2(right) Knut et al. [14]. The samples have been annealed at various temperatures where Knut et al. the samples annealed at 150 and 800 °C show an absorption structure very similar to what was observed for the 5% Co-doped samples. Magnetization data supports the notion that these samples consists of AFM coupled Co^{2+} clusters (~60 atoms). The other samples show additional structure in the absorption which is especially clear for the 600 °C sample. These are due to a formation of a spinel phase in the sample as the sample is annealed between 400 and 600 °C. Due to the small dimensions of the particles comprising these secondary phases they can be difficult to detect by XRD. The magnetic moment per Co atom is found to decrease for these samples which is consistent with a low spin state found in the spinel structure. The same

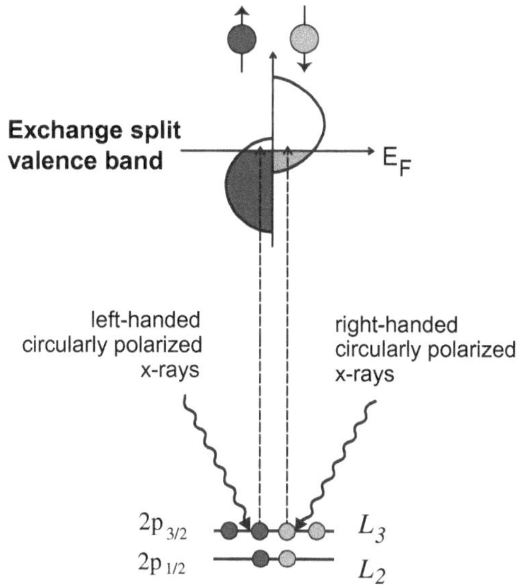

FIGURE 3 Right-handed circularly polarized light excites more spin up electrons than left-handed at the L_3 edge. This enables us to measure the difference in occupied spin up and spin down 3d states with XMCD.

set of samples are also discussed later in connection to X-ray photoelectron spectroscopy in Section II.

Even though the above mentioned samples do not exhibit a ferromagnetic behavior, it has frequently been found in other studies that a ferromagnetic state with varying transition temperatures can be obtained. The most common methods to study magnetization are by Superconducting Quantum Interference Device (SQUID) and vibrating sample magnetometry (VSM). Unfortunately these methods offer no information about the origin of the magnetic response. X-ray magnetic circular dichroism (XMCD), which is more like a branch of XAS rather than a separate technique, does not only give elemental but in many cases also chemical specificity to the magnetic response.

X-ray absorption has different cross sections depending on the polarization of the light. In the case of circularly polarized light, see Figure 3, the cross-section is sensitive to both the magnitude and direction of magnetization in the sample. The difference in absorption between right- and left-handed circular polarized light is called XMCD. Circularly polarized light carries an angular momentum which is either in the direction of motion or opposite to it, σ^+ $(+\hbar)$ and σ^- $(-\hbar)$ respectively. Usually the angular momentum from a photon can only be transferred to the electron orbital momentum and thus conserving the spin of the electron. For electrons which experience a strong spin–orbit coupling, as for 2p electrons in transition metals, the orbital and spin quantum numbers

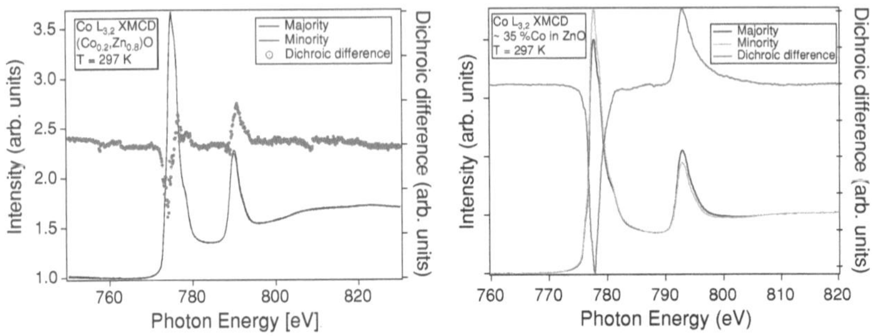

FIGURE 4 Using the XMCD technique the ferromagnetism of the Co-doped ZnO has been studied. (left) The dichroic signature is indicative of Co with oxygen in the nearest neighbor shell. (right) The XMCD signal here has a metallic signature, indicating formation of Co structures with largely metallic character. (Karis et al., unpublished data.)

are no longer good quantum numbers and the spin is not necessarily conserved. For right polarized light at the L_3 edge 62.5% of the excited electrons are spin up while the same holds for left polarized light but now for spin down electrons. Since the absorption cross-section is proportional to the number of empty 3d states and magnetic materials have more unoccupied states in one spin direction than the other; it follows that the absorption of circularly polarized X-rays are sensitive to the magnetization of the sample. Even though this method works for all atoms which exhibit a net magnetic moment, it has proven to be particularly useful in connection to $L_{2,3}$ absorption edges of transition metals. In this case it is possible to obtain both spin and orbital magnetic moments from XMCD data by using magneto-optical sum rules [6,7]. The magnetic moments are obtained as Bohr magnetons per atom and therefore no prior knowledge of the samples, as the thickness or concentration of magnetic ions, is necessary.

The XMCD data in Figure 4(left) was recorded for a sample with nominal Co doping content of 20% in ZnO. We begin by considering the actual absorption data recorded with opposite magnetization of the sample. These spectra show signatures of multiplet structure, manifested as fine structure in the absorption lines. This fine structure is most clearly visible for the L_3 line, appearing at lower photon energies, due to the smaller lifetime broadening of this level. The XMCD data furthermore reveals that among the Co atoms probed within the probing depth of the technique we find a small fraction of the Co atoms coupled ferromagnetically to each other. These atoms exhibit multiplet structure that is most clearly manifested in the dichroic difference spectrum. Multiplet features are not consistent with metallic Co and are likely from a Co^{2+} electronic state that should give rise to a magnetic moment of 3 Bohr magnetons (μ_B) according to Hund's rules. This particular sample was found to exhibit a very low magnetic moment in the SQUID measurement ($\sim$0.2 μ_B). Superparamagnetic AFM clusters blocked at room temperature, with a net magnetic moment due to uncompensated spins, could explain both the small magnetic moments and the FM signal from Co^{2+}. Now turning our attention to the dichroic signal in Figure 4(right), we immediately observe clear differences relative to the result presented in Figure 4(left). No apparent multiplet structure is visible in this dichroic difference. Instead we find a difference signal that is typically observed for metallic systems [9]. We conclude that the Co atoms in this case occupy sites where the net coupling leads to the formation of an electronic structure in the d-band which is nearly identical to Co metal, thus strongly suggesting that the XMCD signal for the latter samples is due to metallic Co clusters.

All spectra presented so far were obtained using total electron yield (TEY). In practice this method measures the total amount of electrons emitted from the sample, which is proportional to the X-ray absorption. The probing depth of total electron yield is ~10 nm [11] which in most cases is only a fraction of the sample thickness. An alternative route is to instead measure the photon emission, usually called fluorescence yield (FY), which has the advantage of a larger probing depth ~100 nm [11]. This is well illustrated by Opel et al. [8] which found ferromagnetic signatures from $Co_{0.05}Zn_{0.95}O$ samples by magnetometry. The XMCD obtained from both TEY and FY is presented in Figure 5(e) obtained at room temperature in an applied magnetic field of 4 T for the Co L_3 edge. There is a clear multiplet structure in the TEY channel suggesting that most of the Co is in a 2+ oxidation state where the black spectrum is a multiplet calculation of Co^{2+}. The FY shows no multiplet structure and is almost identical to the green spectrum which is a reference of metallic Co. The spectra obtained at 10 K, Figure 5(f), has more structure in the FY channel compared to room temperature due to a stronger magnetic response from Co^{2+} at low temperatures but the metallic character is still clear. From these measurements it can be concluded that metallic Co clusters are formed in the bulk which cannot be studied using total electron yield. Care must be taken when performing measurements in fluorescence yield since a direct correspondence to X-ray absorption is not obtained due to self absorption effect which makes this method unsuitable for extracting magnetization data using magneto-optical sum rules.

The usefulness of XMCD for dilute magnetic semiconductors goes further than determining the chemical state and magnetic moment of the magnetic dopant. Even though a large amount of reports from SQUID measurements suggested ferromagnetic DMS, the lack of XMCD data supporting these statements indicated that the ferromagnetism was not of intrinsic character. The high sensitivity of SQUID magnetometry makes it possible to measure very weak magnetic response, which is usually the case for these types of systems, but this also makes the measurements sensitive to external magnetic pollution. It has been suggested that FM responses can originate from intrinsic defects. By doping ZnO with 2% Cu and varying the amount of oxygen vacancies by controlling the oxygen partial pressure during samples growth, it was shown by Herng et al. [10] that Cu^{2+} substituting Zn results in a paramagnetic response while ferromagnetism is observed for Cu atoms in the

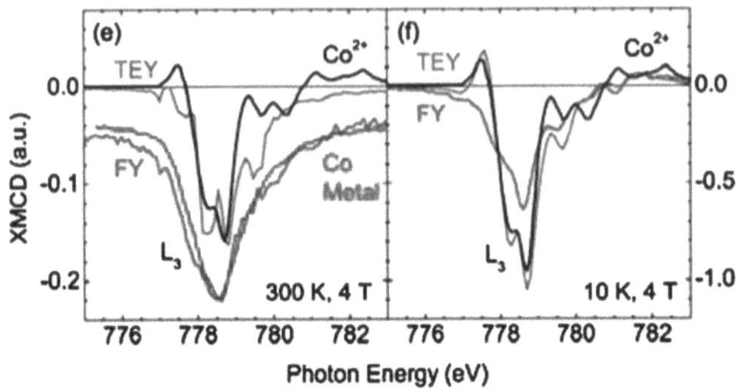

FIGURE 5 XMCD using both the fluorescence (FY) and total electron yield (TEY) in an applied field of 4 T. The TEY exhibits multiplet structure while the FY shows a signature of Co metal. Measurements were performed at both 300 and 10 K, (e) and (f) respectively. From Ref. [8].

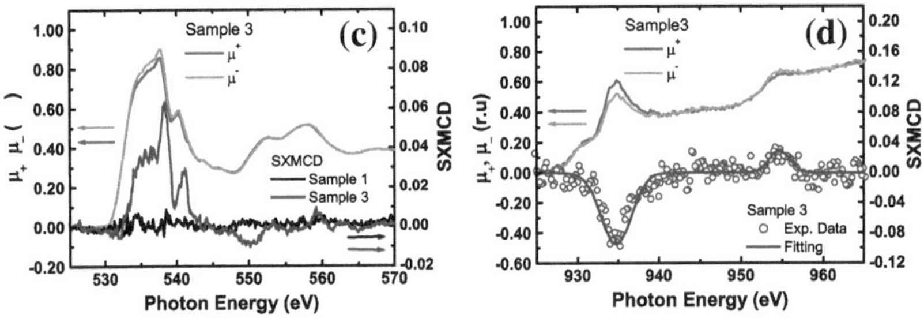

FIGURE 6 XMCD of (c) oxygen K edge and (d) Cu $L_{2,3}$ edges. There is a distinct dichroic difference for both oxygen and copper where the sign difference suggests an anti-ferromagnetic alignment between them. From Ref. [10].

vicinity of oxygen vacancies. Figure 6(c) and (d) shows the dichroic signature from both the O K edge and Cu $L_{2,3}$ edges, respectively. The relative sign difference between oxygen and Cu suggests an anti-ferromagnetic alignment. The multiplet structure indicates that the ferromagnetically active component has Cu^{1+} character which, together with the anti-ferromagnetic alignment between oxygen and copper, supports the notion of oxygen vacancies mediating the FM interaction between Cu atoms.

II. X-RAY PHOTOELECTRON SPECTROSCOPY AND RESONANT PHOTOELECTRON SPECTROSCOPY

Unlike the X-ray absorption process, where the quantity of emitted photons or electrons leaving the sample are measured, X-ray photoelectron spectroscopy

(XPS) relies on information obtained from the kinetic energy of emitted electrons. Electron emission induced by light absorption is a consequence of the photoelectric effect and was first described by A. Einstein, for which he was awarded the 1921 Nobel prize in physics. If the photon energy exceeds the binding energy of an electron and the work function combined, electrons will be emitted from the sample with the kinetic energy described by

$$E_{\mathrm{k}} = \hbar\omega - W - \mathrm{BE}, \qquad (3)$$

where E_{k} is the kinetic energy of the emitted electron, $\hbar\omega$ is the photon energy, W is the work function and BE is the binding energy of the electron. In metals the Fermi level (E_{F}) is often clearly visible which makes it possible to calibrate the binding energy in reference to E_{F} even if W is not known. This is often used for semiconducting samples by measuring an Au foil in electrical contact with the sample since the E_{F} for the metal and semiconductor will align.

The photoelectron emission process is illustrated in Figure 7 where the photoelectron leaves the sample after absorbing a photon with energy $\hbar\omega$, leaving behind an ionized atom [12]. The photoelectrons are registered by an electron energy analyzer which provides a spectra of electron intensity vs. kinetic energy. As illustrated in Figure 7(left), the obtained spectra will show broadened spectral lines both due to the finite lifetime of the ionized atom as well as instrumental broadening from the finite resolution of the electron energy analyzer and the energy distribution of incident X-rays after monochromatization. In a simplified one electron picture the photoelectron emission process of core-level electrons would only provide information on the elemental composition of the sample since the core electrons are very localized and does not take part in any chemical bonding. Fortunately the process is more complicated. Due to initial and final state effects (screening and correlation) a single core level will exhibit chemical shifts and sometimes show additional peaks in the photoelectron spectra which are called satellites. These effects are very sensitive to the chemical environment [13] and can be seen in Figure 7(right) where we show the Co 2p states for substituted Co^{2+} in ZnO and Co^{3+} in $ZnCo_2O_4$. The main lines of this spin–orbit split core level constitutes a well screen final state in which an electron has been transferred from the ligand valence band to the Co 3d state. The satellite structures are due to less screened states, these two types of final states are denoted as $2p^53d^8L^{-1}$ and $2p^53d^7$, respectively, where L^{-1} is a hole in the ligand valence band. In addition to the different spectral shapes there is also a chemical shift between the main lines. Therefore this method is often very efficient for determining the chemical valency of atomic species.

There are basically two different regions which are studied using photoelectron spectroscopy, core-levels and the valence band. Most of the properties (magnetic, optical, electronic transport, heat transport, catalytic) which are of interest in a solid are strongly connected to the valence band structure. The main difference is that valence band electrons are not as localized

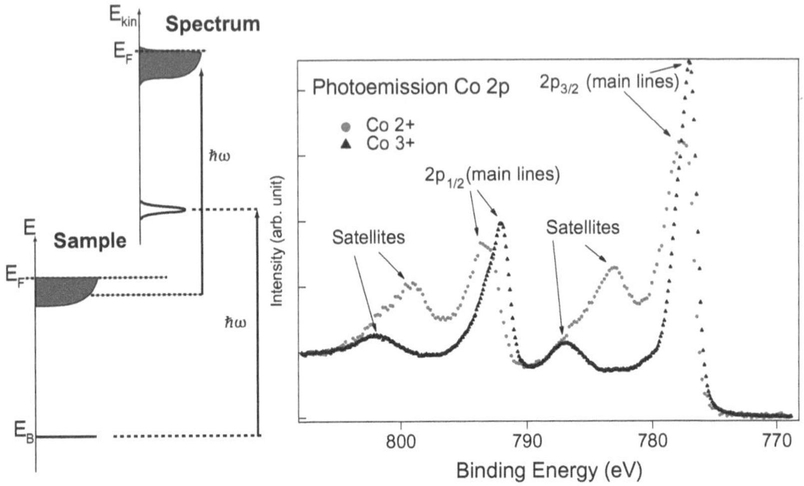

FIGURE 7 (left) Illustration of X-ray photoelectron spectroscopy of core and valence band electrons. The electron binding energy is mapped by measuring the kinetic energy of photoelectrons. (right) Core-level spectra of Co for different valence configurations. (Knut et al., unpublished data.)

as core-level electrons and are able to hybridize, meaning that the valence band states of atoms mix. The valence band and in particular the position of the 3d states of the magnetic atom is important when trying to identify the mechanism which could result in ferromagnetic properties and for comparison to theoretical results. In Figure 8 the valence bands of 5% and 15% doped ZnO are shown, top left and right respectively. The 5% Co-doped samples show additional states at the valence band edge compared to the non-doped ZnO sample. These are confirmed to be Co 3d states using resonant photoelectron spectroscopy. Additional Al co-doping which is expected to have an impact on the Fermi level since it acts as an n-dopant in ZnO, showed an impact on the Co 3d states clearly visible in Figure 8(left, b), but due to high intrinsic n-doping from oxygen vacancies the Fermi level is pinned to the conduction band and is unaffected by Al co-doping. The 15% Co-doped samples which have been annealed to different temperatures exhibit states deeper in the ZnO gap for all annealing temperatures except after 800 °C. These additional states are analyzed further using resonant photoelectron spectroscopy.

The surface sensitivity of XPS is valuable for many types of experiments but it can also be problematic when bulk properties are sought. Often samples needs to be prepared in situ to minimize contributions from surface contamination. If the sample is sensitive to oxidation it needs to have a protective capping layer or the top layers will not be representative for the bulk properties, in either case the top layers need to be removed which is difficult without affecting the sample. It is not possible to study structures that are buried deeper than a few nm using traditional XPS [15–19]. According to the universal curve (see Figure 9), the

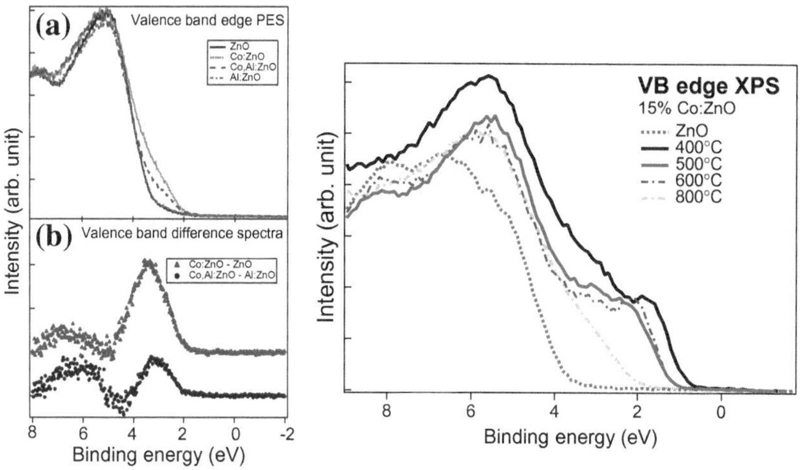

FIGURE 8 (left) (a) Valence band of 5% Co-doped ZnO. (From Ref. [2]) (b) difference between Co-doped and non-doped ZnO. (right) Valence band of 15% Co-doped ZnO. From Ref. [14].

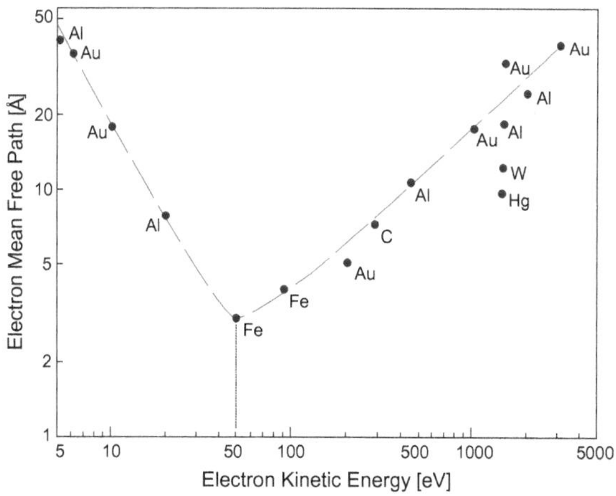

FIGURE 9 The universal curve which illustrates the mean free path (MFP) of electrons in solids as a function of the electron kinetic energy. The MFP is relatively material insensitive and higher kinetic energies increase the probing depth, for energies larger than 50 eV.

mean free path (MFP) of electrons can be increased by using electrons with high kinetic energies and it also indicates that the MFP is relatively material insensitive. Electrons with high kinetic energies can be obtained by using high photon energies. The use of hard X-rays for XPS actually predates the use of soft X-rays in the 1950s due to difficulties in transmission of low energy electrons in the detectors. In 1960s the development of detectors, spectrometers, and soft X-ray sources led to large advances in soft-x photoelectron spectroscopy. During the last decade there has been a large increase in the use of HIKE, where hard X-ray photoelectron spectroscopy (HAXPES) is a common synonym. This is mainly due to advances in the engineering of energy analyzers capable of handling high kinetic electrons with high resolution. Also, new synchrotron sources and monochromators with high resolving power in the hard X-ray region have been necessary to enable the full power of this technique.

Hard X-ray photoelectron spectroscopy can be used for studying core-levels with high binding energy but the main use has been to study core-levels which are available even with soft X-rays but with a much higher bulk sensitivity. The Co 2p core-levels presented in Figure 7 are obtained with HIKE using 3 keV photon energy and hence are good representative of the bulk electronic structure.

A. Resonant Photoelectron Spectroscopy (RPES)

If the valence band photoelectron emission is recorded using a photon energy which corresponds to an absorption edge, we will experience an interesting phenomenon called resonant photoelectron emission. In this case the photon can be absorbed either by a direct photoelectron event of a valence band electron or by exciting a core electron into an empty conduction band state. In the latter case, the atom may relax by filling the core hole with an electron from the valence band and emit an electron from the valence band.

For a Co atom excited with a photon energy corresponding to the L_3 edge we hence have the following possible events

$$2p^63d^7 + \hbar\omega \rightarrow 2p^63d^6 + e^- \quad \text{(photoelectron)}, \tag{4}$$
$$2p^63d^7 + \hbar\omega \rightarrow 2p^53d^8 \rightarrow 2p^63d^6 + e^- \text{(absorption and autoionization)} \tag{5}$$

in which we observe that the final state is identical for both direct photoelectron emission and for the autoionization [20]. The cross-section for absorption is generally much higher than for direct photoelectron emission of valence band electrons at these energies. Therefore the total cross section for valence band states corresponding to atoms which have an absorption edge at the used photon energy is strongly enhanced. It should be noted that this is a simplified picture of the resonant process. Since the initial and final states are identical, the two different processes cannot be separated and only a complete quantum mechanical treatment, taking the possibility of interference into account, can correctly describe the resonant spectra [21].

Often it is sufficient to perform the experiment at two different photon energies, on and off resonance, which are marked by dashed lines in Figure 2. A time-consuming but very useful RPES experiment consists of measuring the valence band over the complete absorption edge energy range. This is illustrated in Figure 10(left) where the valence band of 15% Co-doped ZnO has been

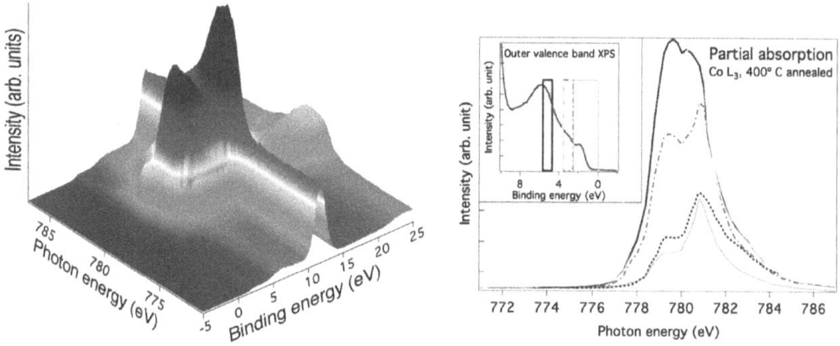

FIGURE 10 (left) Valence band RPES spectrum of the 400 °C sample. The valence band was recorded with photon energies ranging across the Co L_3 edge. (right) Partial absorption of the Co L_3 edge. The absorption was constructed by integration over specific BE ranges from the RPES spectrum. The inset shows the VB XPS and the boxes indicate which BE ranges have been integrated. From Ref. [14].

obtained for energies ranging over the Co L_3 absorption edge. The peak at about 11 eV binding energy are due to Zn 3d states and the strong intensity increases close to the valence band edge are due to Co 3d states.

From the RPES shown in Figure 10(left) for the 400 °C sample we obtained (see figure caption for more details) the partial absorption spectra shown in Figure 10(right). This method can resolve the Co L_3 absorption for different chemical species if the valence electrons are separated in binding energy. This is the case for substituted Co found at the valence band edge and the states found in the band gap. The XAS of the Co_3O_4 spinel (black dotted) and the partial absorption of the states located at low binding energies (solid orange) are very similar while the partial absorption of the states located at higher binding energies (solid black) is very similar to the XAS found in Figure 2(left). This identifies the location of the Co 3d states in the valence band depending on the chemical state. X-ray diffraction (XRD) shows no indication of Co_3O_4 or $ZnCo_2O_4$ for the 400 °C sample, while spectroscopy suggests a large amount of these secondary phases. This might be due to a small crystal size of the secondary phases rendering them invisible to XRD or that the samples contain a large amount of Co 3+ which has not relaxed into the spinel structure.

Similarly, RPES performed on 5% Fe doped ZnO reveals several chemically different Fe states. The excitation energies for Fe^{2+} and Fe^{3+} in XAS show peaks around 708 and 710 eV, respectively, see Figure 11(right). Interestingly it was found that the valency of the Fe atoms is sensitive to X-ray exposure, where a reduction occurs. This can be explained by the formation of oxygen vacancies on the surface of the Fe doped ZnO. Therefore, several different Fe states were identified with very different resonant features in the valence band as shown in Figure 11(left). The plotted resonant features have been obtained by subtracting off resonant valence band spectra from the measured resonant

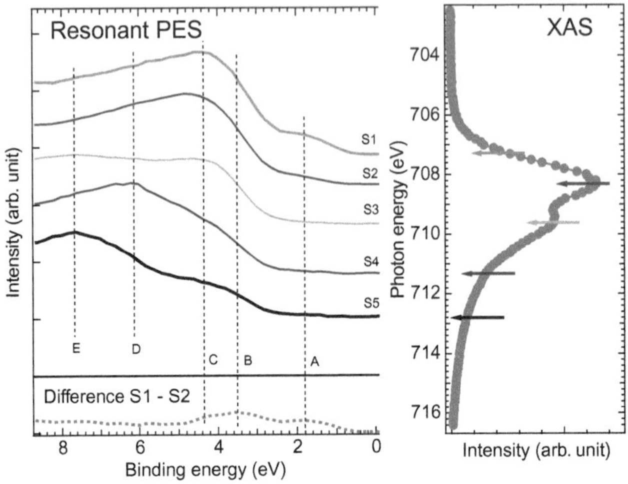

FIGURE 11 (left) Valence band RPES spectra of the 5% Fe doped ZnO. The valence band was recorded with photon energies ranging across the Fe L_3 edge. (right) X-ray absorption spectrum of the Fe L_3 edge. (Knut et al., unpublished.)

spectra. As Fe substitutes Zn^{2+} it will obtain a tetrahedral coordination which is energetically unfavorable for Fe^{2+} and therefore the bulk valency appears to be mainly of 3+ valency. The magnetic properties suggest that the Fe arrange in small anti-ferromagnetic clusters (~5 atoms) and hence this is not a potential magnetic semiconductor but the sensitivity to light makes this material more interesting for its photocatalytic properties.

III. X-RAY EMISSION SPECTROSCOPY

Techniques using photon excitation and photons as detected species have not been very common for investigations of DMS systems. Some examples are provided in Refs. [22–30]. Soft X-ray emission (SXES or XES) and the related resonant inelastic X-ray scattering (RIXS) is potentially a very powerful technique for investigations of electronic structure as it offers many of the advantages of resonant photoelectron spectroscopy presented above in Section II with the additional advantage that the problem of inelastic background from electrons is virtually absent in the XES spectrum and we therefore devote this subsection to explain some of the key aspects of the technique. One of the central problems in studies of electronic structure in general is to study how the electronic structure of a doped candidate DMS system is different from its constituent part, i.e., to investigate the consequences of hybridization.

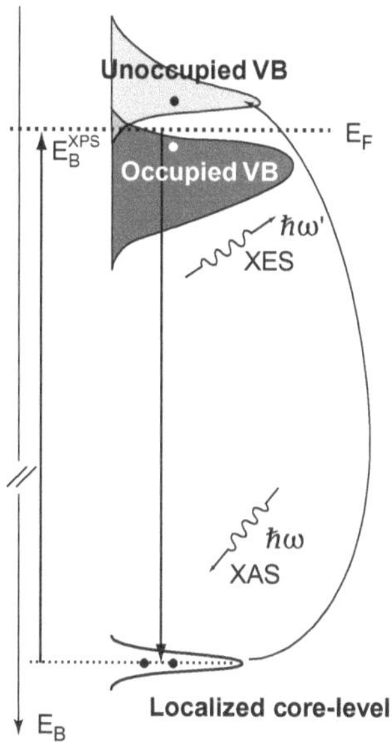

FIGURE 12 The relationship of XP, XA, and XE energies for a well-screened system (see text).

In studies of well-screened systems, like a metal, the XE process is conveniently described as a two-step process. Within this type of description we consider excitation and decay (emission) as separate events. This is schematically illustrated in Figure 12. Upon absorption of an X-ray photon an electron is transferred from a core level into a previously unoccupied level. In order to avoid "initial state" satellites in the XE spectrum it is necessary to use threshold excitation. In the subsequent decay, an electron from the valence band fills the core hole under emission of an X-ray photon. As follows from Figure 12, the high energy cut-off in the emission spectrum will correspond to emission from E_F. In general, the XE final state will correspond to a single valence hole state as in UPS, and we may write down the following energy relation

$$-E_B^{UPS} = \hbar\omega^{XES} - E_B^{XPS}. \quad (6)$$

The relation above allows us to relate the XE emission energy to the UPS binding energy scale via the XPS binding energy.

The real power of XES in studies of electronic structure lies in the fact that it constitutes a local probe, due to the presence of a core hole in the initial state of the decay. This allows us to resolve atomic contributions to the valence band much in the same manner as for RPES discussed above. Since the XE process also follows the dipole selection rules, it is possible to further decompose the contributions into local symmetry resolved components (σ, π and δ contributions). A disadvantage is the requirement of advanced instrumentation and access to synchrotron radiation [31], a caveat that is shared with essentially all methods in this Chapter.

IV. EXTENDED X-RAY ABSORPTION FINE STRUCTURE (EXAFS)

In this last section we discuss the usefulness of structural probes for investigations of systems where small amounts of impurity phases are a possible cause of physical properties measured by global methods like magnetization measurements or even transport. Instead of focussing on conventional in-house methodologies like X-ray diffraction we will explore the consequences of relying on a *local* structural probe like extended X-ray absorption fine structure (EXAFS), a technique closely related to X-ray absorption spectroscopy. The section is organized as follows. In the first subsection we briefly discuss the underlying physics for the EXAFS process which will clarify the distinction to XRD in terms of a local environment around a particular species and the global picture that emerges from X-ray diffraction performed with X-ray sources not tuned to enhance any particular elemental contribution to the scattering. After this introduction to the method we proceed and survey the application of EXAFS in investigations of DMS systems. We do not argue that our coverage is complete but rather selected to illustrate certain points that we would bring to the readers' attention. It should also be clear that the views expressed here are entirely interpretations of the authors of this survey and often not a conclusion expressed by the authors of the original work.

A. Electron Scattering and the Extended X-ray Absorption Fine Structure

When a bound electron is excited into previously unoccupied states it can be viewed as propagating semi-bound photoelectron with kinetic energy $E_{\mathrm{K}} \sim \hbar\omega - E_{\mathrm{T}}$ and a wave vector $k = \frac{2m}{\hbar}\sqrt{E_{\mathrm{K}}}$, where $\hbar\omega$ is the photon energy of the exciting radiation and E_{T} is the 'threshold energy', i.e., the onset of the absorption. Strictly speaking this is an approximation that works well for more than some tens of eV above E_{T}, which is where the fine structure is considered anyway. Due to the presence of surrounding atoms the final state wave function is modified. A way to conceptualize this phenomenon is to view it as a scattering of spherical waves emitted from the absorbing atom and scattering by the surrounding leading to constructive and destructive interference of the primary and scattered waves. This is illustrated in the cartoon in Figure 13. In general, one has to consider many scattering paths to model the spectrum which is usually done within the realm of multiple scattering formalism implemented in some software package for EXAFS analysis (see, e.g., [32–39]). From the experimentally determined X-ray absorption signal $\mu(E)$ one obtains the purely

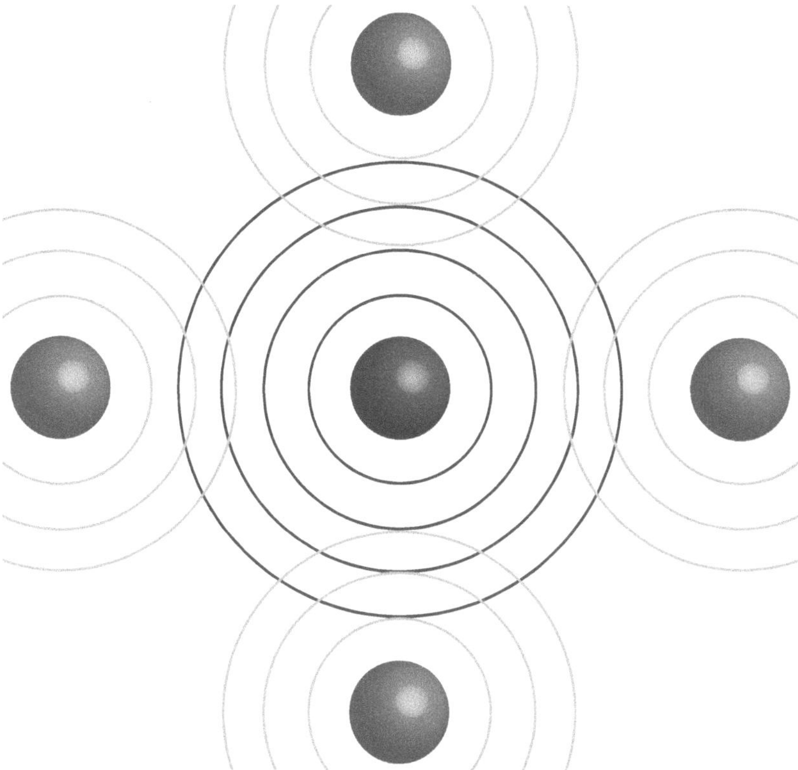

FIGURE 13 The excitation of a photoelectron creates propagating spherical waves that scatter off the surrounding. The interference of primary and secondary waves results in a modulation of the photelectron cross sections which is a signature of the *local* environment around the emitter.

oscillatory part of the absorption coefficient (normalized) $\chi(E) = \frac{\mu(E)-\mu_0(E)}{\mu_0(E)}$. $\mu_0(E)$ is a smooth atomic-like contribution to the absorption, obtained by fitting or calculations. Using the relation for k above we can obtain $\chi(k)$ which can be described as a superposition of terms like $\sin(2kR_j + \phi_j)/R_j^2$, where R_j is the scattering path length of path j and ϕ_j is a path dependent phase shift. The Fourier transform of $\chi(k)$ will contain structural information in real space.

An important point to remember when comparing EXAFS to other methods for structural investigations, is that while diffraction by X-rays or neutrons reflect a global order, EXAFS reflects the local order around the emitting atom. One can therefore obtain structural information even for amorphous systems using EXAFS. The complementarity of EXAFS and diffraction methods have been beautifully illustrated in the work by Mikkelsen and Boyce [40] (see Figure 14).

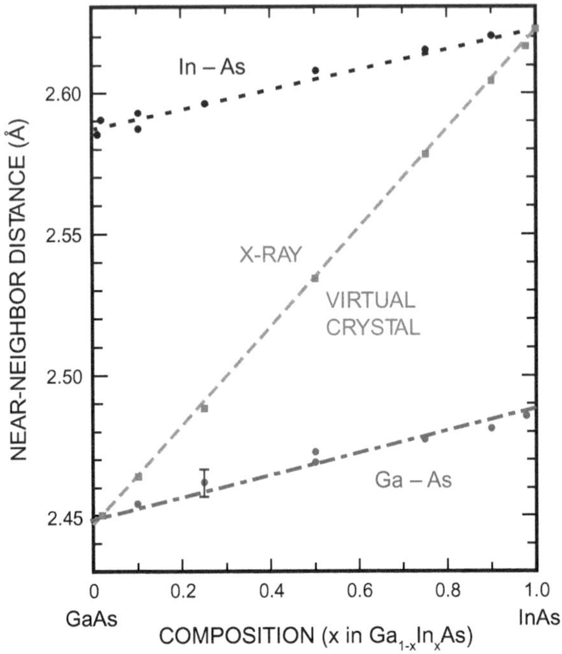

FIGURE 14 Mikkelsen and Boyce compared X-ray diffraction data reflecting global order with the local structure as mapped by the EXAFS data. While the X-ray data is found to be well described by the virtual crystal approximation (Vegard's law), the variation of Ga–As and In–As distances is found to be much smaller. A weighted average of the two nearest neighbor distances agrees well with Vegard's law. The figure is adopted from [40].

The physical properties of DMS systems are considered to be strongly dependent on the local coordination, environment, bond length, bond angle, valence, and the site symmetry of the magnetic ion as well as the host lattice. Hence, EXAFS would be ideally suited for the study of such systems. A potential caveat when considering what information can be obtained from a local probe like EXAFS is that though we are selective to a local environment around one particular atomic species, the signal will describe an average of all such environments present in the system under investigation. If the majority of a dopant is in a particular site, where it does not contribute to a particular physical property, and a small fraction of dopants forms an impurity phase with a particular property associated to it, the EXAFS signal will reflect the statistical average of all sites and only a careful analysis might reveal the existence of an impurity phase. A EXAFS signal obtained for diluted systems like DMS might appear to correspond to a perfectly substituted dopant just because the contribution from the impurity phase is masked by the dominant signal.

B. Applications of EXAFS in DMS Systems

There are not so many investigations on DMS systems using EXAFS over the last decade [41–56]. As expected, considering the caveats in applying EXAFS to a diluted system discussed in Section A, the reports can roughly be separated into three categories; reports of ferromagnetic (or ferrimagnetic) properties correlated to EXAFS spectra interpreted as indications of purely substitutional occupancy of magnetic ions or at least absence of clustering [46–49,51,53,55,56], paramagnetic properties with corresponding interpretations of EXAFS data [52] and indications of magnetic order (ferromagnetic, ferrimagnetic, or anti-ferromagnetic) correlated with an indication of impurity phases in the EXAFS signal [42,54]. There are also studies that mainly concern structural evolution in DMS system as a function of doping or post-growth treatment [41].

In the following we will scrutinize results presented by Shi et al. [55] representing the first category (argued FM and no indication of clustering from XRD nor EXAFS) and the work by Farley et al. [54], which reports on the presence of impurity phases as evidenced by a combination of EXAFS, X-ray absorption, and magnetometry.

Figure 15 from Ref. [55] shows the Co K-edge EXAFS $k^3\chi(k)$ oscillation functions for sol–gel produced $Zn_{1-x}Co_xO$ and sputter-deposited $Co_yZn_{1-y}O$ samples. Shi et al. also provide reference spectra for the Co K-edge of Co metal, CoO, the spinel Co_3O_4, and Zn K-edge EXAFS of wurtzite ZnO powder

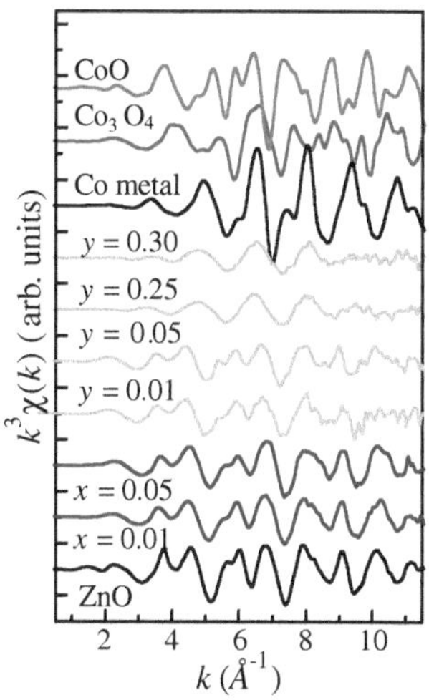

FIGURE 15 The figure shows the Co K-edge EXAFS ($k^3\chi(k)$) for sol–gel produced $Zn_{1-x}Co_xO$ and sputter-deposited $Co_yZn_{1-y}O$ samples. Reference spectra for the Co K-edge of Co metal, CoO, the spinel Co_3O_4, and Zn K-edge EXAFS of wurtzite ZnO powder (bottom curve labeled ZnO) are also provided. From [55].

(bottom curve labeled ZnO). The authors argue that the EXAFS spectra of CoO and Co_3O_4 are different from those of $Zn_{1-x}Co_xO$ and $Co_yZn_{1-y}O$ samples, and that this indicates the absence of CoO or Co_3O_4 phases in the samples. The authors furthermore point out the similarity of the EXAFS spectra of the $Zn_{0.99}Co_{0.01}O$, $Zn_{0.95}Co_{0.05}O$, $Co_{0.01}Zn_{0.99}O$, and $Co_{0.05}Zn_{0.95}O$ samples in Figure 15 to the presented data of the Zn K-edge obtained for ZnO powder. This is interpreted as evidence that the doped Co ions are located at the Zn substitutional sites in both $Zn_{1-x}Co_xO$ ($x \leqslant 0.05$) and $Co_yZn_{1-y}O$ ($y \leqslant 0.05$) samples. The finding that doped ZnO samples appear phase pure by most structural methods up to a level of about 5% TM doping is consistent with several reports [2,57]. When further increasing the Co content for the sputter deposited $Co_yZn_{1-y}O$ series, the authors argue that the shape of the EXAFS signal ($y \geqslant 0.25$) begins to resemble that of the cobalt metal, indicating the presence of metallic cobalt phase for these doping levels. Considering the EXAFS results presented by Shi et al. in Ref. [55] we can thus conclude the following: (i) The data indicate phase pure samples produced both by solution chemistry (sol–gel) methods and physical vapor methods (sputtering) for doping lower than 5%. (ii) For high levels of doping the sputter-deposited samples are found to exhibit indications of metallic Co. How does these findings correlate to the magnetic properties of these samples? In Figure 16 the magnetization data from the work by Shi et al. [55] is given. In the top panel the authors have given the M(T) for the sol–gel produced samples. The behavior is typical for

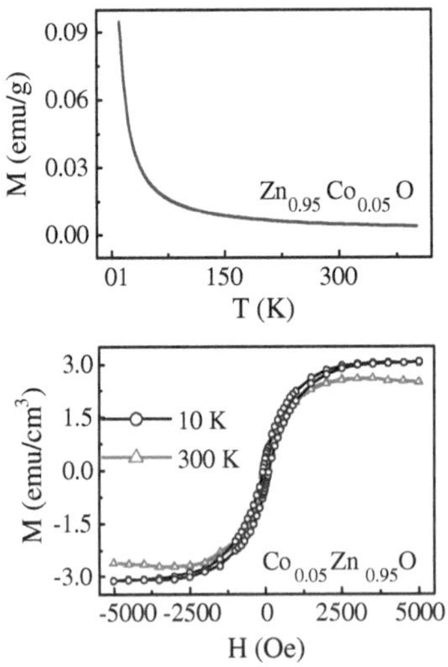

FIGURE 16 Magnetization data obtained Co-doped ZnO obtained by sol–gel top showing an apparent paramagnetic behavior (top) and by sputtering, showing indications of some type of magnetic ordering, respectively. From Ref. [55].

a paramagnetic system. The provided values can be scaled to the same units as the bottom panel by assuming that the sample has the same density as pure ZnO ($\sim$ 5.606 g/cm^3) giving 0.1 emu/g $\simeq$ 0.56 g/cm^3. Converting these units to more easily understood moments per Co atom shows that the provided values of the magnetization correspond to values of less than 0.1 μ_B/Co, i.e., well below what is expected for a Co^{2+} ion which should contribute with 3 μ_B. As the lowest temperature data is not easy to derive from the provided figure, a part of this discrepancy could certainly be a temperature effect. No M(H) data is provided for the sol–gel samples.

The M(H) data for the sputter-deposited samples from Ref. [55] is given in the lower panel of Figure 16. No M(T) data is provided for the sputter-deposited samples. There are clear indications of some type of magnetic order as evidenced by the saturation of the magnetization at moderate applied fields. Shi et al. argue that the bottom panel is evidence of ferromagnetic order in the sputter-deposited samples. Converting the reported values to moments per Co atom give moments of the order of 0.1 μ_B, again much lower than the expected 3 μ_B from a Co^{2+}, suggesting that there is a correlation of the moments but likely not ferromagnetic ordering. In the work by Iuşan et al. [2] a very similar magnetization behavior is reported and is attributed to small clusters of Co atoms with anti-ferromagnetic interactions dominating. A possible alternative interpretation of the results reported in Ref. [55] is thus that the EXAFS data is not sensitive to the relatively small variations that occur in the material for small doping but it is really the nanoscale inhomogeneity of the material that is responsible for the observed magnetic properties. Iuşan et al. argue that the required cluster size is of the order of ten Co atoms which is certainly below the detection limit of most structural methods, though EXAFS would in principle be better suited to detect the variation in local geometry.

We conclude this subsection with the report by Farley et al. [54] that combines all of the experimental methods deriving from the X-ray absorption method accounted for in this chapter to investigate candidate DMS materials based on TM doping of ZnO from sol–gel methods. In the samples used in this study the level of TM doping was 6% for Mn and 3% for Co, i.e., at levels where reports suggest phase pure samples based on X-ray diffraction and related techniques. Figure 17 shows the EXAFS data from Ref. [54]. The top two panels give the k^3 weighted EXAFS for ZnO powder, Mn- and Co-doped films (left) and reference samples (A = CoO; B = Co_3O_4; C = MnO; D = Mn_3O_4; E = Mn_2O_3; F = MnO_2)(right), respectively. The bottom two panels give the corresponding Fourier transforms. The authors report that the coordination in the first metal-oxygen shell of both Co- and Mn-doped films is 4, consistent with a substitution for tetrahedral Td Zn2+ ions in the ZnO structure. Inclusion of the spinels $ZnAl_2O_4$ and $ZnMn_2O_4$ (possible contaminant phases) does not lead to significantly improved fits, though their contribution to the EXAFS cannot be discounted. No indication of anti-ferromagnetic Co_3O_4, another possible contaminant spinel phase, containing Co2+ and Co3+ on tetrahedral

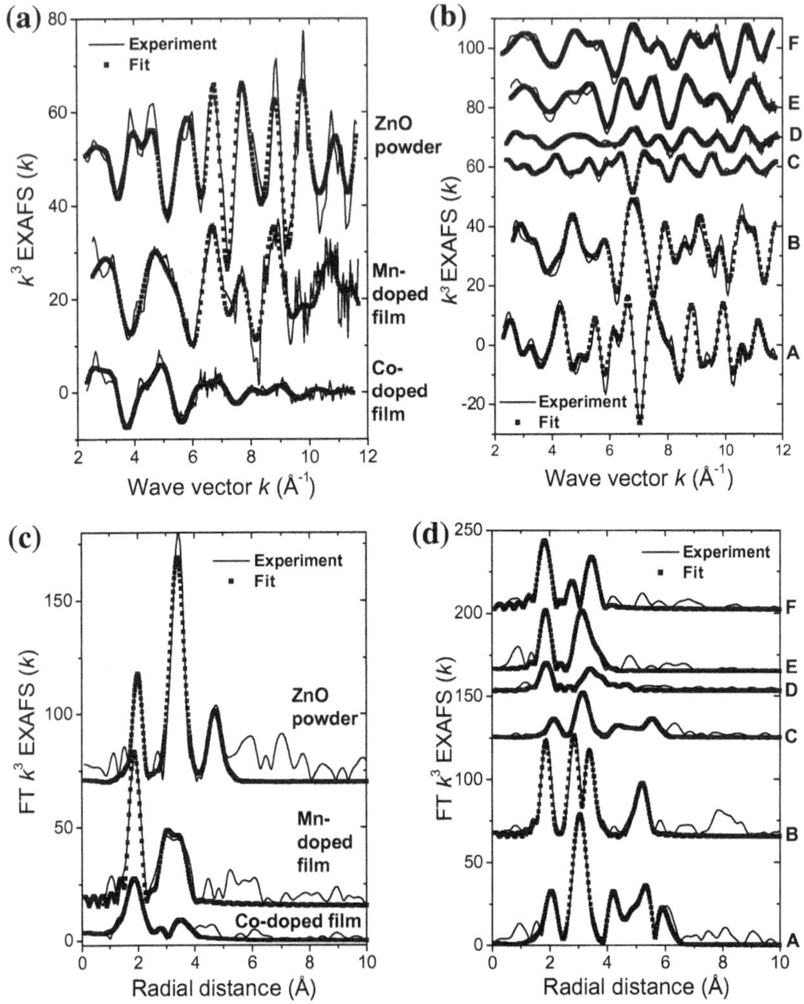

FIGURE 17 (a)–(b) k^3-weighted EXAFS with fits and (c)–(d) corresponding Fourier transforms of Co, Mn and Zn K-edge data of thin films and reference powders: (a) and (c) ZnO powder; Mn-doped film; Co-doped film. (b) and (d) Reference materials: A = CoO; B = Co_3O_4; C = MnO; D = Mn_3O_4; E = Mn_2O_3; F = MnO_2. From Ref. [54].

and octahedral sites, respectively, is found either. Rather the authors argue that both XRD and EXAFS indicate that the samples are phase pure substituted ZnO. However, the characterization of the near edge X-ray absorption structures (XAS), evidence of multiple valence states ($Co^{2+,3+}$, $Mn^{3+,4+}$) consistent with the presence of spinels like Co_3O_4, $ZnCo_2O_4$, or $ZnMn_2O_4$ in the doped films. We note the similarity of these findings to those reported in Ref. [14].

To summarize this subsection we have found a number of investigations of DMS systems using EXAFS for structural investigations. EXAFS has some apparent advantages as it, in principle, allows for investigations of the local structure around the dopant atom. As evidenced by the reports available in the literature we conclude that only combinations of investigations of structural, electronic, and magnetic properties allow for a complete understanding of a complex system like candidate DMS materials.

REFERENCES

[1] Stöhr Joachim. NEXAFS spectroscopy. Heidelberg: Springer-Verlag; 1992.

[2] Iusan Diana, Knut Ronny, Sanyal Biplab, Karis Olof, Eriksson Olle, Coleman Victoria A, et al. Electronic structure and chemical and magnetic interactions in ZnO doped with Co and Al: experiments and ab initio density-functional calculations. Phys Rev B 2008;78(8):085319. doi:10.1103/PhysRevB.78.085319. URL: <http://link.aps.org/abstract/PRB/v78/e085319>.

[3] Kobayashi M, Ishida Y, Hwang JI, Mizokawa T, Fujimori A, Mamiya K, et al. Characterization of magnetic components in the diluted magnetic semiconductor $Zn_{1-x}Co_xO$ by X-ray magnetic circular dichroism. Phys Rev B 2005;72(20):201201. ISSN 0163-1829. URL: <http://dx.doi.org/10.1103/PhysRevB.72.201201>.

[4] Krishnamurthy S, McGuinness C, Dorneles LS, Venkatesan M, Coey JMD, Lunney JG, et al. Soft-X-ray spectroscopic investigation of ferromagnetic Co-doped ZnO. J Appl Phys (USA) 2006;99(8):08M111. ISSN 0021-8979. URL: <http://dx.doi.org/10.1063/1.2165916>.

[5] Okabayashi J, Ono K, Mizuguchi M, Oshima M, Gupta SS, Sarma DD, et al. X-ray absorption spectroscopy of transition-metal doped diluted magnetic semiconductors $Zn_{1-x}M_xO$. J Appl Phys (USA) 2004;95(7):3573–5. ISSN 0021-8979. URL: <http://dx.doi.org/10.1063/1.1652248>.

[6] Carra Paolo, Thole BT, Altarelli Massimo, Wang Xindong. X-ray circular dichroism and local magnetic fields. Phys Rev Lett 1993;70(5):694. doi:10.1103/PhysRevLett.70.694.

[7] Thole BT, Carra P, Sette F, van der Laan G. X-ray circular dichroism as a probe of orbital magnetization. Phys Rev Lett 1992;68(12):1943. doi:10.1103/PhysRevLett.68.1943.

[8] Opel M, Nielsen K-W, Bauer S, Goennenwein STB, Cezar JC, Schmeisser D, et al. Nanosized superparamagnetic precipitates in cobalt-doped ZnO. Eur Phys J B 2008;63(4):437. doi:10.1140/epjb/e2008-00252-4.

[9] Langer J, Hunter-Dunn J, Hahlin A, Karis O, Sellmann R, Arvanitis D, et al. Cap layer influence on the spin reorientation transition in Au/Co/Au. Phys Rev B 2002;66:172401.

[10] Herng TS, Qi D-C, Berlijn T, Yi JB, Yang KS, Dai Y, et al. Room-temperature ferromagnetism of Cu-doped ZnO films probed by soft X-ray magnetic circular dichroism. Phys Rev Lett 2010;105:207201. doi:10.1103/PhysRevLett.105.207201. URL: <http://link.aps.org/doi/10.1103/PhysRevLett.105.207201>.

[11] Naftel SJ, Sham TK. Co $L_{3,2}$-edge and multi-detection channel XAFS studies of Co–Si interactions. J Synchrotron Radiat 1999;6(3):526–8. ISSN 1600-5775. doi:10.1107/S0909049599000989. URL: <http://dx.doi.org/10.1107/S0909049599000989>.

[12] Hüfner S. Photoelectron spectroscopy. Berlin/Heidelberg/New York: Springer-Verlag; 1995.

[13] de Groth F, Kotani Akio. Core level spectroscopy. New York: CRC Press, Taylor and Francis Group; 2008.

[14] Knut R, Wikberg JM, Lashgari K, Coleman VA, Westin G, Svedlindh P, et al. Magnetic and electronic characterization of highly Co-doped ZnO: an annealing study at the solubility limit. Phys Rev B 2010;82(9):094438. doi:10.1103/PhysRevB.82.094438.

[15] Holmström E, Olovsson W, Abrikosov IA, Niklasson AMN, Johansson B, Gorgoi M, et al. Sample preserving deep interface characterization technique. Phys Rev Lett 2006;97:266106. doi:10.1103/PhysRevLett.97.266106. URL: <http://link.aps.org/doi/10.1103/PhysRevLett.97.266106>.

[16] Johansson EMJ, Odelius M, Gorgoi M, Karis O, Ovsyannikov R, Schafers F, et al. Valence electronic structure of ruthenium based complexes probed by photoelectron spectroscopy at high kinetic energy (HIKE) and modeled by DFT calculations. Chem Phys Lett 2008;464(4–6):192–7. ISSN 0009-2614. URL: <http://dx.doi.org/10.1016/j.cplett.2008.09.016>.

[17] Karis O, Svensson S, Rusz J, Oppeneer PM, Gorgoi M, Schafers F, et al. High-kinetic-energy photoemission spectroscopy of Ni at 1s: 6-eV satellite at 4 eV. Phys Rev B 2008;78(23):233105. doi:10.1103/PhysRevB.78.233105. URL: <http://link.aps.org/abstract/PRB/v78/e233105>.

[18] Gorgoi M, Svensson S, Schäfers F, Öhrwall G, Mertin M, Bressler P, et al. The high kinetic energy photoelectron spectroscopy facility at bessy progress and first results. Nucl Instr Meth Phys Res A 2009;601(1–2):48–53. ISSN 0168-9002. doi:10.1016/j.nima.2008.12.244. URL: <http://www.sciencedirect.com/science/article/B6TJM-4VGPHJW-1/2/eefae1c20c5d68dfb3c6eac4bebfe199>. Special issue in honour of Prof. Kai Siegbahn.

[19] Granroth Sari, Knut Ronny, Marcellini Moreno, Andersson Gabriella, Svensson Svante, Karis Olof, et al. Investigation of interface properties of Ni/Cu multilayers by high kinetic energy photoelectron spectroscopy. Phys Rev B 2009;80(9):094104. doi:10.1103/PhysRevB.80.094104. URL: <http://link.aps.org/abstract/PRB/v80/e094104>.

[20] Brühwiler PA, Karis O, Mårtensson N. Charge-transfer dynamics studied using resonant core spectroscopies. Rev Mod Phys 2002;74(3):703. doi:10.1103/RevModPhys.74.703.

[21] Weinelt M, Nilsson A, Magnuson M, Wiell T, Wassdahl N, Karis O, et al. Resonant photoemission at the 2p edges of Ni: resonant raman and interference effects. Phys Rev Lett 1997;78:967.

[22] Bouaine A, Green RJ, Colis S, Bazylewski P, Chang GS, Moewes A, et al. Appearance of ferromagnetism in Co-doped CeO_2 diluted magnetic semiconductors prepared by solid-state reaction. J Phys Chem C 2011;115(5):1556–60. doi:10.1021/jp1088096.

[23] Gautam S, Thakur P, Chae KH, Chang GS, Subramanain M, Jayavel R, et al. Electronic structure of Co-doped ZnO thin films by X-ray absorption and emission spectroscopy. J Korean Phys Soc 2009;55(1):167–72.

[24] Thakur P, Bisogni V, Cezar JC, Brookes NB, Ghiringhelli G, Gautam S, et al. Electronic structure of Cu-doped ZnO thin films by X-ray absorption, magnetic circular dichroism, and resonant inelastic X-ray scattering. J Appl Phys 2010;107(10):103915. doi:10.1063/1.3372758.

[25] Krishnamurthy Satheesh, Kennedy Brian, McGee Fintan, Venkatesan M, Coey JMD, Lunney James G, et al. Resonant soft X-ray emission and X-ray absorption studies on ga1-xmnxn grown by pulsed laser deposition. Phys Status Solidi C: Current Topics In Solid State Physics 2011;8, No 5, 8 (5): 287, 2011. doi:10.1002/pssc.201000810.

[26] Surkova TP, Galakhov VR, Kurmaev EZ. Soft X-ray emission spectra and ferromagnetism in wide-gap doped semiconductors. Low Temp Phys 2009;35(1):79–82. doi:10.1063/1.3064914.

[27] Krishnamurthy S, McGuinness C, Dorneles LS, Venkatesan M, Coey JMD, Lunney JG, et al. Soft-X-ray spectroscopic investigation of ferromagnetic Co-doped ZnO. J Appl Phys 2006;99(8):08M111. doi:10.1063/1.2165916.

[28] Jin J, Chang GS, Zhou YX, Zhang XY, Boukhvalov DW, Kurmaev EZ, et al. X-ray absorption and emission spectroscopic investigation of Mn doped ZnO films. Appl Surf Sci 2011;257(24):10748–51. doi:10.1016/j.apsusc.2011.07.091.

[29] Thakur P, Gautam S, Chae KH, Subramanain M, Jayavel R, Asokan K. X-ray absorption and emission studies of Mn-doped ZnO thin films. J Korean Phys Soc 2009;55(1):177–182.

[30] Guo J-H, Gupta Amita, Sharma Parmanand, Rao KV, Marcus MA, Dong CL, et al. X-ray spectroscopic study of the charge state and local ordering of room-temperature ferromagnetic Mn-doped ZnO. J Phys: Condens Matter 2007;19(17):172202. doi:10.1088/0953-8984/19/17/172202.

[31] Warwick Tony, Padmore Howard, Voronov Dmitriy, Yashchuk Valeriy. A soft X-ray spectrometer using a highly dispersive multilayer grating. In: Sri 2009: the 10th international conference on synchrotron radiation instrumentation, vol. 1234. 2010. p. 971.

[32] Lee PA, Pendry JB. Theory of the extended x-ray absorption fine structure. Phys Rev B 1975;11:2795–811. doi:10.1103/PhysRevB.11.2795. URL: http://link.aps.org/doi/10.1103/PhysRevB.11.2795.

[33] Rehr JJ, Albers RC. Theoretical approaches to X-ray absorption fine structure. Rev Mod Phys 2000;72(3):621–54. doi:10.1103/RevModPhys.72.621.
[34] Ravel B, Newville M. Athena and artemis: interactive graphical data analysis using ifeffit. Phys Scr 2005;T115:1007–10.
[35] Ravel B, Newville M. Athena, Artemis, Hephaestus: data analysis for X-ray absorption spectroscopy using ifeffit. J Synchr Rad 2005;12:537–41. doi:10.1107/S0909049505012719.
[36] Newville M. Ifeffit: interactive XAFS analysis and feff fitting. J Synchr Rad 2001;8.
[37] Zabinsky SI, Rehr JJ, Ankudinov A, Albers RC, Eller MJ. Multiple-scattering calculations of X-ray-absorption spectra. Phys Rev B 1995;52(4):2995–3009.
[38] Rehr JJ, Albers RC, Zabinsky SI. High-order multiple-scattering calculations of X-ray-absorption fine structure. Phys Rev Lett 1992;69:3397–400. doi:10.1103/PhysRevLett.69.3397. URL: http://link.aps.org/doi/10.1103/PhysRevLett.69.3397.
[39] Mustre de Leon J, Rehr JJ, Zabinsky SI, Albers RC. Ab initio curved-wave X-ray-absorption fine structure. Phys Rev B 1991;44:4146–56. doi:10.1103/PhysRevB.44.4146. URL: http://link.aps.org/doi/10.1103/PhysRevB.44.4146.
[40] Mikkelsen JC, Boyce JB. Extended X-ray-absorption fine-structure study of $Ga_{1-x}In_xAs$ random solid solutions. Phys Rev B 1983;28:7130–40. doi:10.1103/PhysRevB.28.7130. URL: http://link.aps.org/doi/10.1103/PhysRevB.28.7130.
[41] Soo YL, Kioseoglou G, Huang S, Kim S, Kao YH, Takatani Y, et al. Studies of impurities in magnetic semiconductors: an example of important XAFS applications. J Synchr Rad 2001;8:874–6. doi:10.1107/S0909049500016708.
[42] Soo YL, Kioseoglou G, Huang S, Kim S, Kao YH, Takatani Y, et al. Local structure around fe in the diluted magnetic semiconductors $Ga_{1-x}Fe_xAs$ studied by X-ray absorption fine structure. Phys Rev B 2001;63(19):195209. doi:10.1103/PhysRevB.63.195209.
[43] Robouch BV, Kisiel A, Konior J. Statistical model for atomic distances and site occupation in zinc-blende diluted magnetic semiconductors (dmss). J Alloys Compd 2002;340(1–2):13–26.
[44] Hashimoto M, Emura S, Asano R, Tanaka H, Teraguchi N, Suzuki A, et al. Local structure of rare-earth-doped diluted magnetic semiconductor gagdn. In: 5th international conference on nitride semiconductors (Icns-5), Proceedings. 2003. p. 922. doi:10.1002/pssc.200303320.
[45] Sikora M, Kapusta C, Maksymowicz L, Lubecka M, Cieciwa B, Szymczak R. et al. Exafs study of indium doped magnetic semiconductor $CdCr_2Se_4$. J Alloys Compd 2004;362(1–2):151–5. doi:10.1016/S0925-8388(03)00576-0.
[46] Soo YL, Kim S, Kao YH, Blattner AJ, Wessels BW, Khalid S, et al. Local structure around Mn atoms in room-temperature ferromagnetic (In,Mn) as thin films probed by extended X-ray absorption fine structure. Appl Phys Lett 2004;84(4):481–3. doi:10.1063/1.1640465. URL http://link.aip.org/link/?APL/84/481/1.
[47] Asahi H, Zhou YK, Hashimoto M, Kim MS, Li XJ, Emura S, et al. Gan-based magnetic semiconductors for nanospintronics. J Phys: Cond Mater 2004;16(48):S5555–62. doi:10.1088/0953-8984/16/48/011.
[48] Li BB, Xiu XQ, Zhang R, Tao ZK, Chen L, Xie ZL, et al. Analysis of local structures around NI atoms doped in ZnO-based diluted magnetic semiconductors by fluorescence exafs. Chinese Phys Lett 2006;23(4):907–10.
[49] Liao Yen-Fa, Hsu Hua-Shu, Huang Yen-Heng, Huang Tzu-Wen, Lin Ming-Zhe, Lee Chih-Hao, et al. Evidence for replacing zinc with cobalt and iron atoms in epitaxial ZnO:CoFe-diluted magnetic semiconductors. J Magn Magn Mater 2006;304(1):E161–3. doi:10.1016/j.jmmm.2006.02.246.
[50] Chattopadhyay S, Key SD, Shibata T, Viswanatha R, Balasubramanian M, Stoupin S, et al. Exafs studies of nanocrystals of $Zn_{1-x}Mn_xO$: a dilute magnetic semiconductor oxide system. X-Ray Absorption Fine Structure-Xafs13 2007;882:937.
[51] Wei Xu, Yingxue Zhou, Xinyi Zhang, Dongliang Chen, Yaning Xie, Tao Liu, et al. Local structures of Mn in dilute magnetic semiconductor ZnMnO. Solid State Commun 2007;141(7):374–7. doi:10.1016/j.ssc.2006.11.027.
[52] Han S-W, Kang H-S, Jiang B-Z, Seo S-Y, Park H-W, Ri H-C. Structural and magnetic properties of ZnMnTe semiconductor. J Magn Magn Mater 2007;310(2):2114–6. doi:10.1016/j.jmmm.2006.10.788.

[53] Li Bin-Bin, Shen Hong-Lie, Zhang Rong, Xiu Xiang-Qiang, Xie Zhi. Structural and magnetic properties of codoped ZnO based diluted magnetic semiconductors. Chinese Phys Lett 2007;24(12):3473–6.

[54] Farley NRS, Edmonds KW, Freeman AA, van der Laan G, Staddon CR, Gregory DH, et al. Magnetic properties of sol-gel-derived doped ZnO as a potential ferromagnetic semiconductor: a synchrotron-based study. New J Phys 2008;10:055012. doi:10.1088/1367-2630/10/5/055012.

[55] Tongfei Shi, Zhenguo Xiao, Zhijun Yin, Xinhua Li, Yuqi Wang, Hongtao He, et al. The role of Zn interstitials in cobalt-doped ZnO diluted magnetic semiconductors. Appl Phys Lett 2010;96(21):211905. doi:10.1063/1.3437082.

[56] Li M, Zhang B, Wang JZ, Shi LQ, Cheng HS, Wang YZ, et al. Exafs and sr-xrd study on Mn occupations in $Zn_{1-x}Mn_xO$ diluted magnetic semiconductors. Nucl Instrum Methods Phys Res Section B: Beam Interact Mater Atoms 2011;269(21):2610–3. ISSN 0168-583X. doi:10.1016/j.nimb.2011.07.094. URL http://www.sciencedirect.com/science/article/pii/S0168583X11007385.

[57] Iuşan D, Kabir M, Grånäs O, Eriksson O, Sanyal B. Microscopic picture of Co clustering in ZnO. Phys Rev B 2009;79(12):125202. ISSN 1098-0121. doi:10.1103/PhysRevB.79.125202.

Chapter 4

Theory of Magnetism in Graphene

Oleg V. Yazyev* and M.I. Katsnelson†

*Institute of Theoretical Physics, Ecole Polytechnique Fédérale de Lausanne (EPFL), CH-1015 Lausanne, Switzerland

†Institute for Molecules and Materials, Radboud University of Nijmegen, Toernooiveld 1, 6525 ED Nijmegen, The Netherlands

Chapter Outline Head

I. INTRODUCTION

Carbon plays unique role in nature, from nucleosynthesis processes in stars to its crucial importance as a basic material substrate of life. It is not surprising therefore that all its properties is a subject of great interest. Magnetic properties of elemental carbon were studied in a context of geology and cosmology [1], biochemistry [2], physics [3], and material sciences [4]. The possibility of intrinsic long-range magnetic order in carbon is intriguing from both basic and

Advanced Functional Materials. http://dx.doi.org/10.1016/B978-0-44-453681-5.00004-2

applied scientific point of view since the properties of sp-electron magnetic materials can be essentially different from conventional itinerant-electron ferromagnets. In particular, it follows from general theoretical considerations that they may be perspective candidates to magnetic semiconductors with the Curie points T_C above room temperature [5]. In principle, such materials may possess a number of other attractive properties such as low density, biocompatibility, plasticity, etc., which stimulates the search for light-element based magnetism [6]. At the same time, usually, very low values of magnetic moments are discussed which makes the problem extremely difficult from the experimental point of view, especially, keeping in mind that everything at our planet is polluted by magnetic iron oxides, essential components of the dust. This issue is therefore one of the most controversial in condensed matter physics and materials science (see, e.g., a detail review of studies of magnetism in CaB_6 in Ref. [5]).

Discovery of graphene and its outstanding electronic and structural properties (for review, see Refs. [7–11]) has given also a huge impact on studies of carbon magnetism. This is a part of a more general problem of the role of many-body effects in graphene. Theory predicted (long ago before the beginning of systematic experimental studies of graphene) that the correlation effects are essential leading to the non-Fermi-liquid behavior [12]. Until recently, it seemed that the whole graphene physics in essentially one-electron, and the many-body effects are not relevant, at least, without quantized magnetic fields. However, a breakthrough improvement of the quality of the samples allowed very recently to avoid the effects of disorder and charge inhomogeneities, which hide the correlation effects, and to observe many-body renormalization of the Fermi velocity predicted theoretically [13]. There are some theoretical arguments that freely suspended undoped graphene is not too far from various correlation-induced instabilities (see, e.g., [14–18]). The issue is completely open now, both theoretically and experimentally.

Magnetism in graphene and related materials, as an essentially many-body phenomenon, is very interesting therefore not only for "applied" reasons mentioned above but also in a context of a general problem about the role of correlation effects in physics of graphene. Here we present a review of current experimental and theoretical status of this problem.

II. EXPERIMENTAL EVIDENCE

The search for carbon-based magnetism has been unsuccessful for a long time even despite the fact that carbon is able to form very diverse and complex molecular structures. While stable organic molecules with localized magnetic moments (radicals) are known for long time, the first reproducible experimental reports of long-range magnetic ordering in p-block compounds were published in 1991 when magnetic ordering was observed in crystalline p-nitrophenyl nitronyl nitroxide (p-NPNN) [19,20] and in a charge transfer complex of C_{60} and tetrakis (dimethylamino) ethylene (TDAE) [21]. Molecular structures of these organic materials made of light elements only (C, H, N and O) are shown in Figure 1(a) and (b). In their molecular crystals the uncompensated electron spins are localized on weakly coupled molecular units. Because of the weak coupling between electron spins, the long-range magnetic order is realized only at low temperatures. The two organic materials mentioned above, p-NPNN and TDAE-C_{60}, are characterized by Curie temperatures of 0.6 and 16 K, respectively. Since 1991 a large number of other organic magnetic materials have been examined. In all cases the temperatures below which long-range magnetic order is established (Curie temperatures, T_C, and Néel temperatures, T_N, in the case of ferromagnetic and antiferromagnetic orderings, respectively) were much lower than room temperature, which renders such materials useless for practical applications.

The bottom-up approach towards chemical synthesis of graphene-like molecular radicals (in other words, nanometer-size magnetic graphene fragments) deserves special attention. The simplest example of such magnetic polyaromatic molecules is triangulane which will be discussed in great detail in the next section. Although triangulane itself has never been isolated, successful synthesis of its chemical derivatives (one of them is shown in Figure 1(c)) has been reported [22,24]. The spin-triplet ground state of these chemical compounds was verified by the electron spin resonance measurements. In principle, this example can be considered as an indirect proof of edge magnetism in graphene systems, at least in finite fragments produced by means of the chemical bottom-up approach. The synthesized triangulane derivatives are reactive molecules, but nevertheless can be handled in common organic solvents and stored for many months at room temperature provided the solution is isolated from atmospheric oxygen [24]. Larger magnetic triangular molecules have not been synthesized so far. Successful synthesis of a rectangular graphene fragment shown in Figure 1(d) has been reported by Konishi et al. [23]. The molecule is equivalent to a short segments of zigzag graphene nanoribbons. Experiments confirm pronounced biradical character (i.e., antiferromagnetic correlation between two localize electron spins) of this chemical structure. For an up-to-date overview of this subject we would refer the reader to a recent article of Morita et al. [25].

The next stimulating experiment was reported in 2001 when ferromagnetism with $T_C \approx 500$ K was observed in rhombohedral C_{60} under high pressure [26]. This observation, however, demonstrates very well the controversial character of the field. Five years later several authors retracted the original publication since the measured content of magnetic impurities was shown to be close to the amount needed to explain the observed magnetization of the samples [27]. In addition, the measured T_C was found to be very similar to the one of cementite Fe_3C. The question of possible high-temperature magnetic ordering in

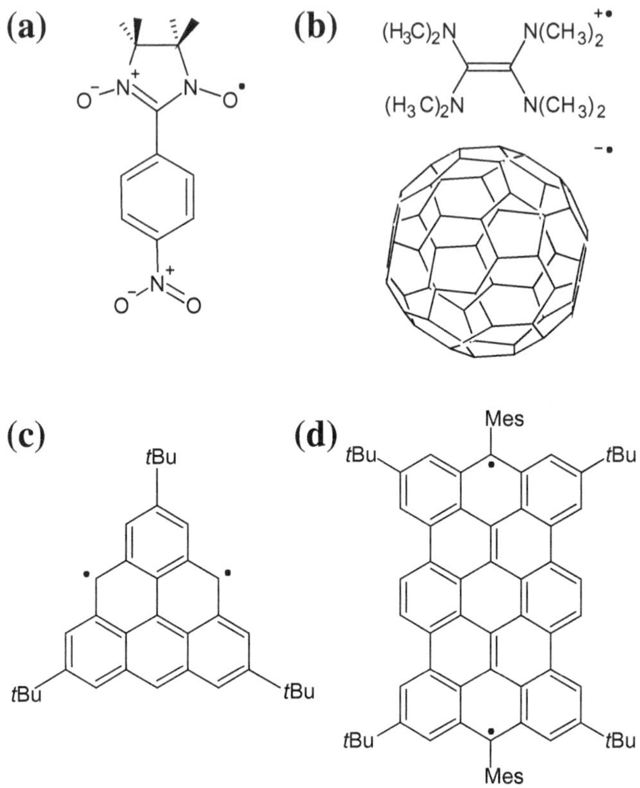

FIGURE 1 Chemical structures of (a) *p*-nitrophenyl nitronyl nitroxide (*p*-NPNN) and (b) the C_{60}-tetrakis (dimethylamino) ethylene (TDAE) charge transfer complex. Each molecular radical unit bears a single uncompensated electron spin. In chemical notation the presence of an unpaired electron is usually denoted by a thick dot. (c) *tert*-Butyl (*t*Bu) derivative of triangulane synthesized by Inoue et al. [22]. Its spin-triplet ground state has been verified by means of electron spin resonance measurements. (d) Chemically synthesized short fragment of graphene with zigzag edges which show pronounced biradical character (Mes = 2,4,6-trimethylphenyl) [23].

C_{60}-based materials remains open. For a recent discussion and references, see Ref. [28].

More recently, room-temperature ferromagnetism was observed in highly oriented pyrolytic graphite (HOPG) irradiated with high-energy (2.25 MeV) protons [29]. Figure 2 shows the magnetization loop for a proton-irradiated sample compared to untreated HOPG. Further experimental investigations revealed that the magnetic order in proton-bombarded graphite has two-dimensional, that is, graphene-like character [30] and originates from the carbon π-electron system rather than from possible d-element impurities [31]. Interestingly, it was shown that the chemical nature of the high-energy particles

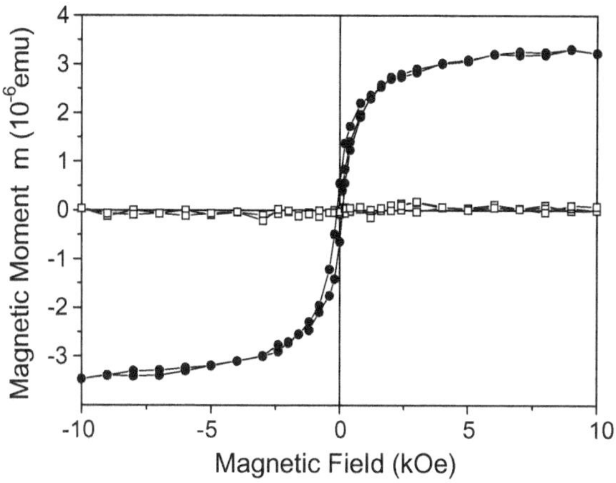

FIGURE 2 Magnetic moment of proton irradiated (filled circles) and untreated (empty squares) graphite as a function applied magnetic field as measured by Esquinazi et al. at $T = 300$ K. The weak hysteresis loop can be recognized. Reprinted from [29]. Copyright 2003 by the American Physical Society.

may play a role in producing magnetic ordering. While proton irradiation leads to the onset of ferromagnetism in irradiated samples, both helium [29] and iron [32] ions show no clear effect. On the other hand, the implantation of carbon ions was also reported to induce ferromagnetism in HOPG [33]. Interestingly, it has also been shown that the magnetism of irradiated graphite samples originates mostly from $\approx$ 10-nm-deep surface layer which is below the penetration depth of high-energy particles [34].

A number of reports have also pointed out that intrinsic graphite samples exhibit weak ferromagnetism [35,36]. Very recently, by using a combination of scanning probe techniques and magnetization measurements, Červenka and colleagues have shown that the intrinsic ferromagnetism of graphite is related to the presence of grain boundaries, intrinsic topological defects of polycrystalline bulk materials [37]. These observations have been rationalized by the fact that two-dimensional grain boundaries can be viewed as arrays of dislocations, and each of them may carry a localized magnetic moment at certain circumstances. The electronic properties of grain boundaries in graphite show strong dependence on the distance between the constituent dislocations (Figure 3).

Several groups reported the observation of room-temperature magnetic hysteresis in graphene samples produced in bulk quantities from graphite using the chemical approaches [38,39]. On the other hand, one report has concluded the absence of ferromagnetism in graphene produced by sonic exfoliation of graphite in different solvents [40].

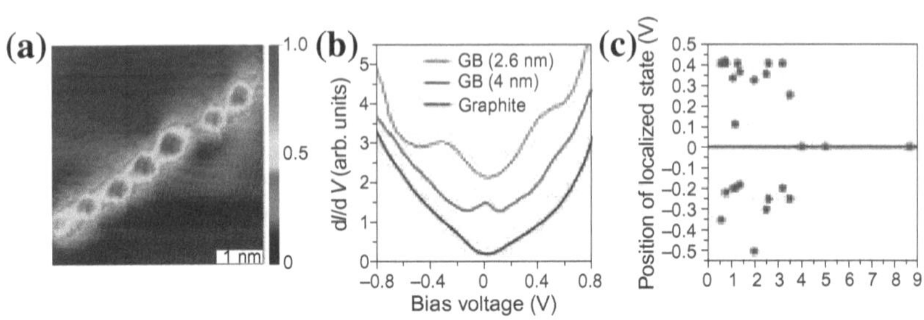

FIGURE 3 (a) Scanning tunneling microscopy image of a grain boundary in highly oriented pyrolitic graphite showing a 1D superlattice with a small periodicity of 1.4 nm. (b) $\mathrm{d}I/\mathrm{d}V$ spectra measured at two grain boundaries and compared to the $\mathrm{d}I/\mathrm{d}V$ spectrum of bare graphite surface. The spectra measured at grain boundary defect show either a pair of bonding/antibonding states or one localized state at the Fermi level. (c) Energy positions of localized states measured for a series different grain boundaries as a function of their superlattice periodicity. Reprinted from [37]. Copyright 2009 by Macmillan Publishers Limited.

Finally, the edge-state magnetism in graphene has recently received an indirect experimental confirmation based on the local probe measurements. The scanning tunneling microscopy/spectroscopy measurements of Tao and co-authors [41] performed on graphene nanoribbons with ultrasmooth low-symmetry (chiral) edges reveal the presence of edge states (Figure 4(a)). Importantly, the edge-localized states appear to be systematically split into a pair of peaks (Figure 4(b)) with the magnitude of splitting showing a clear correlation with the nanoribbon width (similar to the splitting observed at the grain boundaries in highly-oriented pyrolitic graphite [37] shown in Figure 3). This observation provides a strong indication of magnetic ordering fully consistent with the theoretical predictions which will be discussed in the following section.

III. THEORETICAL FRAMEWORK

A. Atomistic models: model Hamiltonians and counting rules

The simplest model Hamiltonian which can be utilized for studying magnetic structures based on graphene is the one-orbital mean-field Hubbard model. This model considers only the π-symmetry electronic states formed by the unhybridized p_z atomic orbitals of carbon atoms. The π-electron states constitute the dominant contribution to all low-energy properties of graphene including majority of magnetic effects. The Hubbard model Hamiltonian can be written as sum of two terms,

$$\mathcal{H} = \mathcal{H}_0 + \mathcal{H}'. \tag{1}$$

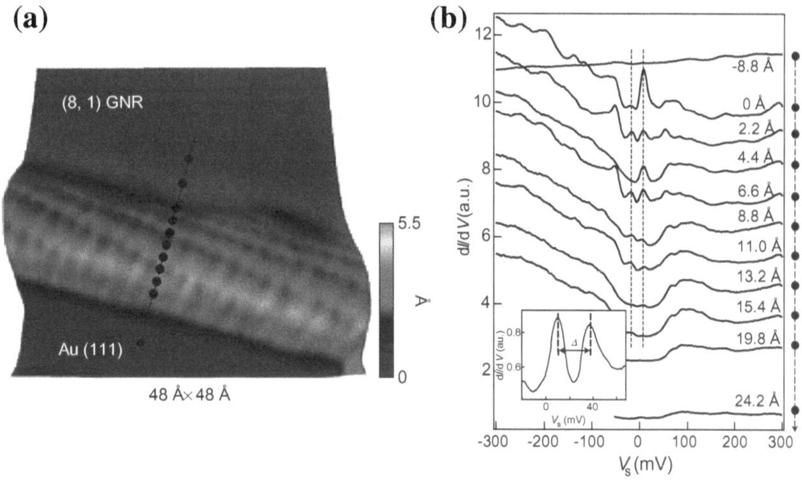

FIGURE 4 (a) Atomically resolved scanning tunneling microscopy (STM) image of the edge of (8, 1) chiral graphene nanoribbon on Au(1 1 1) surface. (b) dI/dV spectra of (8, 1) graphene edge measured at different points (black dots, panel a) along a line perpendicular to the edge at $T = 7$ K. Inset shows a higher resolution dI/dV spectrum for the edge of another graphene nanoribbon of (5, 2) chirality. Reprinted from [41]. Copyright 2011 by Macmillan Publishers Limited.

Here, the first term is the nearest-neighbor tight-binding Hamiltonian

$$\mathcal{H}_0 = -t \sum_{\langle i,j \rangle,\sigma} \left[c^{\dagger}_{i\sigma} c_{j\sigma} + \text{h.c.} \right], \tag{2}$$

in which the operators $c_{i\sigma}$ ($c^{\dagger}_{i\sigma}$) annihilate (create) an electron with spin σ at site (or atom) i, $\langle \cdot, \cdot \rangle$ stands for the pairs of nearest-neighbor atoms, and 'h.c.' is the Hermitian conjugate counterpart. The only empirical parameter in this model is the hopping integral t which effectively defines the energy scale. The often-quoted value of t is 2.7 eV (see, e.g., Ref. [42]). The hopping integral can also be related to the Fermi velocity ν_F close to the Dirac point as $t = 2\nu_F/(\sqrt{3}a)$ ($a = 0.246$ nm is the lattice constant of graphene). The experimental values $\nu_F \approx 10^6$ m/s leads to a larger $t = 3.09$ eV .

From the standpoint of practical calculations, the Hamiltonian matrix is the sparse $N \times N$ matrix for a system with N carbon atoms having sp^2 hybridization. The off-diagonal matrix elements (i, j) and (j, i) are equal to $-t$ if atoms i and j are covalently bonded to each other, and to 0 otherwise. In a neutral graphene system each sp^2 carbon atom contributes one p_z orbital and one π electron. Such π-electron system is called half-filled. Under realistic conditions all graphene systems are expected to be at half-filling or close to it. One important property of the tight-binding model Hamiltonian is electron–hole symmetry, i.e., its spectrum is symmetric with respect to zero energy. In other words, in a neutral graphene system for each eigenvalue $\epsilon < 0$ corresponding to an occupied (bonding) state, there is an unoccupied (anti-bonding) state with $\epsilon^\star = -\epsilon$. The states with $\epsilon = 0$ are called zero-energy states (also referred to as non-bonding or midgap states).

Despites its simplicity, the nearest-neighbor tight-binding model is able to describe accurately a broad range of electronic properties of graphene, carbon nanotubes, and other sp^2 carbon materials. However, studying the electronic structure of magnetic graphene systems requires more explicit treatment of electron–electron interactions. Within the Hubbard model these interactions are introduced through the on-site Coulomb repulsion term

$$\mathcal{H}' = U \sum_i n_{i\uparrow} n_{i\downarrow}, \tag{3}$$

where $n_{i\sigma} = c^\dagger_{i\sigma} c_{i\sigma}$ is the spin-resolved electron density at site i and the empirical parameter $U > 0$ defines the magnitude of the on-site Coulomb repulsion. The Hubbard model considers only the short-range Coulomb repulsion, that is, two electrons interact only if they occupy the p_z atomic orbital of the same atom. Nevertheless, being an effective parameter U can incorporate long-range interactions, to some degree.

Unfortunately, straightforward application of the Hubbard model to many realistic systems is no longer trivial and requires further modification. The mean-field approximation

$$\mathcal{H}'_{\mathrm{mf}} = U \sum_i \left(n_{i\uparrow} \langle n_{i\downarrow} \rangle + \langle n_{i\uparrow} \rangle n_{i\downarrow} - \langle n_{i\uparrow} \rangle \langle n_{i\downarrow} \rangle \right) \tag{4}$$

allows us to significantly reduce the computational demand. Here, a spin-up electron at site i interacts with the mean spin-down electron population $\langle n_{i\downarrow} \rangle$ at the same site, and vice versa. This mean-field model represents a lattice version of the unrestricted Hartree–Fock method [43]. A number of studies employing exact diagonalization, quantum Monte Carlo and functional renormalization group techniques [44–47] verified the validity of mean-field approximation for the relevant values of $U/t \sim 1$ (see below). In other words, under most circumstances magnetic graphene nanostructures cannot be considered as strongly correlated electronic systems.

From the computational point of view, the electron–electron interaction term affects only the diagonal elements of the Hamiltonian matrix. The diagonal elements of the spin-up and spin-down blocks now depend on the unknown $\langle n_{i\downarrow} \rangle$ and $\langle n_{i\uparrow} \rangle$, respectively. The problem can be solved self-consistently starting from some initial values of $\langle n_{i\sigma} \rangle$ which can be chosen randomly. However, one has to keep in mind that in certain cases the broken-symmetry (antiferromagnetic) solutions can be obtained only if the initial guess breaks the spin-spatial symmetry [48]. The process of (1) calculating the matrix elements of the Hamiltonian matrix, (2) its diagonalization, and (3) the computation of updated spin densities is then repeated iteratively until all values of $\langle n_{i\sigma} \rangle$ are converged. The final self-consistent solution provides the spin densities

$$M_i = \frac{\langle n_{i\uparrow} \rangle - \langle n_{i\downarrow} \rangle}{2}, \tag{5}$$

at each atom i and the total spin of the system $S = \sum_i M_i$. Remarkably, for a given graphene structure both local and total spins (magnetic moments) depend exclusively on the dimensionless parameter U/t.

The magnitude of U/t deserves special attention. So far, no straightforward experimental determination of this empirical parameter in graphene-based systems has been reported. The closest material investigated in this context is *trans*-polyacetylene, a one-dimensional sp^2 carbon polymer which can be viewed as a minimum-width zigzag graphene nanoribbon. Magnetic resonance studies of neutral soliton states in this material resulted in the range of values $U \approx 3.0$ eV [49,50]. It has been shown that the results of mean-field Hubbard model calculations correspond closely to the results obtained using first-principles methods based on density functional theory (which are also mean-field approaches) for certain values of U/t [51–53]. In particular, the results of the local-spin-density approximation calculations are best fitted using $U/t \approx 0.9$ while the generalized-gradient-approximation exchange–correlation functionals correspond to $U/t \approx 1.3$. The hybrid functional approaches involving exact exchange lead to larger values of U/t [52], as well as the very recent constrained RPA calculations [18] giving $U/t \approx 3$. One should stress, however, that the latter calculations demonstrate essentially non-local character of the effective Coulomb interaction in graphene remaining the question on the optimal value of U within the Hubbard model open.

In general, $U/t = 1$ can be considered as a reasonable choice for this empirical parameter, and, in general, increasing U/t results in the enhancement of magnetic moments. At $U/t \approx 2.23$ ideal graphene undergoes a Mott–Hubbard transition becoming an antiferromagnetically ordered insulating phase [54]. One should not forget, however, that U/t is an effective parameter; its magnitude can also be influenced by external conditions. For instance, a reduced value of U/t would be appropriate for a graphene system deposited on metallic substrate because of the enhanced screening of Coulomb interactions.

An evident advantage of the simple model Hamiltonians introduced above is the availability of rigorously proved theorems which allow to make important conclusion regarding the electronic structure of graphene systems in a 'counting rule' fashion rather than by performing straightforward numerical diagonalization of the Hamiltonian matrices. The spectrum of the tight-binding Hamiltonian of a honeycomb system (i.e., a graphene system with all faces being hexagons; all carbon atoms are either 2- or 3-fold coordinated) can be analyzed using the so-called benzenoid graph theory [55]. In particular, this theory provide direct access to the number of zero-energy states of the nearest-neighbor tight-binding Hamiltonian. The number of such states is equal to the graph's nullity

$$\eta = 2\alpha - N, \tag{6}$$

where N is the total number of sites and α is the maximum possible number of non-adjacent sites (i.e., the sites which are not the nearest neighbors to each other). The onset of magnetism in the system is determined by the so-called Stoner criterion which refers to the competition of the exchange energy gain and the kinetic-energy penalty associated with the spin-polarization of the system [56]. The gain in exchange energy is due to the exchange splitting of the electronic states subjected to spin-polarization [57]

$$\Delta_S = \epsilon_\uparrow - \epsilon_\downarrow = \frac{U}{2}\sum_i n_i^2, \tag{7}$$

where $\sum_i n_i^2$ is the inverse participation ratio, a measure of the degree of localization of the corresponding electronic state. The kinetic-energy penalty is proportional to the energy of this state. Thus, the zero-energy states undergo spin-polarization at any $U > 0$ irrespective of their degree of localization. One can view spin-polarization as one of the mechanisms for escaping an instability associated with the presence of low-energy electrons in the system. Other mechanisms, such as the Peierls distortion, were shown to be inefficient in the case of graphene nanostructures [52].

Although the benzenoid graph theory is able to predict the occurrence of zero-energy states, it is not clear how the electron spins align in these states. The complementary knowledge is supplied by Lieb's theorem [58] which determines the total spin of any bipartite system described by the Hubbard model. This theorem states that in the case of repulsive electron–electron interactions ($U > 0$), a bipartite system at half-filling has the ground state characterized by the total spin

$$S = \frac{1}{2}|N_A - N_B|, \tag{8}$$

where N_A and N_B are the numbers of sites in sublattices A and B, respectively. The ground state is unique and the theorem holds in all dimensions without the necessity of a periodic lattice structure.

B. Finite graphene molecules

We will now consider possible scenarios of magnetic ordering in different graphene systems. It is natural to start this discussion by addressing the origin of magnetism in finite graphene fragments as a function of their shape and size. In this section we will apply both the counting rules discussed above and perform explicit model Hamiltonian calculations. Three simple examples of nanometer-size graphene fragments are shown in Figure 5. In simplest case, these π-electron systems can be realized in the corresponding all-benzenoid polycyclic aromatic hydrocarbon (PAH) molecules. That is, the edges of these fragments are assumed to be passivated by hydrogen atoms and all carbon atoms are sp^2-hybridized. Current progress in synthesizing such molecules and understanding their properties has recently been reviewed [59].

The hexagonal graphene fragment shown in Figure 5(a) is equivalent to the coronene molecule. For this fragment, the number of sites belonging to the two sublattices is equal, $N_A = N_B = 12$. The number of non-adjacent sites is maximized when all atoms belonging to either of the two sublattices are selected, i.e., $\alpha = 12$. Thus, both the number of zero-energy states η and the total spin S are zero. The tight-binding model predicts a wide band gap of $1.08t \approx 3.0$ eV for this graphene molecule. As expected, the mean-field Hubbard model solution for this fragment does not reveal any magnetism. This is consistent with experimental results.

The second graphene fragment shown in Figure 5(b) has triangular shape. The corresponding hypothetical PAH molecule is called triangulane. The two sublattices of this triangular fragment are no longer equivalent ($N_A = 12$ and $N_B = 10$). The unique choice maximizing the number of non-adjacent sites is achieved by selecting the atoms belonging to the dominant sublattice

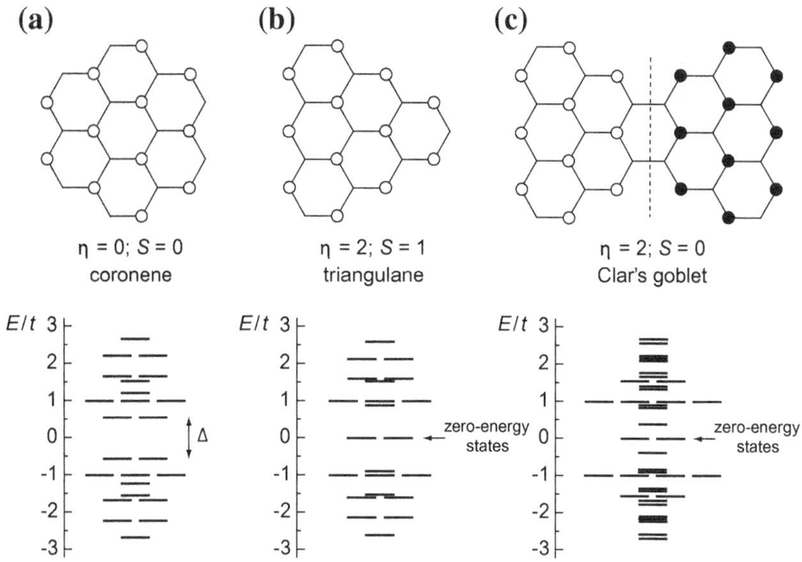

FIGURE 5 Atomic structures and tight-binding energy spectra of three graphene fragments: (a) coronene, (b) triangulane, and (c) a bowtie-shaped fragment ("Clar's goblet"). Non-adjacent sites are labeled by circles. For clarity, empty and filled circles distinguish sublattice A and sublattice B, respectively. The bottom panels show the corresponding tight-binding spectra with zero-energy states explicitly indicated.

A ($\alpha = N_A = 12$). Thus, the benzenoid graph theory predicts the presence of two zero-energy states on sublattice A. Lieb's theorem predicts the $S = 1$ (spin-triplet) ground state at half-filling or, equivalently, a magnetic moment of 2 μ_B per molecule. The two low-energy electrons populate a pair of zero-energy states according to Hund's rule, that is, their spins are oriented parallel to each other. The mean-field Hubbard model results ($U/t = 1.2$) for this system at half-filling are shown in Figure 6(a). One can see that spin-polarization lifts the degeneracy of the zero-energy electronic states and opens an energy gap $\Delta_S = 0.30t \approx 0.8$ eV. The system is stabilized by spin-polarization. Most of the spin-up electron density localized on the atoms in sublattice A (see Figure 6(a)) originates from the two electrons populating the non-bonding states. However, there is also a considerable amount of spin-down density on the atoms in sublattice B which is compensated by an equivalent contribution of the spin-up density in sublattice A. The occurrence of the induced magnetic moments is a manifestation of the spin-polarization effect which is related to the exchange interaction of the fully populated states with the two unpaired electrons.

The bowtie-shaped graphene molecule shown in Figure 6(c) is composed of two triangulane fragments sharing one hexagon. For this system Lieb's theorem predicts the spin-singlet ground state since $N_A = N_B = 19$. However, the choice of the set of atoms which maximizes the number non-adjacent sites is less evident in this case. Figure 6(c) shows such a selection ($\alpha = 20$) which involves the atoms belonging to both sublattice A and sublattice B in the two triangular parts. Hence, there are $\eta = 2 \times 20 - 38 = 2$ zero-energy states as verified by the tight-binding calculation (Figure 6(c)). The zero-energy states are spatially localized in the two parts of the molecule [60]. To satisfy the spin-singlet ground state, the two zero-energy states have to be populated by two electrons with oppositely oriented spins. In other words, the ground electronic configuration breaks spin-spatial symmetry resulting in antiferromagnetic ordering. This result can be verified by mean-field Hubbard model calculations as shown in Figure 6(b). It can be argued that this example violates Hund's rule. However, one has to keep in mind that each of the two non-bonding states is localized within one of the graphene sublattices. That is, there are two electronic sub-bands, each populated by electrons according to Hund's rule. The coupling between the electron spins in these two sub-bands is antiferromagnetic due to the superexchange mechanism [61,62]. The polyaromatic molecule corresponding to the considered bowtie fragment was hypothesized by Eric Clar and named "Clar's goblet" after him [63]. Attempts to synthesize this molecule have failed [64].

The two counting rules can be applied to larger graphene fragments. It was shown that the total spin of triangular fragments with edges cut along the zigzag direction scales linearly with fragment size [51,65,66] as illustrated in Figure 7. The average magnetic moment per carbon atom decays with increasing the system size. The evolution of magnetic properties with increasing size for

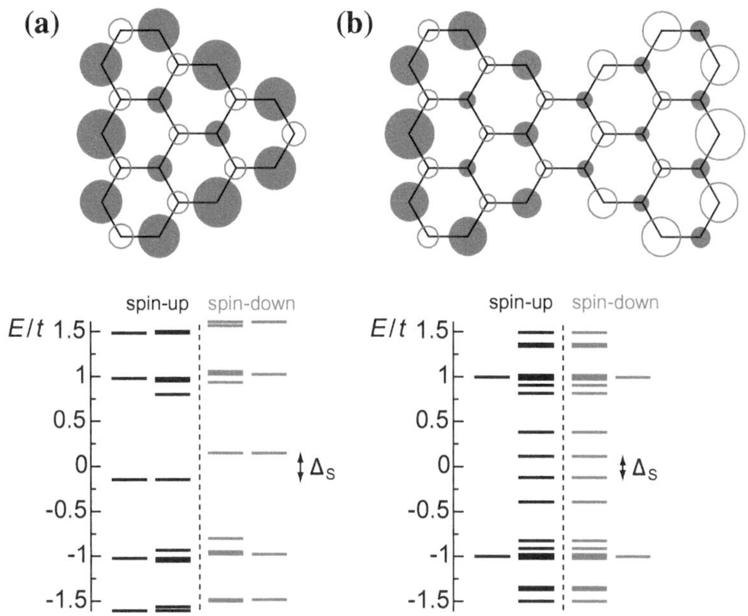

FIGURE 6 Local magnetic moments and spin-resolved energy levels obtained through the mean-field Hubbard model calculations for (a) triangulane and (b) "Clar's goblet" ($U/t = 1.2$). Area of each circle is proportional to the magnitude of the local magnetic moment at each atom. Filled and empty circles correspond to spin-up and spin-down densities, respectively. The bottom panels show the corresponding mean-field Hubbard model spectra for the two spin channels.

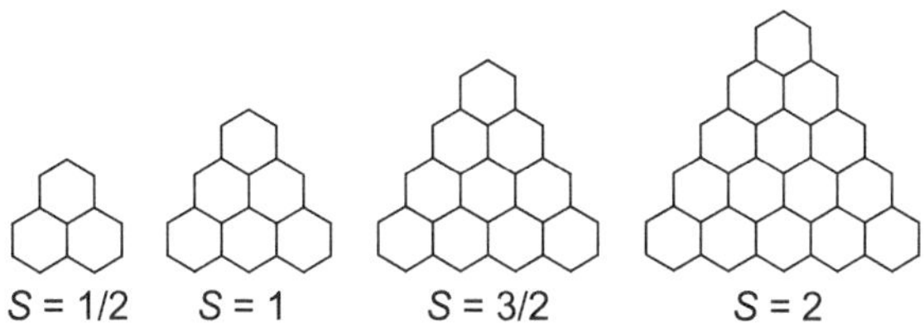

FIGURE 7 Evolution of the total spin of triangular graphene fragments with size. The total spin scales linearly with fragment size but the average magnetic moment per carbon atom decays with increasing the system size.

hexagonal fragments with edges cut along the same zigzag direction is less trivial. It has been shown theoretically that above some critical size the system undergoes a transition into a broken-symmetry antiferromagnetic state [51]. The critical size itself depends strongly on the value of U/t. However, it is easier to explain the origin of this behavior in large systems from the standpoint of edge magnetism covered in the next section.

A few words have to be said about the possibility of realizing in practice the magnetic graphene fragments we have discussed. It is expected that such magnetic systems are more reactive than the non-magnetic polyaromatic molecules. Although triangulane itself has never been isolated, successful synthesis of its chemical derivatives has been reported [22,24]. The spin-triplet ground state of these chemical compounds was verified by the electron spin resonance measurements. In principle, this example can be considered as an indirect proof of edge magnetism in graphene systems, at least in finite fragments produced by means of the chemical bottom-up approach. The synthesized triangulane derivatives are reactive molecules, but nevertheless can be handled in common organic solvents and stored for sufficiently long time at room temperature provided the solution is isolated from atmospheric oxygen [24]. Larger magnetic triangular molecules have not been synthesized so far. The PAH molecule corresponding to the considered bowtie fragment was hypothesized by Eric Clar and named "Clar's goblet" after him [63]. Attempts to synthesize this molecule have failed [64].

The examples shown above illustrate how three different magnetic scenarios can be realized in very simple finite graphene systems. These examples also provide a way for designing nanostructures with predefined magnetic interactions, a highly useful tool for developing novel spintronic devices. The value of this approach has already been demonstrated by the proposal of reconfigurable spintronic logic gates exploiting the strong antiferromagnetic couplings in the bowtie-shaped graphene fragments [60]. Several devices for controlling spin-currents based on triangular graphene fragments have also been described recently [67]. A number of other novel properties predicted for zero-dimensional graphene fragments are potentially interesting from both fundamental and applied points of view [68–71].

C. One-dimensional graphene edges and nanoribbons

Larger graphene nanostructures can be investigated by focusing on the edges. Convenient model for studying the edge effects are graphene nanoribbons, one-dimensional periodic strips of graphene. The basic structural parameters of a graphene nanoribbon are the crystallographic direction of its edges and its width. There are two high-symmetry crystallographic directions in graphene, zigzag, and armchair, which correspond to the directions of $(1,0)$ and $(1,1)$ vectors on graphene lattice. Alternatively, the edge direction can be described by using the so-called chirality angle

$$\theta = \arcsin\sqrt{\frac{3}{4}\left(\frac{m^2}{n^2+nm+m^2}\right)}, \tag{9}$$

where n and m define the translational vector (n, m) oriented along the edge. Edge chiralities of $\theta = 0°$ and $\theta = 30°$ correspond to zigzag and armchair edges, respectively. The structures of armchair and zigzag nanoribbons of ≈ 1.5 nm width are shown in Figure 8.

The electronic structures of armchair and zigzag nanoribbons are remarkably different. Figure 8(a) shows the tight-binding band structure of a ~ 1.5 nm wide armchair nanoribbon. For this particular armchair graphene nanoribbon, introducing a pair of parallel armchair edges opens a gap of $0.27t$. As expected for the nearest-neighbor tight-binding model, the density of states plot is symmetric with respect to $E = 0$ and shows a number of van Hove singularities as a result of quantum confinement in one dimension. In general, the tight-binding model predicts either metallic or semiconducting behavior for armchair nanoribbons [72–76], and the two situations alternate as the nanoribbon's width changes. The band gap of semiconducting graphene nanoribbons decreases with increasing width. In the case of metallic nanoribbons two bands cross the Fermi level at the Γ point.

Within the same model all zigzag graphene nanoribbons are metallic and feature a flat band extending over one-third of the one-dimensional Brillouin

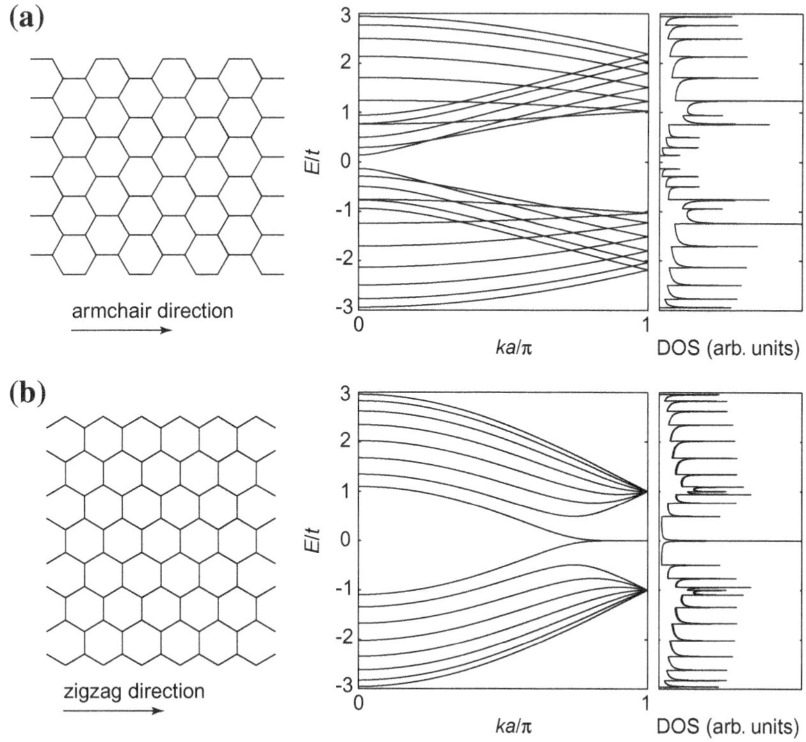

FIGURE 8 Atomic structures, tight-binding band structures and the corresponding density of states plots for (a) armchair and (b) zigzag graphene nanoribbons.

zone at $k \in (2\pi/3a; \pi/a)$ ($a = 0.25$ nm is the unit cell of the zigzag edge) as shown in Figure 8(b) [77]. These peculiar low-energy states are localized at the edge and decay quickly in the bulk region of graphene. High density of low-energy electronic states suggests a possibility of magnetic ordering close to the half-filling once electron–electron interactions are taken into account. The mean-field Hubbard model solution for this system reveals magnetic moments localized at the edges as shown in Figure 9(a) [78]. The ground state is characterized by ferromagnetic ordering along the zigzag edge while the mutual orientation of the magnetic moments localized at the opposite edges is antiparallel. Thus, the net magnetic moment of a zigzag nanoribbon is zero and Lieb's theorem is satisfied ($N_A = N_B$). The band structure and the density of states for the mean-field Hubbard model solution are compared to the tight-binding band structure in Figure 8(b). The introduced electron–electron interactions split the whole flat-band segment turning opening a band gap of $\Delta_z^0 = 0.16t$ (at $U/t = 1$). The high-amplitude peak in the tight-binding density of states splits into two pairs of van Hove singularities. One pair characterized by splitting Δ_z^0 corresponds to the band extrema at $k = 2\pi/3a$ while the larger splitting $\Delta_z^1 = 0.27t$ corresponds to another pair of band extrema at the Brillouin zone boundary $k = \pi/a$. The spin-polarization has almost no effect on the electronic states at higher energies. All van Hove singularities at higher energies are the result of quantum confinement in one dimension. The band structures for the two spin channels are equivalent, but spin-spatial symmetry is broken.

It is interesting to discuss the dependence of splittings Δ_z^0 and Δ_z^1 on the nanoribbon width. Figure 10 reproduces the results of first-principles

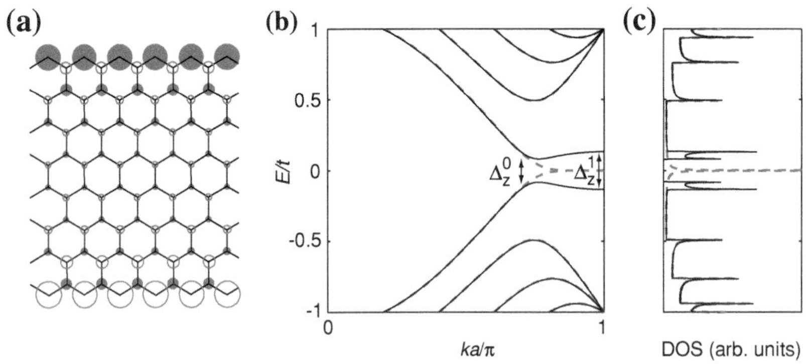

FIGURE 9 (a) Local magnetic moments in a zigzag graphene nanoribbon calculated using the mean-field Hubbard model. Area of each circle is proportional to the magnitude of the local magnetic moment at each atom. Filled and empty circles correspond to spin-up and spin-down densities, respectively. (b) Mean-field Hubbard-model band structure and the corresponding density of states plot (solid lines) compared to the tight-binding model results (dashed lines). The band structures for spin-up and spin-down electrons are equivalent.

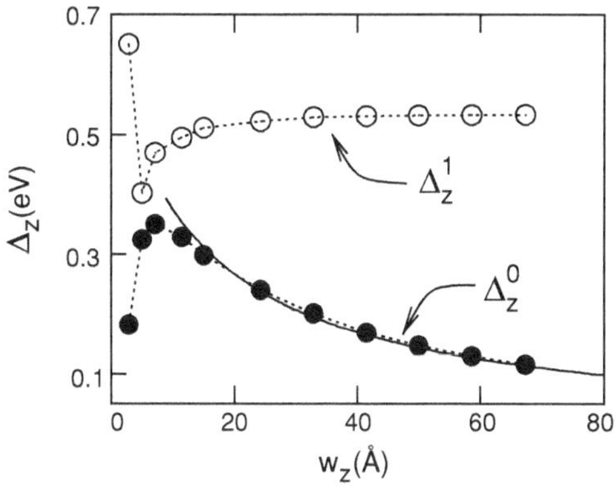

FIGURE 10 The variation of splitting Δ_z^0 and Δ_z^1 calculated from first-principles as a function of the width w_z of zigzag graphene nanoribbons. Reproduced from [73]. Copyright 2006 by the American Physical Society.

calculations for zigzag graphene nanoribbons from Ref. [73]. The states at $k = \pi/a$ are the most strongly localized states at the edge. The corresponding splitting Δ_z^1 thus shows very weak dependence beyond $w = 1$ nm. On the contrary, the states characterized by $k = 2\pi/3a$ penetrate deep into the bulk region. The corresponding splitting Δ_z^0 displays roughly w^{-1}-dependence on the nanoribbon width. These slowly decaying states are also responsible for the antiferromagnetic interaction between the magnetic moments localized at the opposite edges of the nanoribbon. The magnitude of the antiferromagnetic coupling has also been investigated theoretically and showed a w^{-2}-dependence as a function of nanoribbon width w [79]. The inter-edge magnetic coupling strength of ~25 meV per unit cell has been calculated from first principles for a ~1.5 nm wide zigzag nanoribbon [52]. Unlike the antiferromagnetic ground state, a zigzag graphene nanoribbon with ferromagnetic interedge orientation is a metal with two bands crossing the Fermi level at $k \approx 2\pi/3a$. The possibility of switching between the two states was exploited in a theoretical proposal of a graphene-based magnetic sensor [80]. The coupling between the magnetic moments localized at the edges can be controlled by means of either electron or hole doping of the nanoribbons [81,82]. High-doping levels eventually suppress magnetism since the flat band shifts away from the Fermi level, thus eliminating the electronic instability associated with the presence of low-energy electrons [82].

The presence of zero-energy states is actually a generic feature of low-symmetry edges, not only the high-symmetry zigzag edges. Already in the pioneering work of Nakada et al. it was demonstrated numerically that all chiral graphene nanoribbons exhibit such edge states. Later, it was shown analytically that in the limit of infinite nanoribbon width the density of localized zero-energy states per edge unit length

$$\rho(\theta) = \frac{2}{3a_0}\cos\left(\theta + \frac{\pi}{3}\right) \tag{10}$$

changes monotonically with the chirality angle θ [83]. The maximum $\rho = 1/(3a)$ is achieved in the case of zigzag edges while only strictly armchair edges show no edge states. An example of chiral graphene nanoribbon ($\theta = 13.9°$) is shown in Figure 11(a). Periodicity of this structure is defined by the $(3, 1)$ translational vector. The chiral edge is composed of one armchair and two zigzag units. Figure 11(b) shows the magnetic moments per edge length M

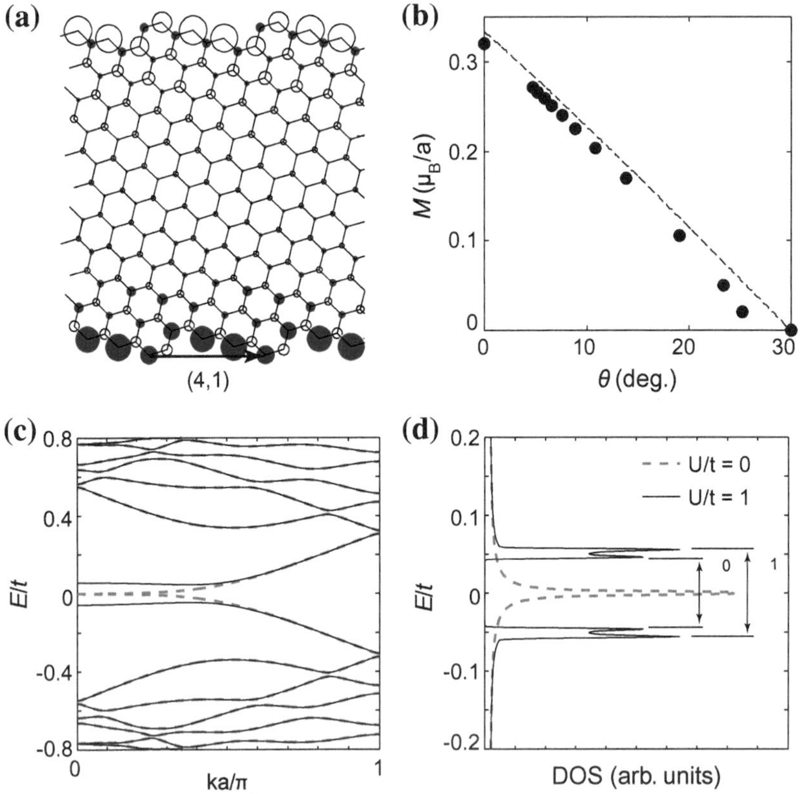

FIGURE 11 (a) An example of chiral graphene nanoribbon ($\theta = 13.9°$) defined by periodicity vector $(3, 1)$. Filled and empty circles correspond to spin-up and spin-down densities, respectively. (b) Magnetic moment per edge unit length M as a function of chirality angle θ (circles) calculated for ~7-nm-wide chiral graphene nanoribbons using the mean-field Hubbard model approach ($U/t = 1$). The dashed line corresponds the edge state density in the limit of infinite width (Expr. (10)). (c) Band structure and (d) density-of-states plot for the nanoribbon shown in panel (a) obtained at $U/t = 0$ (dashed line) and $U/t = 1$ (solid line).

calculated for a series of chiral graphene nanoribbons characterized by different angles θ using the mean-field Hubbard model approach [84]. The calculated values of M follow closely the density of edge states given by expression (10). Only graphene nanoribbons with edges oriented along armchair direction or close to it have non-magnetic ground state. Like in the case of zigzag nanoribbons, chiral nanoribbons feature a zero-energy edge-state flat band when considered in the absence of electron–electron interactions ($U/t = 0$), while at $U/t = 1$ the ground state solution is a semiconducting magnetic state. However, the energy splittings, in particular Δ^1, are reduced in comparison with the corresponding splittings in the case of zigzag graphene nanoribbon of equivalent width. This is illustrated by the band structure and density-of-states plots calculated for the above-mentioned (3, 1) chiral graphene nanoribbon with the edge orientation ($\theta = 13.9°$) situated right in between the high-symmetry zigzag ($\theta = 0°$) and armchair ($\theta = 30°$) directions.

Another illustrative example of the edge-state effects was presented in a tight-binding investigation of a large graphene quantum dot [85]. The local density of states plot (Figure 12(a)) shows the presence of edge states along the entire perimeter of this irregular shape fragment. The density of edge states is diminished only in the regions where the edge orientation is close to the armchair direction. The effects of random atomic-scale disorder at the edge were investigated as well. It turns out that in the case of rough edges the density of zero-energy states decrease, but do not vanish completely (Figure 12(b)). That is, the edge magnetism in graphene is expected to be rather robust with respect to disorder.

It has been realized that the novel magnetic properties of graphene nanostructures are very promising from the point of technological applications, especially in spintronics. The pioneering idea was introduced by Son, Cohen,

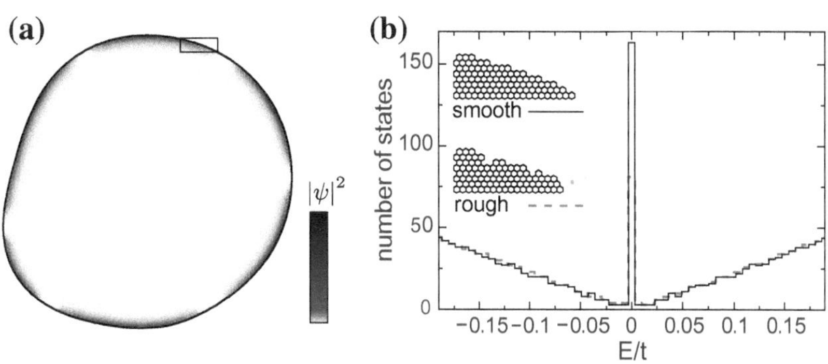

FIGURE 12 (a) The local density of states due to edge states in a graphene quantum dot of irregular shape with smooth boundaries. (b) Number of states per energy interval dE for the quantum dot shown in panel (a) with smooth (solid line) and rough (dashed line) edges. Reproduced from [85]. Copyright 2008 by the American Physical Society.

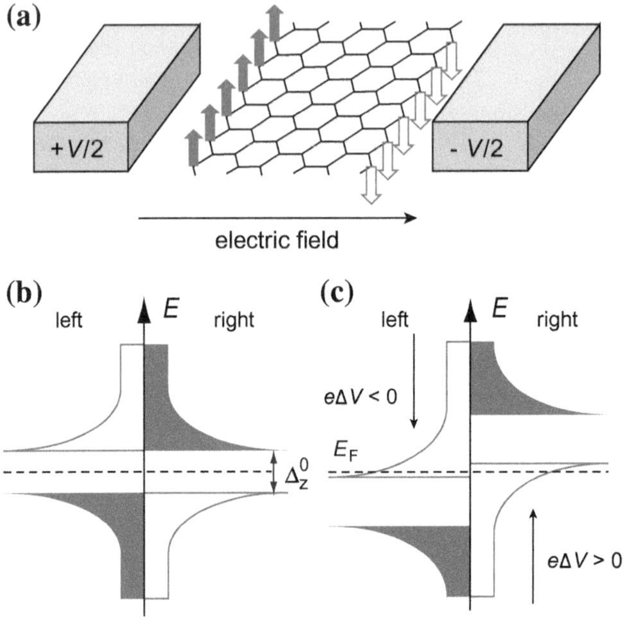

FIGURE 13 Electric-field-induced half-metallicity in zigzag graphene nanoribbons. (a) Electric field is applied across the nanoribbon, from left edge (spin-up) to right edge (spin-down). (b) Schematic representation of the spin-resolved local density of states for the opposite edges at zero applied field. (c) Applied electric field closes the band gap at the Fermi level E_F for spin-down electrons selectively.

and Louie, who have predicted that external electric fields induce half-metallicity in zigzag graphene nanoribbons [86]. The half-metallicity refers to the coexistence of a metallic state for electrons with one spin orientation and an insulating state for electrons with the opposite spin orientation. An electric field is applied across the nanoribbon as shown in Figure 13(a). At zero field the system is characterized by the energy gap Δ_z^0 for the spin-polarized states localized at both edges (Figure 13(b)). An applied electric field breaks the symmetry and closes the gap for one of the spin directions selectively (Figure 13(c)). The critical field required for inducing the half-metallicity was estimated to be $3.0/w$V , where w is the nanoribbon width in Å. The direction of the applied electric field defines the spin channel with metallic conductivity. If realized in practice, this simple device would offer efficient electrical control of spin transport – a highly demanded component of spintronic circuits.

However, magnetic order in the nanometer scale two-dimensional systems is particularly sensible to thermal fluctuations. In particular, the Mermin–Wagner theorem excludes long-range order in one-dimensional magnetic systems (such as the magnetic graphene edges) at any finite temperature [87]. The range of magnetic order is limited by the temperature-dependent spin correlation lengths ξ^{α} ($\alpha = x, y, z$) which define the decay law of the spin correlation

$$\langle \hat{s}_i^{\alpha} \hat{s}_{i+l}^{\alpha} \rangle = \langle \hat{s}_i^{\alpha} \hat{s}_i^{\alpha} \rangle \exp(-l/\xi^{\alpha}), \tag{11}$$

where $\hat{s}_i^{\alpha}$ are the components of magnetic moment unit vector $\hat{\mathbf{s}}_i$ at site i. In principle, the spin correlation length imposes the limitations on the device dimensions. In order to establish this parameter one has to determine the energetics of the spin fluctuations contributing to the breakdown of the ordered ground-state configuration.

The energetics of the transverse and longitudinal spin excitations (Figure 14(a) and (b)) have been explored using density-functional-theory calculations [88]. The magnetic correlation parameters in the presence of spin-wave fluctuations, the dominant type of spin disorder in this case, were obtained with the help of one-dimensional Heisenberg model Hamiltonian

$$H = -a \sum_i \hat{\mathbf{s}}_i \hat{\mathbf{s}}_{i+1} - d \sum_i \hat{s}_i^z \hat{s}_{i+1}^z, \tag{12}$$

where the Heisenberg coupling $a = 2\kappa/a_z^2 = 105$ meV corresponds to the spin-wave stiffness $\kappa = 320$ meV Å^2 calculated from first principles. The estimated small anisotropy parameter $d/a \approx 10^{-4}$ originates from the weak spin–orbit interaction in carbon. This simple model Hamiltonian has known analytic solutions [89]. Figure 14(c) shows the spin correlation lengths calculated for our particular case. Above the crossover temperature $T_x \approx 10$ K,

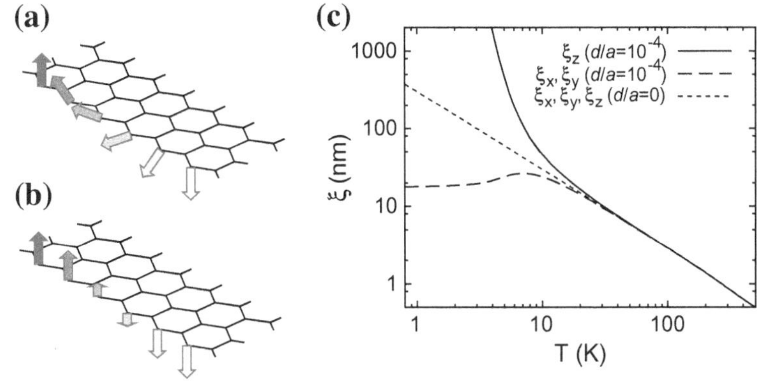

FIGURE 14 Schematic representation of the transverse (a) and the longitudinal (b) spin excitation at zigzag graphene edges. The magnetic moments are shown by arrows. Their directions are represented by orientation and the magnitude is shown by their lengths. (c) Correlation lengths of magnetization vector components orthogonal (ξ_z) and parallel (ξ_x, ξ_y) to the graphene plane as a function of temperature. Reproduced from [88]. Copyright 2008 by the American Physical Society.

weak magnetic anisotropy does not play any role and the spin correlation length $\xi \propto T^{-1}$. However, below T_x the spin correlation length grows exponentially with decreasing temperature. At $T = 300$ K the spin correlation length $\xi \approx 1$ nm.

From a practical point of view, this means that the dimensions of spintronic devices based on the magnetic zigzag edges of graphene and operating at normal temperature conditions are limited to several nanometers. At present, such dimensions are very difficult to achieve, which can be regarded as a pessimistic conclusion. Nevertheless, one has to keep in mind that the spin stiffness predicted for the magnetic graphene edges is still higher than the typical values for traditional magnetic materials. That is, graphene outperforms d-element based magnetic materials, and there is a room for improvement. Achieving control over the magnetic anisotropy d/a could possibly raise the crossover temperature T_x above 300 K and thus significantly extend ξ. Possible approaches for reaching this goal include chemical functionalization of the edges with heavy-element functional groups or coupling graphene to a substrate.

D. Defect-induced magnetism in graphene and graphite

Experimental observations of ferromagnetic ordering in irradiated graphite as well as recent reports of magnetic ordering in chemically processed graphene have already been mentioned in the overview of experimental progress. These results are particularly exciting because of the fact that the induced magnetic ordering is stable at room temperature and well above. Let us now try to understand the origin of magnetism in irradiated graphite. The present section covers the cases of both graphene and graphite which has a three-dimensional crystalline lattice composed of weakly coupled graphene layers.

The basic picture of the radiation damage process in carbon materials is relatively simple. Irradiation of graphite with high-energy particles (e.g., protons) produces several types of point defects. In carbon materials the defects are created as a result of so-called "knock-on collisions" involving the direct transfer of kinetic energy from the high-energy incident particles to the individual atoms [90,91]. If the transferred energy is larger than the displacement threshold T_d, the recoil atom may leave its equilibrium position leading to the formation of a pair of point defects – a vacancy defect and an interstitial. The displacement threshold T_d for carbon atoms in graphitic materials was found to be ~20 eV in a number of studies [92–95]. Creation of defects due to electron stopping, i.e., the process involving possible electronic excitations and ionization of individual atoms, is less important in carbon materials since electronic excitations in metals are delocalized and quench instantly [90,91].

After slowing down, reactive particles may also produce chemisorption defects. In particular, protons are able to bind to individual carbon atoms in graphene lattice resulting in their rehybridization into the sp^3-state. Such

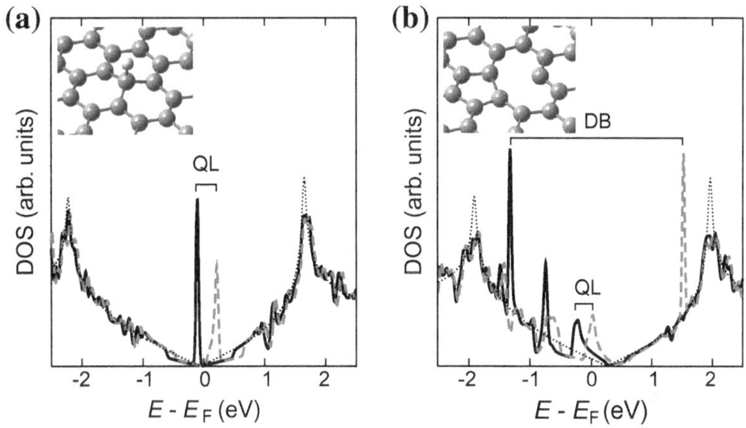

FIGURE 15 Spin-resolved density of states (DOS) for (a) the hydrogen chemisorption defect and (b) the vacancy defect in graphene calculated from first principles. Solid and dashed curves correspond to the majority and minority spins, respectively. Dotted curve shows the reference density of states of the ideal graphene. Exchange-split peaks which correspond to quasi-localized (QL) and dangling-bond (DB) states are labeled. The inset show atomic structures of the defects.

defects are referred to as hydrogen chemisorption defects. From the point of view of single-orbital models both vacancy and hydrogen chemisorption defects (Figure 15, insets) are equivalent – one p_z-orbital is removed from the π-system of graphene. In the first case, the p_z-orbital is eliminated together with the knocked-out carbon atom. The hydrogen chemisorption does not remove the carbon atom from the crystalline lattice, but once rehybridized the atom is unable to contribute its p_z-orbital to the π-electron system. These two types of defects are further referred to as p_z-vacancies.

The defects described above are the primary defects in the radiation-damage process. More complex defects can be produced at later stages of the process. For instance, single-atom vacancies and interstitial may aggregate producing extended defects. Complexes involving two or more different defects can also be formed upon irradiation. Examples are complexes of hydrogen with vacancies and interstitials [96], and intimate Frenkel pairs [97,94]. Radiation damage in graphitic materials may also produce the Stone–Wales defects [98,99].

The single atom p_z-vacancies described above have a particularly profound effect on the electronic structure of graphene. Let us consider a periodically repeated supercell of graphene composed of $2N (N_A = N_B = N)$ carbon atoms. Elimination of one atom from sublattice A introduces a zero-energy state in the complementary sublattice ($\alpha = N_B$; thus $\eta = 2N_B - ((N_A - 1) + N_B) = 1$). Such zero-energy states extending over large distances are called quasi-localized states since they show a power-law decay [100,101]. The quasi-localized states have been observed in a large number of scanning tunneling microscopy (STM) studies of graphite as triangular $\sqrt{3} \times \sqrt{3}R30^\circ$ superstructures extending over a few nanometers and localized around point defects [102–104]. For the single-defect model we have adopted, Lieb's theorem predicts a magnetic moment of $|(N_A - 1) + N_B| = 1\ \mu_B$ per supercell, that is, the presence of a defect induces ferromagnetic ordering.

This result has been widely confirmed using both first-principles [96,105–107] and mean-field Hubbard model [108,57] calculations. Figure 15 shows the spin-resolved density-of-states plots for hydrogen chemisorption and vacancy defects obtained using first-principles calculations [106]. In the first case, the sharp peak close to the Fermi level corresponds to the quasi-localized state band induced by the chemisorbed hydrogen atom. The peak is fully split by exchange and the system is characterized by a magnetic moment of 1 μ_B irrespective of defect concentration. The distribution of spin density around the defective site clearly shows a $\sqrt{3} \times \sqrt{3}R30^\circ$ superstructure [106]. The case of vacancy defect is somewhat more complicated. In addition to the quasi-localized state, there is also a localized non-bonding state due to the presence of one σ-symmetry dangling bond (Figure 15(b), inset). The dangling-bond state shows a very strong exchange splitting and contributes 1 μ_B to the total magnetic moment of the defect. However, the magnetic moment due to the quasi-localized state is partially suppressed in this case due to the self-doping effect related to the structural reconstruction of the vacancy [106]. The overall magnetic moment per vacancy defect varies from 1.12 to 1.53 μ_B for defect concentrations ranging from 20% to 0.5%.

Magnetic moments due to dangling bonds can also be contributed by other types of defects, e.g., the bridge-configuration interstitial defect in graphene [109]. However, one has to keep in mind that magnetic ordering due to only localized magnetic moments in graphene-based system is improbable at high temperatures. The Ruderman–Kittel–Kasuya–Yoshida interaction is weak in this case due to the semi-metallic electronic structure of graphene [110,111]. On the other hand, magnetic ordering due to the quasi-localized states can be considered as it inerant magnetism without excluding a possible contribution of dangling-bond magnetic moments to the net magnetic moment of a defective carbon system.

The system with one defect placed in a periodically repeated supercell is only a first-approximation model of disordered graphene for two reasons. First, all defects are located in the same sublattice of the graphene layer. Second, the defects form an ordered periodic superlattice. A more realistic description of disorder can be achieved by constructing models with defects randomly distributed in a large enough supercell [112]. Such models allow defects to occupy both sublattices at arbitrary concentrations and eliminate any short-range order in the spatial arrangement of defects. Larger supercells are needed for building the disordered models which makes first-principles calculations impractical. However, such system can still be treated using the mean-field Hubbard model calculations.

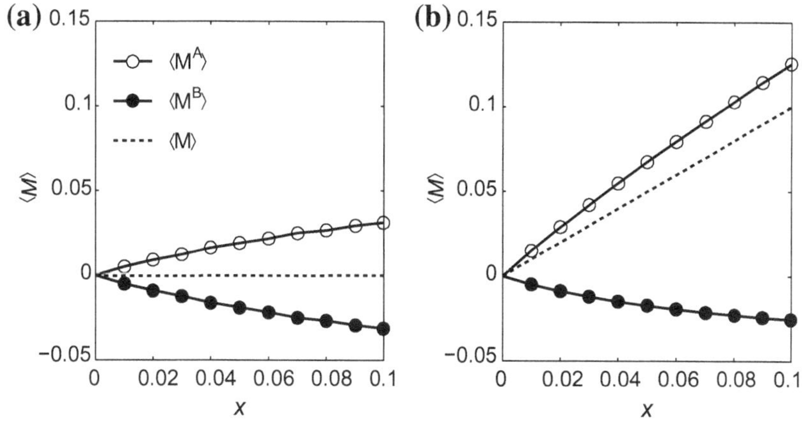

FIGURE 16 Average magnetic moments for the atoms in sublattices A and B as a function of defect concentration x. The defects are either (a) distributed equally between the two sublattices or (b) belong to sublattice B only. Net magnetic moments per carbon atom (dotted line) are shown.

Figure 16 shows the mean magnetic moment $\langle M^A \rangle$ and $\langle M^B \rangle$ per carbon atom in sublattice A and sublattice B as a function of defect concentration x [112]. The resulting values have been averaged over many random placements of defects in the simulation supercell for two different scenarios: defects equally distributed over the two sublattices ($N_A^d = N_B^d$, Figure 16(a)) and defects occupying to sublattice B only ($N_A^d = 0$, Figure 16(b)). In the first case, the magnetic moments in the two sublattices compensate each other. The overall magnetic ordering is of *antiferromagnetic* character. When defects populate only one sublattice the system exhibits *ferromagnetic* ordering. The net magnetic moment per carbon atom $\langle M \rangle = (\langle M^A \rangle + \langle M^B \rangle)/2 = x/2$ scales linearly with the defect concentration. Both numerical results are in full agreement with Lieb's theorem.

Experimental observations of ferromagnetism in irradiated graphite point to the conclusion that sublattices of individual graphene layers in bulk graphite are populated by defects differently: that is, there must be a mechanism which makes the sublattices of graphene inequivalent. Such an intrinsic discriminating mechanism was ascribed to the stacking order of graphite layers in bulk graphite [112]. The lowest-energy ABA stacking order of individual graphene sheets in graphite breaks the equivalence of the two sublattices as shown in Figure 17(a). In fact, only local ABA stacking order is required to discriminate between the two sublattices of the middle sheet. The mechanism can be demonstrated for the case of hydrogen chemisorption defects. First-principles calculations show that the configuration which involves hydrogen chemisorbed on sublattice B is 0.16 eV lower in energy than on sublattice A (see Figure 17(b) and (c). This energy difference is sufficient to trigger a considerable difference

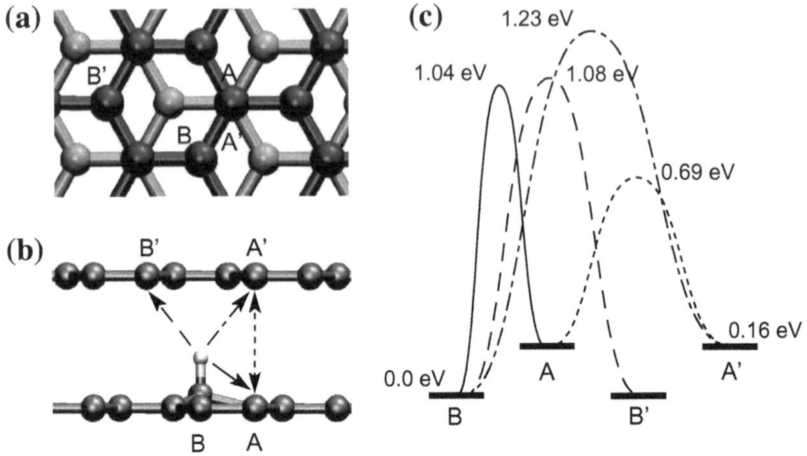

FIGURE 17 (a) Inequivalent carbon atoms (A and B) in ABA graphite. (b) Possible pathways for the diffusion of chemisorbed hydrogen in graphite. (c) Schematic representation of the potential energy surface for the in-plane diffusion of hydrogen in graphite showing the relative energies for the local minima and transition states. Reproduced from [112]. Copyright 2008 by the American Physical Society.

in equilibrium populations of the two sublattices. The energy barrier for the hopping of chemisorbed hydrogen atoms is relatively small ($\sim$1 eV) [113] to allow for thermally activated diffusion at mild temperatures.

Similar discriminating mechanisms may also exist for the other types of defects created by irradiation, e.g., for vacancies. Cross-sections for momentum transfer during knock-on collisions with high-energy incident particles are likely to be very similar for both A and B carbon atoms in graphite. However, the stacking order may have a strong influence on the recombination of interstitial and vacancy defects close to equilibrium conditions. It was also shown that instantaneous recombination of low-energy recoil atoms in graphite is significantly more probable for atoms in position A [94]. That is, more vacancies in sublattice B are left assuming an equal number of knock-on collisions involving the atoms of both types. These results allow us to conclude that the most probable physical picture of magnetic order in irradiated graphite is ferrimagnetism. The magnetic moment induced by defects in sublattice A is larger than the one induced in sublattice B.

The question of magnetic ordering in defective graphene and graphite at finite temperatures remains largely unaddressed. Similar to the one-dimensional system discussed above, an isotropic two-dimensional system cannot develop long-range magnetic ordering at any finite temperature [87]. However, the introduction of a small magnetic anisotropy $d/a \sim 10^{-3}$ already leads to very high transition temperatures [114,30]. Weak magnetic coupling between the individual layers in graphite also produces a pronounced effect on the magnetic transition temperature [115].

E. Theory of *sp*-electron magnetism in narrow impurity bands beyond the mean-field approximation

Defect-induced magnetism in *sp*-electron systems has some general features dramatically distinguishing it from more traditional itinerant d-electron magnetism. Here we will discuss these differences following Ref. [5]. We have shown that, actually, the mean-field consideration of the previous sections works in this situation quite well, contrary to the d-electron case where it fails completely, at least, to describe finite-temperature effects [116].

First of all, even at zero temperature, the renormalization of the effective Stoner parameter due to multiple electron–electron scattering (T-matrix renormalization [117]) should be taken into account. For the case of magnetism in narrow bands $W \ll U$ (W is the effective bandwidth) it decreases the effective value of U to the values of the order of bandwidth. As a result, the effective Stoner criterion $\alpha = I_{\text{eff}} N(E_F) > 1$ (I_{eff} is the effective Stoner parameter and $N(E_F)$ is the density of states at the Fermi level is typically satisfied at the border, and the product is rarely larger than, say, 1.2 [5]. The case of magnetism in a narrow defect bands is different, and the T-matrix renormalization turns out to be not essential. To demonstrate this, let us start with a general multi-band, non-local Hamiltonian for interacting electrons on a crystal lattice:

$$\begin{aligned} \mathcal{H} &= \mathcal{H}_0 + \mathcal{H}', \\ \mathcal{H}_0 &= \sum_{\lambda\lambda'\sigma} t_{\lambda\lambda'} c^{+}_{\lambda\sigma} c_{\lambda'\sigma}, \\ \mathcal{H}' &= \frac{1}{2} \sum_{\{\lambda_{im}\}\sigma\sigma'} \left\langle \lambda_1\lambda_2 \left| v \right| \lambda'_1\lambda'_2 \right\rangle c^{+}_{\lambda_1\sigma} c^{+}_{\lambda_2\sigma'} c_{\lambda'_2\sigma'} c_{\lambda'_1\sigma}, \end{aligned} \tag{13}$$

where $\lambda = im$ are the site number (i) and orbital (m) quantum numbers, $\sigma = \uparrow, \downarrow$ is the spin projection, c^{+}, c are the Fermi creation and annihilation operators, H_t is the hopping Hamiltonian, and the Coulomb matrix elements are defined in the standard way

$$\left\langle 12 \left| v \right| 34 \right\rangle = \int d\mathbf{r}\, d\mathbf{r}' \psi_1^*(\mathbf{r}) \psi_2^*(\mathbf{r}') v\left(\mathbf{r} - \mathbf{r}'\right) \psi_3(\mathbf{r}) \psi_4(\mathbf{r}'), \tag{14}$$

where we define for briefness $\lambda_1 \equiv 1$, etc. Following to [118] let us take into account the ladder (T-matrix) renormalization of the effective interaction:

$$\begin{aligned} \left\langle 13 \left| T^{\sigma\sigma'}(i\Omega) \right| 24 \right\rangle = \left\langle 13 \left| v \right| 24 \right\rangle - \frac{1}{\beta} \sum_{\omega} \sum_{5678} \left\langle 13 \left| v \right| 57 \right\rangle \\ \times G^{\sigma}_{56}(i\omega) G^{\sigma'}_{78}(i\Omega - i\omega) \left\langle 68 \left| T^{\sigma\sigma'}(i\Omega) \right| 24 \right\rangle, \end{aligned} \tag{15}$$

where $\omega = (2n+1)\pi T$ are the Matsubara frequencies for temperature $T \equiv \beta^{-1}(n = 0, \pm 1, \ldots)$. Using the spectral representation for the Green's function

$$G^{\sigma}_{56}(i\omega) = \int_{-\infty}^{\infty} \mathrm{d}x \frac{\rho_{56}(x)}{i\omega - x}, \tag{16}$$

substituting Eq. (16) into Eq. (15) and calculating the sum over Matsubara frequencies by the usual way [119] one can rewrite Eq. (15) in terms of the spectral density matrix $\widehat{\rho}(x)$:

$$\langle 13\,|T(E)|\,24\rangle = \langle 13\,|v|\,24\rangle + \sum_{5678} \langle 13\,|v|\,57\rangle \langle 57\,|P(E)|\,68\rangle \langle 68\,|T(E)|\,24\rangle\,,$$
$$\langle 57\,|P(E)|\,68\rangle = \int_{-\infty}^{\infty} \mathrm{d}x \int_{-\infty}^{\infty} \mathrm{d}y \frac{1 - f(x) - f(y)}{E - x - y} \rho_{56}(x)\rho_{78}(y), \tag{17}$$

where E is the *real* energy. We neglect here and further spin-polarization since the effective exchange parameter I_{eff} in the Stoner criterion should be calculated in the paramagnetic phase.

If we take into account, for simplicity, only on-site Coulomb interaction (the Hubbard approximation) than the T-matrix turns out to be also diagonal in site indices and the matrix equation (17) holds assuming that $1, 2, \ldots, 8$ label only orbital indices and $\widehat{\rho}(x)$ is local (on-site) spectral density.

Let us consider now the case when the energy spectrum is of type of that shown in Figure 15(a), with a broad valence band (region I) and narrow band of quasi-localized (QL) states (region II). We assume that the magnetism is related with the latter ones only. To estimate different contributions to the function $\widehat{P}(E)$ one should take into account that

$$\int_I \int_I \frac{\mathrm{d}x\,\mathrm{d}y}{x+y} \rho(x)\rho(y) \sim \frac{1}{W} Z^2_{\mathrm{band}},$$
$$\int_{II} \int_{II} \frac{\mathrm{d}x\,\mathrm{d}y}{x+y} \rho(x)\rho(y) \sim \frac{1}{W_{\mathrm{imp}}} Z^2_{\mathrm{imp}},$$
$$\int_I \int_{II} \frac{\mathrm{d}x\,\mathrm{d}y}{x+y} \rho(x)\rho(y) \sim \frac{1}{W} \ln\left(\frac{W}{W_{\mathrm{imp}}}\right) Z_{\mathrm{band}} Z_{\mathrm{imp}}, \tag{18}$$

where W is a total bandwidth and W_{imp} is a characteristic width of the region of quasi-localized states (impurity band),

$$Z_{\mathrm{imp}} = \int_{II} \mathrm{d}x \rho(x),$$
$$Z_{\mathrm{band}} = \int_I \mathrm{d}x \rho(x) = 1 - Z_{\mathrm{imp}} \tag{19}$$

are total spectral weights of the impurity QL states and main bands, correspondingly.

To proceed further, one has to specify our model for the impurity band. Let us consider for simplicity the Slater–Koster model of single impurity, with non-degenerate valence band with the band energy $\epsilon(\mathbf{k})$, $\mathbf{k}$ being the quasimomentum, and on-site only impurity potential V. Then the impurity site Green's function reads [120]

$$G_{00}(E) = \left[F^{-1}(E) - V\right]^{-1},$$
$$F(E) = \sum_{\mathbf{k}} \frac{1}{E - \epsilon(\mathbf{k})}. \tag{20}$$

The energy of the impurity localized state E_0 and its spectral weight are determined by the equations

$$V F(E_0) = 1,$$
$$Z_{\text{imp}} = \left| \frac{d}{dE} \left[F^{-1}(E) - V\right] \right|_{E=E_0} = \frac{1}{V^2 \left|F'(E_0)\right|}. \tag{21}$$

For the case of strong scattering centers (such as vacancies or hydrogen impurities) [121–124] one has

$$Z_{\text{imp}} \propto \frac{1}{\left|\ln |E_0/W|\right|} \tag{22}$$

and one can see that, indeed, the T-matrix renormalization of the Stoner parameter is irrelevant. This justifies the use of the mean-field criterion of ferromagnetism at $T = 0$.

The most interesting prediction of the theory [5] is that the *sp*-electron ferromagnetism in the impurity band can lead to much higher Curie temperatures than those for conventional magnetic semiconductors. The point is the ratio of typical energies of Stoner (electron–hole) and spin-fluctuation (magnon) energies. Typically, the latter is much smaller. For example, for the weak itinerant ferromagnets $\alpha \to 1$ the energies of the Stoner excitations is proportional to $(\alpha - 1)^{1/2}$, in comparison with the energies of spin fluctuations $(\alpha - 1)^{3/4}$, so, the latter determines the Curie temperature T_C [116]. In the opposite case of saturated itinerant electron ferromagnets [125–127] a typical magnon energy can be estimated as $\langle H_{\text{kin}} \rangle / M$ where $\langle H_{\text{kin}} \rangle$ is the average hopping energy and M is the average magnetic moment per site. For conventional magnetic semiconductors with small concentration of charge carriers n per site $\langle H_{\text{kin}} \rangle \approx n|t|$ but M is of order of one (magnetic moments are provided by localized electrons). For *sp*-magnets $M \propto n$ as well, and typical magnon energies are of order of $|t|$, that is, much higher than the Fermi energy [5]. Therefore, spins are very rigid and spin rotations are energetically unfavorable, and T_C is determined by the energy of Stoner excitations which can be easily above room temperature.

IV. CONCLUDING REMARKS

Theoretical arguments discussed in this paper show convincingly that defect-induced ferromagnetism of *sp*-electrons *is* possible. It is confirmed by some rigorous statements such as Lieb theorem, by results of density functional calculations, by mean-field analysis within the Hubbard model and by general consideration of the last section demonstrating that the mean-field theory is probably reliable. Experimental search of this magnetism in graphene and other carbon-based material is a very complicated issue. Local probes such as STM should be especially useful here. Unfortunately, up to now there were no attempts to use the spin-polarized STM [128], probably, the most powerful local probe of magnetism. Such experiments are highly desirable. If the existence of *sp*-electron magnetism is confirmed it will have enormous importance for spintronics and other applications.

Acknowledgment

O.V.Y. acknowledges support of the Swiss National Science Foundation (Grant No. PP00P2_133552).

REFERENCES

[1] Coey JMD, Venkatesan M, Fitzgerald CB, Douvalis AP, Sanders IS. Nature 2002;420:156.
[2] Coronado E, Galán-Mascarós JR, Gómes-Garsía CJ, Laukhin V. Nature 2000;408:447.
[3] Odom TW, Huang JL, Cheung CL, Lieber CM. Science 2000;290:1549.
[4] Krasheninnikov AV, Banhart F. Nature 2004;431:672.
[5] Edwards DM, Katsnelson MI. J Phys: Condens Matter 2006;18:7209.
[6] Makarova T, Palacio F. Carbon based magnetism: an overview of the magnetism of metal free carbon-based compounds and materials. Amsterdam: Elsevier; 2006.
[7] Geim AK, Novoselov KS. Nature Mater 2007;6:183.
[8] Katsnelson MI. Mater Today 2007;10:20.
[9] Castro Neto AH, Guinea F, Peres NMR, Novoselov KS, Geim AK. Rev Mod Phys 2009;81:109.
[10] Geim AK. Science 2009;324:1530.
[11] Vozmediano MAH, Katsnelson MI, Guinea F. Phys Rep 2010;496:109.
[12] González J, Guinea F, Vozmediano MAH. Nucl Phys B 1994;424:595.
[13] Elias DC, Gorbachev RV, Mayorov AS, Morozov SV, Zhukov AA, Blake P, et al. 2011 arXiv:11041396.
[14] Herbut IF. Phys Rev Lett 2006;97:146401.
[15] Honerkamp C. Phys Rev Lett 2008;100:146404.
[16] Drut JE, Lähde TA. Phys Rev Lett 2009;102:026802.
[17] Meng ZY, Lang TC, Wessel S, Assaad FF, Muramatsu A. Nature 2010;464 :847.
[18] Wehling TO, Sasioglu E, Friedrich C, Lichtenstein AI, Katsnelson MI, Bluegel S. Phys Rev Lett 2011;106:236805.
[19] Takahashi M, Turek P, Nakazawa Y, Tamura M, Nozawa K, Shiomi D, et al. Phys Rev Lett 1991;67:746.
[20] Tamura M, Nakazawa Y, Shiomi D, Nozawa K, Hosokoshi Y, Ishikawa M, et al. Chem Phys Lett 1991;186:401.

[21] Allemand P-M, Khemani KC, Koch A, Wudl F, Holczer K, Donovan S, et al. Science 1991;253:301.
[22] Inoue J, Fukui K, Kubo T, Nakazawa S, Sato K, Shiomi D, et al. J Am Chem Soc 2001;123:12702.
[23] Konishi A, Hirao Y, Nakano M, Shimizu A, Botek E, Champagne B, et al. J Am Chem Soc 2010;132:11021.
[24] Allinson G, Bushby RJ, Paillaud J-L, Thornton-Pett M. J Chem Soc, Perkin Trans 1995;1:385.
[25] Morita Y, Suzuki S, Sato K, Takui T. Nature Chem 2011;3: 197.
[26] Makarova TL, Sundqvist B, Hohne R, Esquinazi P, Kopelevich Y, Scharff P, et al. Nature (London) 2001;413:716.
[27] Makarova TL, Sundqvist B, Höhne R, Esquinazi P, Kopelevich Y, et al. Nature (London) 2006;440:707.
[28] Boukhvalov DW, Katsnelson MI. Eur Phys J B 2009;68:529.
[29] Esquinazi P, Spemann D, Höhne R, Setzer A, Han K-H, Butz T. Phys Rev Lett 2003;91:227201.
[30] Barzola-Quiquia J, Esquinazi P, Rothermel M, Spemann D, Butz T, Garcia N. Phys Rev B 2007;76:161403.
[31] Ohldag H, Tyliszczak T, Höhne R, Spemann D, Esquinazi P, Ungureanu M, et al. Phys Rev Lett 2007;98:187204.
[32] Barzola-Quiquia J, Höhne R, Rothermel M, Setzer A, Esquinazi P, Heera V. Eur Phys J B 2008;61:127.
[33] Xia H, Li W, Song Y, Yang X, Liu X, Zhao M, et al. Adv Mater 2008;20:4679.
[34] Ohldag H, Esquinazi P, Arenholz E, Spemann D, Rothermel M, Setzer A, et al. New J Phys 2010;12:123012.
[35] Kopelevich Y, Esquinazi P, Torres JHS, Moehlecke S. J Low Temp Phys 2000;119:691.
[36] Esquinazi P, Setzer A, Höhne R, Semmelhack C, Kopelevich Y, Spemann D, et al. Phys Rev B 2002;66:024429.
[37] Červenka J, Katsnelson MI, Flipse CFJ. Nature Phys 2009;5:840.
[38] Wang Y, Huang Y, Song Y, Zhang X, Ma Y, Liang J, et al. Nano Lett 2009;9:220.
[39] Matte HSSR, Subrahmanyam KS, Rao CNR. J Phys Chem C 2009;113:9982.
[40] Sepioni M, Nair RR, Rablen S, Narayanan J, Tuna F, Winpenny R, et al. Phys Rev Lett 2010;105:207205.
[41] Tao C, Jiao L, Yazyev OV, Chen Y-C, Feng J, Zhang X, et al. Nature Phys 2011;7:616.
[42] Reich S, Maultzsch J, Thomsen C, Ordejón P. Phys Rev B 2002;66:035412.
[43] Szabo A, Ostlund NS. Modern quantum chemistry: introduction to advanced electronic structure theory. New York: Macmillian; 1982.
[44] Feldner H, Honecker A, Cabra D, Wessel S, Meng ZY, Assaad FF. Phys Rev B 2010;81:115416.
[45] Kinza M, Ortloff J, Honerkamp C. Phys Rev B 2010;82:155430.
[46] Ijäs M, Harju A. Phys Rev B 2010;82:235111.
[47] Feldner H, Meng ZY, Lang TC, Assaad FF, Wessel S, Honecker A. Phys Rev Lett 2011;106:226401.
[48] Yazyev OV, Wang WL, Meng S, Kaxiras E. Nano Lett 2008;8:766.
[49] Kuroda S, Bando H, Shirakawa H. Solid State Commun 1984;52:893.
[50] Thomann H, Dalton LR, Grabowski M, Clarke TC. Phys Rev B 1985;31:3141.
[51] Fernández-Rossier J, Palacios JJ. Phys Rev Lett 2007;99:177204.
[52] Pisani L, Chan JA, Montanari B, Harrison NM. Phys Rev B 2007;75:064418.
[53] Gunlycke D, Areshkin DA, Li J, Mintmire JW, White CT. Nano Lett 2007;7:3608.
[54] Sorella S, Tosatti E. Europhys Lett 1992;19:699.
[55] Fajtlowicz S, John PE, Sachs H. Croat Chem Acta 2005;78:195.
[56] Mohn P. Magnetism in the solid state: an introduction. Springer Series in solid-state sciences. Berlin: Springer-Verlag; 2003.
[57] Palacios JJ, Fernández-Rossier J, Brey L. Phys Rev B 2008;77:195428.
[58] Lieb EH. Phys Rev Lett 1989;62:1201.
[59] Wu J, Pisula W, Mullen K. Chem Rev 2007;107:718.
[60] Wang WL, Yazyev OV, Meng S, Kaxiras E. Phys Rev Lett 2009;102:157201.
[61] Kramers HA. Physica 1934;1:182.

[62] Anderson PW. Phys Rev 1950;79:350.
[63] Clar E. The aromatic sextet. London: Wiley; 1972.
[64] Clar E, Mackay CC. Tetrahedron 1972;28:6041.
[65] Wang WL, Meng S, Kaxiras E. Nano Lett 2008;8:241.
[66] Ezawa M. Phys Rev B 2007;76:245415.
[67] Ezawa M. Eur Phys J B 2009;67:543.
[68] Akola J, Heiskanen HP, Manninen M. Phys Rev B 2008;77:193410.
[69] Ezawa M. Phys Rev B 2008;77:155411.
[70] Ezawa M. Phys Rev B 2009;79:241407.
[71] Agapito LA, Kioussis N, Kaxiras E. Phys Rev B 2010;82:201411.
[72] Ezawa M. Phys Rev B 2006;73:045432.
[73] Son Y-W, Cohen ML, Louie SG. Phys Rev Lett 2006;97:216803.
[74] Barone V, Hod O, Scuseria GE. Nano Lett 2006;6:2748.
[75] Brey L, Fertig HA. Phys Rev B 2006;73:235411.
[76] Peres NMR, Castro Neto AH, Guinea F. Phys Rev B 2006;73:195411.
[77] Nakada K, Fujita M, Dresselhaus G, Dresselhaus MS. Phys Rev B 1996;54:17954.
[78] Fujita M, Wakabayashi K, Nakada K, Kusakabe K. J Phys Soc Jpn 1996;65:1920.
[79] Jung J, Pereg-Barnea T, MacDonald AH. Phys Rev Lett 2009;102:227205.
[80] Muñoz-Rojas F, Fernández-Rossier J, Palacios JJ. Phys Rev Lett 2009;102:136810.
[81] Sawada K, Ishii F, Saito M, Okada S, Kawai T. Nano Lett 2009;9:269.
[82] Jung J, MacDonald AH. Phys Rev B 2009;79:235433.
[83] Akhmerov AR, Beenakker CWJ. Phys Rev B 2008;77:085423.
[84] Yazyev OV, Capaz RB, Louie SG. Phys Rev B 2011;84:115406.
[85] Wimmer M, Akhmerov AR, Guinea F. Phys Rev B 2010;82:045409.
[86] Y-Son W, Cohen ML, Louie SG. Nature (London) 2006;444:347.
[87] Mermin ND, Wagner H. Phys Rev Lett 1966;17:1133.
[88] Yazyev OV, Katsnelson MI. Phys Rev Lett 2008;100:047209.
[89] Joyce GS. Phys Rev Lett 1967;19:581.
[90] Banhart F. Rep Prog Phys 1999;62:1181.
[91] Krasheninnikov AV, Banhart F. Nature Mater 2007;6:723.
[92] Crespi VH, Chopra NG, Cohen ML, Zettl A, Louie SG. Phys Rev B 1996;54:5927.
[93] Smith BW, Luzzi DE. J Appl Phys 2001;90:3509.
[94] Yazyev OV, Tavernelli I, Rothlisberger U, Helm L. Phys Rev B 2007;75:115418.
[95] Zobelli A, Gloter A, Ewels CP, Seifert G, Colliex C. Phys Rev B 2007;75:245402.
[96] Lehtinen PO, Foster AS, Ma Y, Krasheninnikov A, Nieminen RM. Phys Rev Lett 2004;93:187202.
[97] Ewels CP, Telling RH, El-Barbary AA, Heggie MI, Briddon PR. Phys Rev Lett 2003;91:025505.
[98] Kaxiras E, Pandey KC. Phys Rev Lett 1988;61:2693.
[99] Stone AJ, Wales DJ. Chem Phys Lett 1986;128:501.
[100] Pereira VM, Guinea F, Lopes dos Santos JMB, Peres NMR, Castro Neto AH. Phys Rev Lett 2006;96:036801.
[101] Huang W-M, Tang J-M, Lin H-H. Phys Rev B 2009;80:121404.
[102] Mizes HA, Foster JS. Science 1989;244:559.
[103] Kelly KF, Halas NJ. Surf Sci 1998;416:L1085.
[104] Ruffieux P, Gröning O, Schwaller P, Schlapbach L, Gröning P. Phys Rev Lett 2000;84:4910.
[105] Duplock EJ, Scheffler M, Lindan PJD. Phys Rev Lett 2004;92:225502.
[106] Yazyev OV, Helm L. Phys Rev B 2007;75:125408.
[107] Boukhvalov DW, Katsnelson MI, Lichtenstein AI. Phys Rev B 2008;77:035427.
[108] Kumazaki H, Hirashima DS. J Phys Soc Jpn 2007;76:064713.
[109] Lehtinen P, Foster AS, Ayuela A, Krasheninnikov A, Nordlund K, Nieminen RM. Phys Rev Lett 2003;91:017202.
[110] Vozmediano MAH, López-Sancho MP, Stauber T, Guinea F. Phys Rev B 2005;72:155121.
[111] Dugaev VK, Litvinov VI, Barnas J. Phys Rev B 2006;74:224438.
[112] Yazyev OV. Phys Rev Lett 2008;101:037203.
[113] Wehling TO, Katsnelson MI, Lichtenstein AI. Phys Rev B 2009;80:085428.
[114] Irkhin VY, Katanin AA, Katsnelson MI. Phys Rev B 1999;60:1082.
[115] Pisani L, Montanari B, Harrison NM. New J Phys 2008;10:033002.

[116] Moriya T. Spin fluctuations in itinerant electron magnetism. Berlin: Springer; 1985.
[117] Kanamori J. Prog Theor Phys 1963;30:275.
[118] Galitski VM. Zh Eksp Teor Fiz 1958;34:151.
[119] Mahan GD. Many-particle physics. New York: Plenum; 1990.
[120] Vonsovsky SV, Katsnelson MI. Quantum solid-state physics. Berlin: Springer; 1989.
[121] Peres NMR, Guinea F, Castro Neto AH. Phys Rev B 2006;73:125411.
[122] Wehling TO, Katsnelson MI, Lichtenstein AI. Chem Phys Lett 2009;476:125.
[123] Wehling TO, Yuan S, Lichtenstein AI, Geim AK, Katsnelson MI. Phys Rev Lett 2010;105:056802.
[124] Yuan S, De Raedt H, Katsnelson MI. Phys Rev B 2010;82:115448.
[125] Edwards DM. Phys Lett A 1967;24:350.
[126] Irkhin VY, Katsnelson MI. J Phys Chem C 1985;18:4173.
[127] Irkhin VY, Katsnelson MI. J Phys: Condens Matter 1990;2:7151.
[128] Wiesendanger R, Güntherodt H-J, Güntherodt G, Gambino RJ, Ruf R. Phys Rev Lett 1990;65:247.

Chapter 5

Synthesis of Graphene and Derivatives

Helena Grennberg* **and Ulf Jansson**†

**Department of Chemistry – BMC, Uppsala University, Sweden*

† Department of Chemistry – Ångström Laboratory, Uppsala University, Sweden

Chapter Outline Head

I. INTRODUCTION

Experimental research on graphene and graphene-related carbon materials was initiated in the early 00'ties, gained speed with the first experimental study by Novoselov et al. [1] and is still, experimentally as well as from a theory

Advanced Functional Materials. http://dx.doi.org/10.1016/B978-0-44-453681-5.00005-4

perspective, an expanding field. Potential applications as well as the theoretical basis for the interest in this material are discussed elsewhere in this book. For any application of the material, (i) scalable synthetic methods and (ii) transfer of the graphene product from the synthetic environment to a solid substrate suitable for characterization or device fabrication are primary requirement. In the present chapter, three of the main synthetic routes fulfilling both requirements are discussed: Exfoliation from graphite, molecular synthesis, and CVD growth on metal substrates. The discussion covers selected parts of the literature up to the end of 2011. Synthesis from SiC [2] where the graphene is strongly coupled to the growth substrate and hence not yet fulfilling the second requirement is not discussed. Methods for characterization of as-produced graphene with regard to flake thickness and presence of defects have been reviewed by for example Allen et al. [3].

According to an early definition, developed for graphite intercalation compounds [4], graphene is a one-atom thick sheet of hexagonally arranged sp^2 hybridized carbon atoms. Despite the strict definition, it has become established practice to use the term graphene also for flakes of more than one layer, of any sheet size from around square micrometer and up and also for carbon sheets with significant deviations from the all-carbon and all-sp^2 criteria such as graphite oxide (GO) and reduced graphite oxide (RGO). In this chapter, we have used "graphene" for materials prepared via non-oxidative routes. Depending on synthetic method the product can be single- layer or with a distribution of thicknesses. In case of oxidation–reduction routes, the resulting material is denoted RGO.

II. SYNTHESIS BY SCALABLE EXFOLIATION FROM GRAPHITE

Exfoliation from graphite has been the most popular route to graphene flakes. Graphene prepared by micromechanical exfoliation using sticky tape [1] is still the experimental benchmark and thus the method is widely used despite being low-yielding and time-consuming. In contrast, routes where flakes upon exfoliation are dispersed in a solvent can be fast, readily scalable and hence with potential of being low-cost. Current strategies include chemical exfoliation via graphite oxides, and more recently developed direct processes involving ultrasonication or supercritical conditions. Compared to graphene from micromechanical cleavage, molecular synthesis or by gas-phase routes, graphene prepared by exfoliation is less homogenous both regarding size and thickness and to the presence of process-induced defects. The carbon source can be graphite powder, requiring fractionation of exfoliated graphene from bulk graphite and exfoliated multilayer flakes, or molecularly intercalated graphite foil where surplus graphite is more easily removed from the suspension. Due to the high sheet-to-sheet contact area, the intermolecular forces to overcome are larger than for unbundling/exfoliation of structurally related materials such as carbon nanotubes, and therefore often more forcing and potentially damaging reaction conditions are used.

A. Graphite Properties

The source of graphite has a profound influence on the size and thickness of the exfoliated flakes. Wu et al. subjected five types of graphite to identical reaction conditions: oxidation, thermal expansion followed by dispersion, and reduction using H_2 [5]. It was found that both the grain size and the crystallinity of the starting graphite materials play important roles in the number of RGO layers produced (Figure 1). Artificial graphite, with a small lateral grain size and low crystallinity, gave a significantly higher proportion of flakes with 1–3 layers than was the case for HOPG, with larger grains. The reagent–graphite interactions ultimately leading to exfoliation take place at the edges of the graphite layers and in the interlayer space. For a starting material with lower crystallinity, the proportion of edges is higher, the diffusion routes for intercalation of reagents shorter and the efficiency of exfoliation thus higher. For exfoliation by the oxidation–expansion–dispersion–reduction route, it was thus possible to tune the number of layers of the exfoliated flakes by choice of starting material.

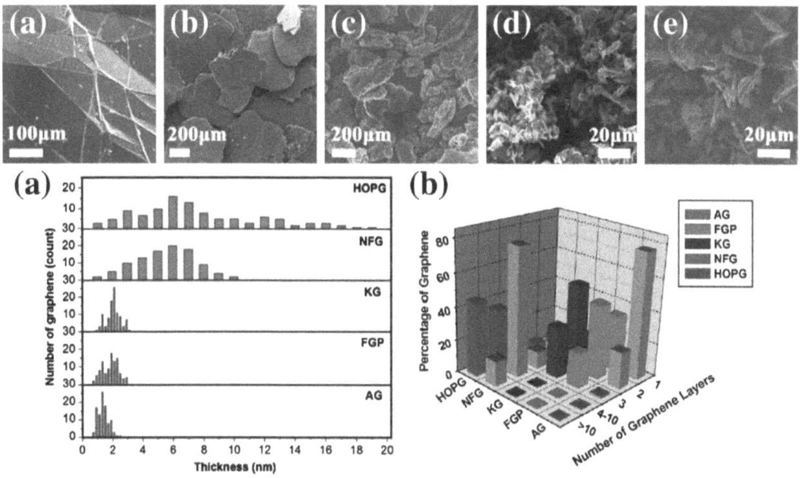

FIGURE 1 Top row: SEM images of (a) highly oriented pyrolytic graphite (HOPG), (b) natural flake graphite (NFG), (c) Kish graphite (KG), (d) flake graphite powder (FGP) and (e) artificial graphite (AG). Lower row: Statistical distributions of the thickness and the number of layers of flakes prepared from different starting graphites. (a) Histograms of the thickness distribution, obtained from 100 flakes for each sample. (b) Histogram of the distribution of the number of graphene layers, obtained from the corresponding thickness distribution in (a). Reproduced from Ref. [5].

B. Exfoliation via Oxidized Graphite

The most commonly employed large-scale exfoliations have been those via graphite oxide (GO), using combinations of oxidizing agents in strongly acidic media followed by a reductive treatment, as pioneered by Stankovich et al. [6] and extensively reviewed by Ruoff et al. [7] Most oxidation–exfoliation protocols to form GO are adaptations of the method by Hummers and Offeman [8] where the graphite is treated with potassium permanganate in concentrated sulphuric acid or the much older Brodie [9] or Staudenmaier [10] methods that use mixtures of potassium chlorate and nitric acid. Graphite oxide can also be prepared using the strong acids nitric, sulphuric, or perchloric acid without additional co-oxidants [11]. It is of importance to safe working practice to avoid accidents in the exothermic oxidation–expansion stage as well as during workup of the reaction mixtures and while drying the GO.

In the oxidation process, oxygen-containing functional groups are introduced at reactive edge sites as well as in the layers. As the reaction proceeds, the interlayer forces are weakened, eventually leading to exfoliation of highly oxidized sheets of hydrophilic non-conductive GO easily processable in water. Although hydrophilic, GO can be transferred to organic solvents by the use of cationic surfactants [12].

Using expanded graphite and solvothermal conditions, exfoliated flakes with a lower content of oxygen-containing functional groups are obtained. Treating graphite with oleum (SO_3 in H_2SO_4) results in intercalated graphite, from which partially oxidized graphene flakes with properties close to pristine graphene can be exfoliated [13].

By reductive treatments, a material with appearance and properties resembling that of pristine graphene can be obtained. Reduced GO (RGO) is black, hydrophobic, conducting and has a significantly lower relative content of oxygen than GO. The most established protocols utilize hydrazine [6,7], a reagent known to transform ketones to methylene units. Due to toxic and explosive properties, the use of hydrazine is restricted or forbidden in many parts of the world. Naturally, the hazards are the same everywhere, and although studies employing hydrazine are among our examples we recommend use of alternatives regardless of the local situation concerning restrictions. Less harmful yet efficient alternatives to hydrazine have been reported, such as H_2 [5], ascorbic acid [14], $NaBH_4$ [15], hydroquinone [16] and metallic iron [17]. Thermal treatment also reduces the O/C ratio [18], as does solvothermal treatment with the amide solvent NMP [19].

Each graphite–GO–RGO route may give material with different properties, differences due to factors such as source of graphite [5] (Figure 1) and slight differences in reaction conditions for oxidation as well as reduction [20]. In order to improve this general route to graphene, knowledge of the chemical outcome from oxidation to GO and subsequent treatments to obtain RGO is essential [21]. Several structural hypotheses for GO have been put forward,

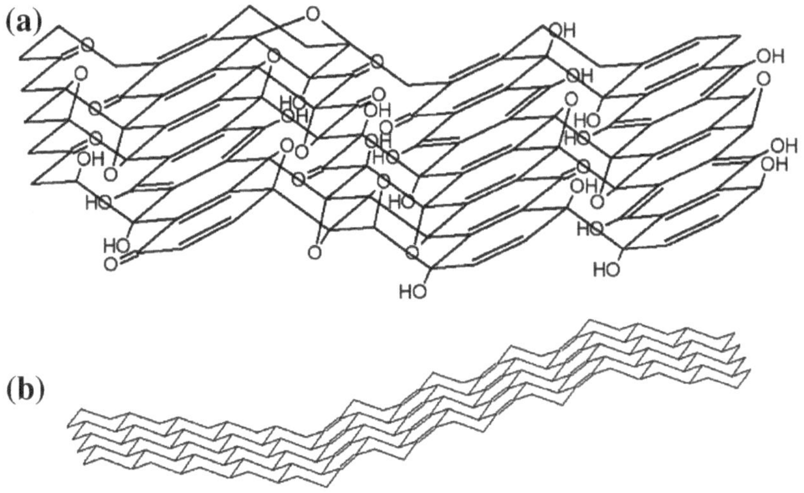

FIGURE 2 The structural model of graphite oxide proposed by Szabó et al. (a) Structure including the possible functional groups. (b) Carbon skeleton view. Reproduced from Ref. [22].

where the one proposed by Szabo et al. [22] (Figure 2) being the model taking most experimental evidence into account. Still the exact details concerning regarding identity, proportion, and position of the introduced oxygen-containing functional groups and the extent of oxidation-induced carbon lattice damage is unresolved.

No reduction of GO has yet produced oxygen-free fully sp^2-hybridized graphene; still RGO materials with conductivities in the same range as micromechanically exfoliated graphene have been reported [7,20]. For a RGO graphene material prepared via Hummers oxidation to GO followed by a thermal expansion, dispersion in the amide solvent NMP and reduced using H_2, Wu et al. [5] obtained linear current–voltage characteristics and conductivities in individual flakes similar to what was reported by Novoselov et al. [1], and in the same range as reported by Wang et al. for graphene flakes prepared by a less oxidizing first step followed by solvothermal reduction with hydrazine in DMF [23]. Zheng et al. reported that treatment of graphite with a combination of nitric and sulphuric acids followed by exfoliation under solvothermal conditions in the presence of a long-chain aliphatic amine that has the threefold role of neutralizing base, solvent, and surfactant, yield dispersions of large mainly monolayer graphene. The electrical transport characteristics of flakes measured between Au/Ti electrodes on a SiO_2/Si substrate were close to those of pristine graphene (Figure 3) [24].

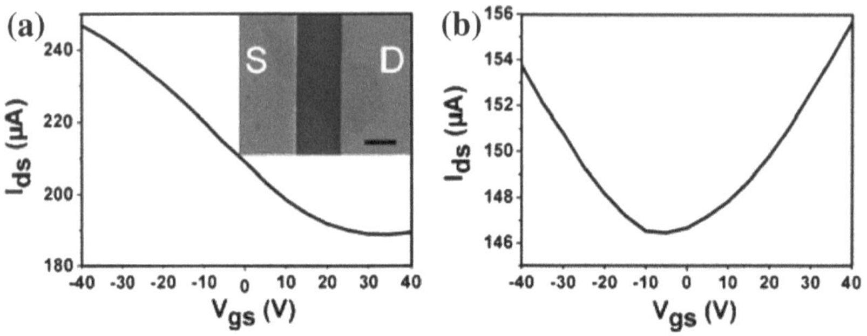

FIGURE 3 Electrical transport characteristics of exfoliated sheets. (a) Current–gate voltage (I_{ds}–V_{gs}) curves for a device measured in air. Inset: SEM image (scale bar: 2 mm) of source (S) and drain (D) electrodes bridged by one sheet. (b) Current–gate voltage curves for a device measured in nitrogen. Reproduced from Ref. [24].

C. Direct Exfoliation

Exfoliation of graphene flakes from graphite can be achieved without involvement of acidic and oxidizing agents. Graphite–potassium intercalation compounds react with protic solvents such as ethanol, acidic (alfa) positions of amides or trace water present. In the exothermic reaction, H_2 and the corresponding base of solvent, with K^+ as counter ion is formed. The combined effect of local heating and evolution of gas results in very efficient exfoliation of graphene, however with limited stability: Graphene from treatment in ethanol formed scrolls [25] whereas the flakes dispersed into N-methylpyrolidinone were air-sensitive [26].

Ultrasonication, commonly utilized for cleaning and to create dispersions is efficient also for preparation of graphene dispersions from graphite without need for any oxidative, acidic, or reductive reagents. The acoustic cavitation; formation and collapse of bubbles in the solvent; result in very high local pressures and temperatures [27], which constitutes a good alternative to expansion–exfoliation caused by exothermic chemical reactions of intercalated species [28]. Transfer of energy to the graphite by cavitation is dependent on the boiling point of the solvent. A solvent with high-boiling point will form fewer cavities that collapse with larger temperature/pressure pulses than what is the case for a solvent with a lower boiling point [29]. Also of importance is the stability of the resulting suspensions, which relies on the nature of the intermolecular bond between the exfoliated graphene flake and the solvent (or surfactant) molecules. In a survey of sonication of graphite powder in a range of organic solvents followed by fractionation by centrifugation in order to separate thin flakes from graphite, Hernandez et al. found that high-boiling amide solvents DMF and NMP gave excellent results concerning yield and quality of few-layer and monolayer graphene [30]. Sonication in the presence of water-soluble surfactants such as sodium dodecylphenylsuphonate [31], sodium cholate [32], or tetrapotassium coronene tetracarboxylate [33] gives, after centrifugation, aqueous dispersions of graphene with a high proportion of few-layer flakes covered with ionic surfactant molecules.

Not all solvents are stable under sonication conditions. In particular, amide solvents such as DMF and NMP may undergo sonication-induced decomposition into radical species [34] with ability to add covalently with introduction of defects to the graphene. Sonication of carbon nanotubes in a chlorinated aromatic solvent has been reported to result in doping of the

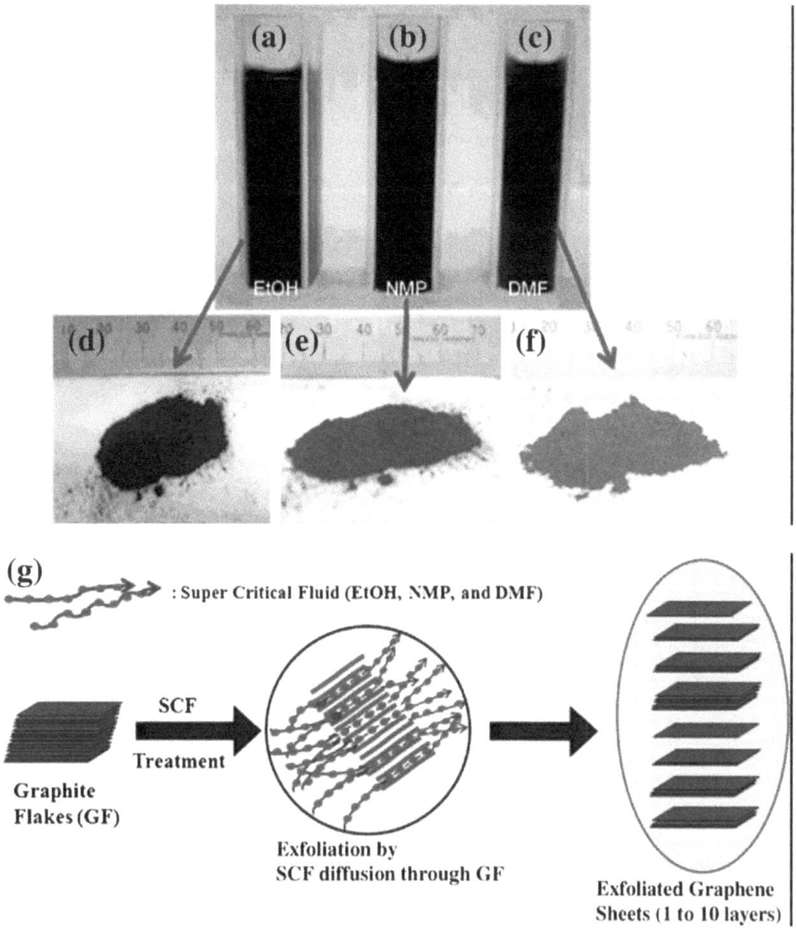

FIGURE 4 Photographs of exfoliated graphene sheets dispersed in (a) ethanol, (b) NMP and (c) DMF and approximately 1 g of their respective dried graphene powders (d–f); (g) scheme showing the SCF exfoliation of graphite crystals to graphene. Reproduced from Ref. [42].

tubes by chlorine [35] and chloroform may decompose to various compounds, including HCl and dichlorocarbene [36], the latter has been reported to react with soluble carbon nanotubes [37]. Moreover, extended random sonication enhance aggregation of suspended graphene [38]. The sonication time and power, and hence the risk of process-induced damage can be significantly decreased if the graphite is pre-treated by intercalation of, for example, ICl [39] or Br_2 [40].

Use of supercritical conditions constitutes a promising further development towards facile preparation of high-quality graphene. Supercritical CO_2 was used by Pu et al. to intercalate graphite powder [41]. On depressurizing, the volume of the confined gas molecules increases, leading to graphite expansion and exfoliation of high-quality graphene flakes into aqueous sodium dodecylsulfate. Rangappa et al. employed supercritical extraction with amide solvents (DMF and NMP) to prepare dispersions with high graphene content (Figure 4). About 90–95% of the exfoliated sheets were <8 layers and the characterized flakes exhibited very good conduction and electron-carrier capacity [42].

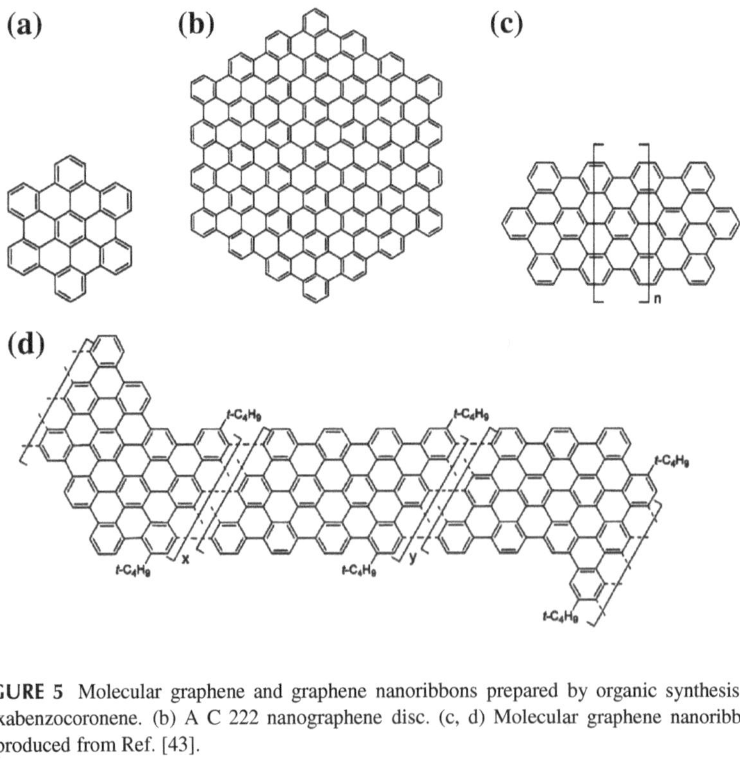

FIGURE 5 Molecular graphene and graphene nanoribbons prepared by organic synthesis. (a) Hexabenzocoronene. (b) A C 222 nanographene disc. (c, d) Molecular graphene nanoribbons. Reproduced from Ref. [43].

III. MOLECULAR SYNTHESIS

By molecular synthesis, almost any molecule imaginable can be prepared with excellent selectivity starting from smaller precursor molecules [43,44]. Already in 1958, Clar et al. reported the synthesis of hexabenzocoronene, a C6-symmetric, all-sp^2 PAH with the molecular formula $C_{42}H_{18}$ [45] (Figure 5, structure (a). Giant molecular graphenes of different shapes have since then been prepared, with C6-symmetric insoluble $C_{222}H_{42}$ (Figure 5, structure b) being one of the larger ones produced with a reasonable isolated yield [46]. Further size extension of the molecular graphenes becomes synthetically highly demanding due to competing side-reactions and limited solubility of the products. With solubility-enhancing branched alkane edge groups as part of the design, soluble molecular nanoribbons more than 10 nm long, with well-defined edge structures and masses of 20000 u were possible to prepare in good to excellent yields [47] (Figure 5c and d). These achievements clearly demonstrate that molecular synthesis should be the preferred route for preparation of large quantities of identical nanographene entities, with the structure of edges and general shape decided already at the synthesis design stage.

IV. GRAPHENE BY CVD

Chemical vapour deposition (CVD) is a well-known technique to deposit thin films using gaseous molecules as reactants. An advantage with the CVD technique is the compatibility with large-scale film growth enabling a rather inexpensive production on large-area substrates. Use of CVD in graphene growth can be separated into two types of studies; graphene synthesis by decomposition of hydrocarbons under ultra-high vacuum (UHV) conditions and more conventional CVD at higher pressures. Several studies have recently demonstrated that the pressure is a critical growth parameter [48,49]. The two pressure ranges will therefore be discussed separately below.

A. Growth of Graphene Under UHV Conditions

Surface scientists have for a long time studied the interaction between carbon and well-defined surfaces of transition metal single crystals. Early, even before the pioneering work by Novoselov and Geim, it was realized that this interaction could give rise to monolayers of graphite, i.e., graphene. In principle, two different routes to graphene on transition metal surfaces were identified; segregation of dissolved carbon from the bulk of the metal to the surface and decomposition of hydrocarbons [50]. In a typical UHV-based growth experiment using hydrocarbons, the metal substrate is cleaned by etching followed by an in situ high-temperature annealing giving a relatively clean and ordered surface. The substrate is then exposed to a hydrocarbon gas usually at pressures ranging from 10^{-8} to 10^{-4} Torr at a given growth temperature. Examples of hydrocarbons used to form graphene include methane, propene, and ethylene but also larger molecules such as toluene and benzene have been used. In most cases, the formation of graphene is confirmed by analytical techniques such as scanning tunneling microscopy (STM) or low-energy electron diffraction (LEED) mounted directly on the UHV chamber.

Hitherto, a large number of studies have been published on graphene formation on metal surfaces at UHV conditions. Although, the growth process is carried out in a completely different pressure range than conventional CVD, this type of studies can give important information on graphene growth also at higher pressures. An interesting observation is that the best metal substrates for graphene growth are elements of transition metals with a relative low-solid solubility of carbon such as Ni, Pt, Ir, Ru and Cu. Also these elements typically form only metastable carbides (Cu forms no carbide). As will be discussed more in detail below, the solubility of carbon at different temperatures plays an important role in the quality of the final graphene film in conventional CVD. Furthermore, the UHV-based graphene growth seems to be most favorable on close-packed surfaces such as Ni(1 1 1) [51,52], Cu(1 1 1) [53,54], Pt(1 1 1) [55], Ir(1 1 1) [56,57], Rh(1 1 1) [58], and Ru(0 0 0 1) [59]. Typically, the graphene has a small misfit to these close-packed surfaces. The lattice constant of bulk graphite is 2.46 Å which should be compared to 2.49 Å for Ni(1 1 1), 2.69 Å for Rh(1 1 1), 2.71 Å for Ru(0 0 0 1), 2.72 Å for Ir(1 1 1) and 2.77 Å for Pt(1 1 1). On Ni(1 1 1) this difference is so small that graphene can grow directly with a 1×1 structure. On the other surfaces, the graphene forms domains with large coincidence lattices or incommensurate structures. Wang et al. recently demonstrated that graphene deposited from ethylene at 1100 K on Rh(1 1 1) form a 12×12 unit cell on a 11×11 substrate cell. The graphene forms domains with sizes larger than 100 nm only limited by substrate defects such as surface steps [58]. Graphene can also form domains with different orientation towards the substrate. One example is Pt(1 1 1) where graphene forms domains with three different rotations and periodicities towards the substrate [55]. This behavior can also be dependent on the growth temperature. Coraux et al. have shown that graphene deposited from ethylene on Ir(1 1 1) at 1320 K forms perfect graphene over areas larger than 2500 Å × 1200 Å [56]. The graphene formed dislocation-free structures growing also over step edges. Reducing the temperature to 1120 K showed a different growth behavior with domains exhibiting some rotational misorientations. STM images of the domain boundaries show edge dislocations with pairs of pentagons and heptagons (Figure 6). This observation illustrates that temperature is an important parameter in graphene growth under UHV conditions. As will be discussed below, it seems to be the most important factor in conventional CVD at higher pressures.

The growth mechanism of graphene on single-crystal surfaces is probably a rather complex process depending on several parameters. Loginova et al. have

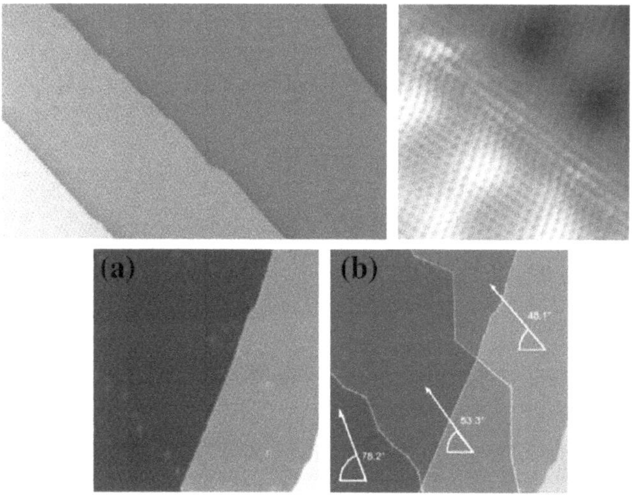

FIGURE 6 STM images of graphene on Ir(1 1 1) deposited in UHV system using ethylene as carbon source. Top left: Moiré structure of graphene deposited at 1320 K showing a coherent domain covering several terraces. Image size 2500 Å × 2500 Å. Top left: Detail of graphene deposited at 1320 K showing growth over step edge. Image size 50 Å × 50 Å. Bottom left: Moiré structure of graphene deposited at 1120 K showing several domains. Domain size 1080 Å × 1080 Å. Bottom right: Color-coded map of graphene deposited at 1120 K showing the domain structures with rotational angles. Reproduced from Ref. [56].

shown that growth rate of graphene deposited from ethylene at Ir(1 1 1) and Ru(0 0 0 1) has a nonlinear relationship with the C-adatom concentration [57]. Moreover, they observed several rotational variants of graphene on Ir(1 1 1) and found that they have different nucleation and growth behavior. For example, on Ir(1 1 1), the "majority orientation" nucleated first followed by the "minority orientations" nucleating on the edges of the initially formed graphene. The minority orientations also exhibited a larger growth rate. Furthermore, their results suggest that the majority orientation on Ru(0 0 0 1), and presumably Ir(1 1 1), only grow downhill the substrate surface steps while the minority orientations on Ir(1 1 1) can grow both downhill and uphill the Ir surface steps [57]. This complex behavior also on well-defined single-crystal surfaces suggests that the graphene growth on less-defined substrates with many orientations and surface defects must be a rather complicated process.

Another important factor in graphene growth on transition metal surfaces is the graphene–substrate interactions. Studies on graphene deposited from hydrocarbons on different metal substrates in UHV systems have shown strong substrate dependence. For example, graphene on Ni(1 1 1) shows carbon–metal bond lengths as small as 2.1 Å [60]. This is considerably less than expected for a purely physisorbed graphene layer which should exhibit distances to the substrate similar to that in graphite (about 3.3 Å). This indicates a strong interaction between graphene and the Ni surface which clearly can affect not only the quality of the graphene layer but also possibilities to transfer the graphene from the metal to another substrate such as SiO_2 [61]. In contrast, Pt(1 1 1) and Ir(1 1 1) exhibit carbon–metal distances as large as 3.7 Å suggesting a much weaker bond to the metal surface.

B. CVD Growth of Graphene at Higher Pressures

Although the growth of graphene by decomposition on well-defined surfaces in UHV systems can be described as a CVD process, it is very far from a large-scale production process. This requires more conventional CVD reactors with the potential to use large-area substrates with a reasonable turnaround time. These processes are therefore carried out at higher pressures with the hydrocarbon in an inert carrier gas using polycrystalline substrates. Also, it can involve more complex gas mixtures including, for example, hydrogen and dopant molecules.

It is rather well-known in the CVD community that Ni metal easily catalyses the formation of well-ordered graphite [62], and it was therefore not surprising that this metal was the first to be used for CVD growth of graphene [63–65]. The mechanism for graphene growth has two separate stages (i) decomposition of the carbon source with deposition of carbon atoms on the Ni surface and (ii) segregation of C from the bulk of the Ni substrate. The heated Ni surface catalyzes the decomposition of carbon-containing molecules and this process will stop when the entire surface is covered by graphene. However, the bulk of the metal substrate allows a limited solubility of carbon (<0.1 at.%) but the maximum solubility decreases rapidly as the substrate is cooled, leading to a surface segregation of carbon which can form additional graphene and hence cause multilayer growth. This is considered to be a main reason for the difficulties to grow graphene with a well-defined thickness over larger areas on polycrystalline Ni substrates. The thickness control can be improved by, for example, rapid thermal CVD where the cooling time is reduced [65].

The high solubility of carbon in Ni is the main reason for the search for an alternative substrate for graphene growth. An excellent choice is Cu which has a very low carbon solid solubility but also a low price. During recent years, numerous papers on graphene formation by CVD on Cu have been published.[48,66,67] Other substrates such as Pt and Ir has also been used to produce graphene with CVD [68] but the high cost of these metals will most likely limit their use for large-scale industrial production of graphene.

A large number of different carbon precursors have been used in CVD of graphene on Cu. This includes hydrocarbons such as methane and ethylene but also alcohols such as ethanol and propanol [69]. The presence of oxygen in the alcohol did not have detrimental effects on the graphene quality. In fact, it is likely that graphene can be synthesized from many different types of carbon-containing precursors. In a rather unusual study, Ruan et al. recently have demonstrated that graphene with good quality can be grown on Cu in a CVD chamber using, for example, cookies, chocolate, grass, and cockroach legs as carbon sources [70]. An interesting conclusion in this study is that the vaporized species from these rather unconventional carbon sources most likely contain high amounts of other elements such as oxygen. The fact that the graphene is of good quality suggests that the purity of the hydrocarbon source is of lesser importance than, i.e., the growth substrate. It is not, however, without importance since studies with a range of more conventional precursors under similar conditions have shown that the graphene quality is affected by a change of precursor [71].

An important parameter in CVD of graphene on Cu is the substrate quality. Earlier studies in UHV on single-crystalline substrates suggest that the most closed-packed surface, i.e., Cu(1 1 1) is particularly favourable. Recent studies have shown that this is the case also in CVD of graphene on Cu [67,72]. On a polycrystalline Cu foil, the highest quality of graphene was obtained on the Cu(1 1 1) grains. This grain orientation also has the lowest surface energy which means that different types of substrate annealing processes increasing the fraction of Cu(1 1 1) orientated grains and their size can be an important factor to optimize the quality of graphene deposited with CVD. An alternative route to graphene growth can also be to use thin Cu films deposited on another substrate. Hu et al. have recently demonstrated that large-area single-layer graphene can be obtained by atmospheric pressure CVD on an epitaxial Cu(1 1 1) film sputtered on a Al_2O_3(0 0 0 1) substrate [73]. Most likely, however, the growth of an epitaxial Cu(1 1 1) film also require a single-crystalline substrate which limits the use of this approach to available substrate sizes.

The quality of the graphene is also dependent on the substrate pretreatment. The Cu substrate can be cleaned by, for example, acid treatments but this can give rise to a surface roughness that may have a detrimental effect on the graphene quality [48]. Ultrasonic cleaning combined with substrate annealing prior to the CVD process may improve the quality of the graphene. As will be discussed below an important part in the formation of graphene is the decomposition of the carbon-containing species on the surface and the subsequent formation of carbon atoms. These reactions are more favorable on surface steps, kinks, and other "sharp" structures on the surface and can contribute to multi-layer growth. By a proper annealing process, the density of such sites can be reduced and create smoother surfaces more suitable for monolayer growth [48].

The CVD process is also strongly dependent on other growth parameters. As discussed above, studies in UHV systems on single crystals have demonstrated that temperature is important. Studies with CVD at higher pressures have indeed confirmed that the temperature seems to be the single most important parameter on Cu substrates [73,74]. Regmi et al. have studied graphene formation from Cu methane at 600–1000 °C and found no graphene below 850 °C. At this temperature, the graphene formed 1–5 μm large domains but with a fraction of about 25% multi-layer graphene [74]. With increased temperature, the domain size increased and the fraction of multi-layer growth was drastically reduced. The total pressure is another potentially important parameter. At lower pressures (<10 Torr), large amounts of mono-layer graphene are formed on Cu at 1000 °C using methane as a carbon source [74]. When the pressure is increased above 10 Torr, the amount of multi-layer graphene increases dramatically and leads at 20 Torr to significant amounts of non-graphitic carbon. It should be noted, however, that there are studies on atmospheric pressure CVD claiming high-quality graphene on Cu [73]. A reduced quality at higher total pressures is logical since the concentration of C atoms available for graphene growth should increase with pressure and favor growth of multi-layer structures. However, most likely the important parameter here is partial pressure of carbon-containing species and not total pressure. Regmi et al. observed reduced quality of the graphene for methane partial pressures above 7.5 Torr (using a CH_4/H_2 ratio of 3:1) [74]. In comparison, the atmospheric pressure studies by Hu et al. used a much more diluted $CH_4/H_2/Ar$ mixture with only 0.4% methane corresponding to a partial pressure of 3 Torr CH_4 [73]. Consequently, future studies separating between total pressure and partial pressures are required to determine the influence of pressure on graphene growth. Also Hu et al. has studied the influence of $CH_4 : H_2$ ratio on the quality and only observed a weak influence, with lower ratios (<0.2) clearly favoring monolayer growth [73].

V. TRANSFER/DEPOSITION

Applications of graphene require high-quality material on a range of substrates, most of which are not identical to those used for epitaxial growth from, i.e., SiC[2] or by CVD on metals (previous paragraph). Graphene grown using the latter approach can be separated from the growth substrates and transferred flake by flake to the substrate of the intended application by solvent-assisted protocols, only to a limited extent currently overlapping with the protocols employed for deposition of graphene prepared by exfoliation or by molecular synthesis.

A. Transfer of CVD-Produced Graphene to Other Substrates

CVD on metals such as Cu is an excellent method to produce high-quality graphene, but the as-grown graphene cannot usually be used directly due to the conducting substrate. A transfer method must therefore be used where the graphene is moved from the metallic substrate to, for example, a dielectric material such as SiO_2. The transferred graphene can then be patterned and/or functionalized for further use in special devices.

Kim et al. demonstrated in 2009 a method to transfer graphene grown on 300 nm Ni film sputtered onto a SiO_2/Si substrate (Figure 7) [64]. After the growth step, a polydimethylsiloxane (PDMS) stamp is attached to the graphene

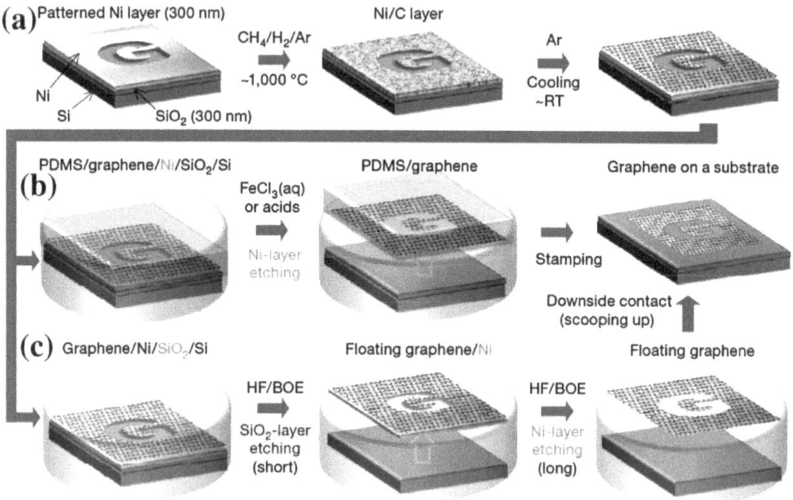

FIGURE 7 Processes for transfer of graphene. The graphene is deposited by CVD on a $Ni/SiO_2/Si$ substrate. The Ni film is 300 nm thick. (a) Gas-phase graphene synthesis on a Ni substrate. (b) Etching of the metal substrate using a $FeCl_3$ solution and transfer using a PDMS stamp, (c) etching using a BOE or hydrogen fluoride solution. Reproduced from Ref. [64].

surface and the Ni layer is etched away using a $FeCl_3$ solution (route b). An alternative route is to remove the graphene layer in, for example, an HF solution (route c).

Reina et al. used a similar $Ni/SiO_2/Si$ substrate and prepared graphene by CVD using methane as carbon source [63]. A support material of polymethyl methacrylate (PMMA) was then spin-coated onto the graphene followed by an etching in an HCl solution. This results in a free-standing PMMA/graphene film which can be placed on a target substrate followed by removal of the PMMA using acetone. An alternative to use a polymer-coated graphene structure is to dissociate the graphene by removing the substrate in an etch process producing a graphene layer floating on the surface of the etch solution (e.g., route c in Figure 7). This has limitations since the graphene consists of small domains with grain boundaries which easily can break up and form smaller fragments. On Ni, the graphene domain overlaps at the boundaries or form larger areas with multilayers making it possible to obtain reasonably coherent graphene layers floating on the surface.

Several related methods for transfer of graphene prepared by CVD on Cu substrates have been developed. Li et al. treated graphene/Cu coated with PDMS and PMMA in iron nitrate solutions [66]. The nitrate solution completely dissolves the Cu substrate overnight leaving a polymer/graphene film which allows transfer of graphene to, for example, a SiO_2 substrate. Also other etching agents have been used to dissolve Cu from graphene. One example is ammonium persulfate [75,76]. For large-scale applications, high-throughput methods for transfer of very large area graphene to suitable substrates are required. A promising step has recently been taken by Bae et al. who reported a roll-to-roll process which has been used to produce as large as 30 inch graphene layers on a PET substrate [77]. Graphene is grown on Cu foil rolls. The foil is then covered by a polymer support and the support/graphene/Cu is in a continuous process rolled through an etching bath removing the Cu. The remaining graphene/support foil can then be rolled onto a target substrate. This can be repeated several times giving several layers of graphene on the target substrate.

A very important issue in all transfer processes is the influence of contaminants on the properties of the graphene. Pirkle et al. have shown that polymer residues can dope the graphene and that a proper cleaning procedure is required [75]. Also etching can lead to damages as well as a chemical doping of the graphene. Some examples of wet-chemical doping have been demonstrated by Bae et al. [77].

B. Deposition of Exfoliated Graphene from Dispersion

Graphene produced as dispersions need to be deposited on solid substrates prior to any characterization or device fabrication. The current toolbox contains variations of "drop" and "dip" methods. Ideally, the deposition process is innocent with respect to the structure and properties of the graphene. In reality, deposition or graphene in a controlled and reproducible fashion requires multi-parameter optimization. Not only is the general method for deposition of importance, also the surface properties of the intended substrate, and the volatility, graphene-solvating and surface wetting properties of the solvent for the dispersions have profound effects on folding and aggregation of the flakes.

(i) Drop Deposition

Drop deposition will deliver all components in the dispersions to the surface. Evaporation of solvent will lead to an accumulation of the dispersed material at the outer perimeter of each drop, leaving "coffee-rings" of material [39]. The uneven distribution of material, including trapped solvent, is more prominent for graphene deposited from solvents with low volatility such as water or DMF, and less problematic in volatile solvent such as chloroform or toluene [40]. Spin- and spray coating are scalable versions of dip-deposition. Recent examples include spin coating of surfactant-stabilized RGO mixed with a conducting polymer onto polyetehenetereftalate (PET) [78], and, using an airbrush, spray coating of aqueous GO-hydrazine dispersions onto preheated quartz substrates, with formation of RGO simultaneous to deposition [79]. In both cases, the resulting films are more than 80% transparent to visible light over a large area (sq microns).

(ii) Dip Deposition

Dipping of a substrate into graphene dispersion will, in contrast to drop depositions, capture only material with an affinity for the substrate surface, or, as demonstrated for carbon nanotube dispersions, a region of a patterned surface [80]. Depending on the concentration of flakes, the coverage will range from individual flakes to quite continuous films. Efficient dip-deposition, favoring monolayers has been achieved using Langmuir–Blodgett methodology. Cote et al. deposited flat GO sheets onto hydrophilic substrates such as silicon wafers pre-treated with aqueous ammonium hydroxide and H_2O_2, glass, quartz, or mica [81]. By adjusting the surface pressure, the coverage of deposited GO can be controlled, from isolated sheets at low pressure, over close-packed monolayers in the medium stage to overpacked edge-folded layers (Figure 8). Reduction of the deposited GO by hydrazine vapour resulted in 95% transparent RGO films with sheet resistance in the 10^7 (ohm) range, similar to what has been determined for RGO reduced in solution. Also graphene prepared by exfoliation of oleum-intercalated graphite into DMF, can be deposited using Langmuir–Blodgett methodology, as was demonstrated by Li et al. [13]. The transparency of a monolayer film deposited on quartz was 90% over a large

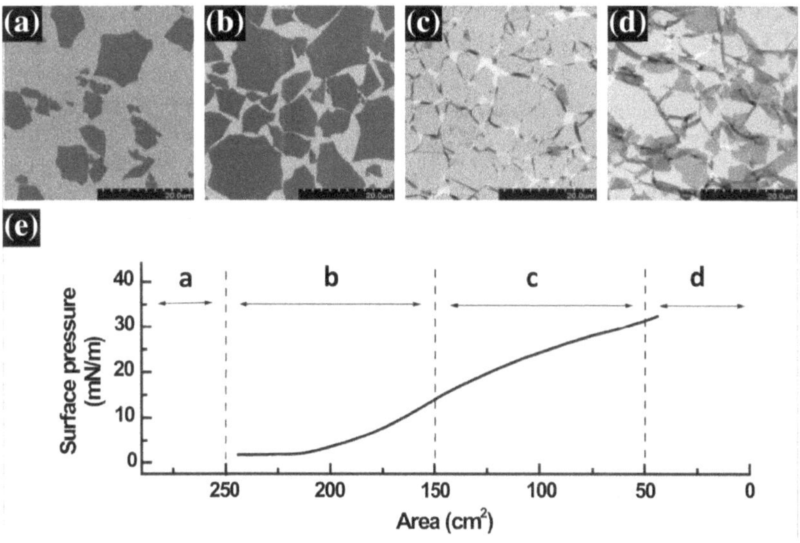

FIGURE 8 SEM images of the collected graphite oxide monolayers collected on a silicon wafer at different regions of the isotherm. The packing density was continuously tuned: (a) dilute monolayer of isolated flat sheets, (b) monolayer of close-packed GO, (c) overpacked monolayer with sheets folded at interconnecting edges, and (d) over packed monolayer with folded and partially overlapped sheets interlocking with each other. (e) Isothermal surface pressure/area plot showing the corresponding regions a–d at which the monolayers were collected. Scale bars in a–d represent 20 μm. Reproduced from Ref. [81].

range of wavelengths with sheet resistance in the 10^5 (ohm) range, reflecting the milder conditions for exfoliation compared to more oxidative routes via GO.

VI. FURTHER MODIFICATIONS BY FUNCTIONALIZATION

Graphene and derivatives thereof chemically belongs to the diverse group of polycyclic aromatic hydrocarbons, in which naftalene, antracene, and benzpyrenes are the smallest planar members. The chemical properties of a series of progressively larger aromatic hydrocarbon compounds are similar, but due to the significantly higher bulk:edge ratio, graphene reactivity *per carbon* is considerably lower than the smaller planar members of the group. The chemical reactivity of graphene is also considerably lower than the structurally related carbon nanotubes and fullerenes possessing permanent curvature and, for all fullerenes, non-hexagonal arrangement of the carbon atoms which considerably increases the reactivity. A perfectly planar hexagonal lattice would be practically non-reactive, but most graphene contains lattice defects [82]. Deviations from perfect graphene such as vacancies, Stone–Wales defects, or 5–7 seams in grain boundaries constitute regions of increased chemical reactivity. Wrinkles and folds, with the curvature resembling that of carbon nanotubes and fullerenes, are also regions with a substantially higher chemical reactivity and thus serve as initiation points for covalent functionalization reactions that may modify local as well as bulk properties [83,84].

A. Covalent Functionalization

(i) Oxidation

Treatment of graphene with oxidizing agents will introduce oxygen-containing functional groups at particularly reactive sites. The more forcing the conditions, the more GO-like, i.e., insulating and hydrophilic, the resulting material will become. Aqueous dispersions of oxidized graphene are brown, orange, or yellow depending on the degree of reaction. Partially oxidized graphene and GO has been further modified by secondary functionalization, where in particular reactions targeting carboxyl units have been employed. The properties and chemistry of GO has been extensively reviewed [7,85].

Flourination of graphene, formally an oxidation, has been achieved by reaction with gaseous F_2 at elevated temperature [86], by exposure to XeF_2 [87], or treatment of deposited graphene with fluorine-based plasma [88]. Also exfoliation of preformed graphite fluoride in the presence of sulfolane yields sheets of fluorinate graphene [89]. The product, as expected from a reaction transforming sp^2 carbon to sp^3, is a hydrophobic insulator with mechanical properties similar to graphene [90].

(ii) Reduction

When exposed to cold hydrogen plasma, graphene undergoes a reversible transformation from conductor to insulator [91]. Graphene on SiO_2, with only one side exposed to the reagent was less reduced than graphene mounted on a holey membrane, where access of both sides allow the formation of "graphane", with a relatively larger D:G ratio than for the partially reduced graphene on SiO_2 where only one side is available for the reagent. The graphene Raman signatures were almost completely recovered after annealing, an observation supporting the general reversibility of the applied reduction process. The one-side reactivity and addition pattern may be different for graphene on other substrates than SiO_2 [92].

(iii) Polar Reagents

Treatment of unsaturated compounds with aqueous HCl, a strong acid with net non-oxidizing properties, add water to reactive localized C=C sites (both C sp^2). The initial H–C–C–OH (both C sp^3), in which formally one carbon has been reduced and one oxidized, may rearrange with formation of other monooxygen functional groups [93]. Nanosheets of graphene was found to react with incorporation of oxygen into the carbon sheet plane, and the acid-treated material exhibited a lower resistivity than the controls. Aqueous HCl has profound effects on the properties of graphene, with an increased conductivity [94,95].

Organic polar reagents such as aryldiazonium salts, aryl azides, or azomethine ylides, well established for functionalization of carbon nanotubes, are reactive towards graphene as well. The structurally versatile reagents results in arylation [96], aziridination [97], and pyrrolidination [98], respectively. As for all covalent functionalization reactions, grafting of the organic groups introduce sp^3 sites in the carbon lattice.

B. Non-Covalent and Supramolecular Functionalization

Non-covalent interaction between graphene and smaller molecules are of fundamental importance for the formation of stable dispersions in a solvent, either via direct graphene–solvent interaction or via graphene–surfactant + surfactant–solvent interactions. Although solvatization depends on non-covalent interactions, it is not categorized as functionalization. In contrast, non-covalent assemblies with in particular optical properties significantly different from those of the individual components are categorized as being supramolecular, and the interactions leading to the assembly as a functionalization. The graphene–molecule interaction can, depending on the structure of the molecule as well as the amount and functional group identity of defects in the graphene, range from van der Waals over dipole–dipole to ionic. Use of the extended pi system to non-covalently attach aromatic molecules

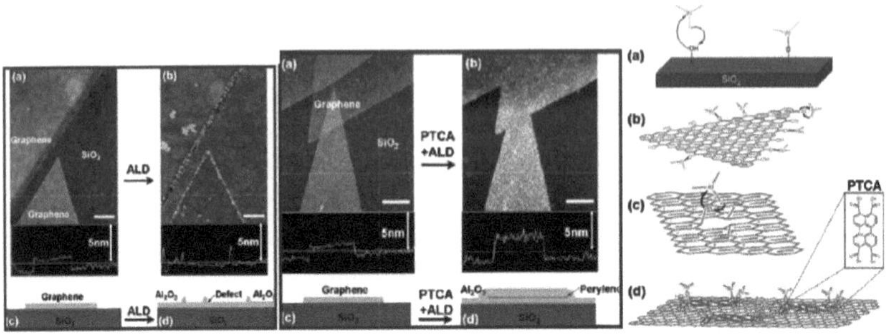

FIGURE 9 Atomic layer deposition of aluminum oxide with trimethylaluminum as Al precursor. Left and middle: AFM images (a) before and (b) after ALD. Left panel: Graphene sheet on SiO_2, no pretreatment. Middle panel: Graphene sheet on SiO_2, pretreatment with perylene tetracarboxylic acid. Right panel. Schematics of ALD on (a) bare SiO_2 substrate, (b) a single layer graphene sheet with edges, (c) graphene containing a defect site, and (d) perylene tetracarboxylic acid (PTCA)-coated graphene. Reproduced from Ref. [100].

in a supramolecular fashion is well established for carbon nanotubes and has been demonstrated adaptable also to graphene. For example, an electron-poor carboxycoronene with characteristic optical properties was employed as a surfactant in a direct exfoliation of graphene into water [33]. Both the long-term dispersability of the non-polar graphene in a polar phase and the differences in absorption and emission indicative of significant charge-transfer interactions are proof of successful functionalization. Both electron-donating and electron-withdrawing molecules may attach and hence influence the pi system of the graphene [99]. Compared to the often prominent changes of the electronic properties observed for covalently functionalized graphene, supramolecular functionalization inflicts only minor changes and will thus be of use for fine-tuning of graphene electronic and surface chemistry properties in devices where an intact graphene sp^2 system is of importance. Supramolecular functionalization has been demonstrated for graphene on SiO_2 (Figure 9). The carboxyperylene derivative preferentially adheres to graphene, leaving the SiO_2 largely uncovered. The non-covalently attached molecules provide homogenously distributed nucleation sites for ALD growth of Al_2O_3, rendering possible a mild route towards layered graphene-containing structures [100].

REFERENCES

[1] Novoselov KS, Geim AK, Morozov SV, Jiang D, Zhang Y, Dubonos SV, et al. Science 2004;306:5269.
[2] (a) Tzalenchuk A, Lara-Avila S, Cedergren K, Syvaejaervi M, Yakimova R, Kazakova O, et al. Solid State Commun 2011;151:1094; (b) Riedl C, Coletti C, Starke U. J Phys D 2010;43:374009.
[3] Allen MJ, Tung VC, Kaner RB. Chem Rev 2010;110:132.

[4] Boehm HP, Setton R, Stumpp E. Carbon 1986;24:241.
[5] Wu Z-S, Ren W, Gao L, Liu B, Jiang C, Cheng H-M. Carbon 2009;47:493.
[6] Stankovich S, Dikin DA, Piner RD, Kohlhaas KA, Kleinhammes A, Jia Y, et al. Carbon 2007;45:1558.
[7] (a) Park S, Ruoff RS. Nature Nanotech 2009;4:217; (b) Dreyer DR, Park S, Bielawski CW, Ruoff RS. Chem Soc Rev 2010;39:228; (c) Zhu Y, Murali S, Cai W, Li X, Won Suk J, Potts JR, et al. Adv Mater 2010;22:3906.
[8] Hummers WS, Offeman REJ. J Am Chem Soc 1958;80:1339.
[9] Brodie BC. Ann Chim Phys 1860;59:466.
[10] Staudenmaier L. Ber Dtsch Chem Ges 1898;31:1481.
[11] Boehm HP, Eckel W, Scholz Z. Anorg Allg Chem 1967;353:236.
[12] Liang Y, Wu D, Feng X, Müllen K. Adv Mater 2009;21:1679.
[13] Li X, Zhang G, Bai X, Sun X, Wang X, Wang E, et al. Nature Nanotech 2008;3:538.
[14] Dua V, Surwade SP, Ammu S, Agnihotra SR, Jain S, Roberts KE , et al. Angew Chem Int Ed 2010;49:2154.
[15] Shin H-J, Kim KK, Benayad A, Yoon S-M, Park HK, Jung I-S, et al. Adv Funct Mater 2009;19:1987.
[16] Wang GX, Yang J, Park J, Gou XL, Wang B, Liu H, et al. J Phys Chem C 2008;112:8192.
[17] Fan Z-J, Kai W, Yan J, Wei T, Zhi L-J, Feng J, et al. ACS Nano 2011;5:191.
[18] McAllister MJ, Li JL, Adamson DH, Schniepp HC , Abdala AA , Liu J, et al. Chem Mater 2007;19:4396.
[19] Dubin S, Gilje S, Wang K, Tung VC, Cha K, Hall AS, et al. ACS Nano 2010;4:3845.
[20] Inagaki M, Kim YA, Endo M. J Mater Chem 2011;21:3280.
[21] An SJ , Zhu Y , Lee SH , Stoller MD , Emilsson T, Park S, et al. J Phys Chem Lett 2010;1:1259.
[22] Szabo T, Berkesi O, Forgo P, Josepovits K, Sanakis Y, Petridis D, et al. Chem Mater 2006;18:2740.
[23] Wang H, Robinson JT, Li X, Dai H. J Am Chem Soc 2009;131:9910.
[24] Zheng J, Di C, Liu Y, Liu H, Guo Y, Du C, et al. Chem Commun 2010;46:5728.
[25] Viculis LM, Mack JJ, Kaner RB. Science 2003;299:1361.
[26] Valles C, Drummond C, Saadaoui H, Furtado CA, He M, Roubeau O, et al. J Am Chem Soc 2008;130:15802.
[27] (a) Suslick KS, Didenko Y, Fang MM, Hyeon T, Kolbeck KJ, McNamara WB, et al. Phil Trans Roy Soc London Ser A 1999;357:1751; (b) Suslick KS, Price GJ. Ann Rev Mater Sci 1999;29:295.
[28] Rangappa D, Sone K, Wang M, Gautam UK, Golberg D, Itoh H, et al. Chem Eur J 2010;16:6488; (b) Cravotto G, Cintas P. Chem Eur J 2010;16:5246; (c) Blake P, Brimicombe PD, Nair RR, Booth TJ, Jiang D, Schedin F, et al. Nano Lett 2008;8:1704; (d) Hamilton CE, Lomeda JR, Sun ZZ, Tour JM, Barron AR. Nano Lett 2009;9:460.
[29] (a) Kelkar MA, Gogate PR, Pandit AB. Ultrason Sonochem 2006;13:523; (b) Vichare NP, Senthilkumar P, Moholkar VS, Gogate PR, Pandit AB. Ind Eng Chem Res 2000;39:1480.
[30] Hernandez Y, Nicolosi V, Lotya M, Blighe FM, Sun Z, De S, et al. Nature Nanotech 2008;3:563.
[31] Lotya M, Hernandez Y, King PJ, Smith RJ, Nicolosi V, Karlsson LS, et al. J Am Chem Soc 2009;131:3611.
[32] Green AA, Hersam MC. Nano Lett 2009;9:4031.
[33] Ghosh A, Rao KV, George SJ, Rao CNR. Chem Eur J 2010;16:2700.
[34] (a) Misik V, Kirschenbaum LJ, Riesz P. J Phys Chem 1995;99:5970; (b) Misik V, Riesz P. J Phys Chem 1994;98:1634.
[35] Moonoosawmy KR, Kruse R. J Am Chem Soc 2008;130:13417.
[36] Semeluk GP, Bernstein RB. J Am Chem Soc 1957;79:46.
[37] Hu H, Zhao B, Hamon MA, Kamaras K, Itkis ME, Haddon RC. J Am Chem Soc 2003;125:14893.
[38] Zhang, Xiao J, Meng X, Monroe C, Huang Y, Zuo J-M. Phys Rev Lett 2010;104:166805.
[39] Shih C-J, Vijayaraghavan A, Krishnan R, Sharma R, Han J-H, Ham M-H, et al. Nature Nanotech 2011;6:439.
[40] (a) Widenkvist E, Boukhvalov DW, Rubino S, Akhtar S, Lu J, Quinlan RA, et al. J Phys D 2009;42:112003; (b) Widenkvist E, Yang W, Akhtar S, Palmgren P, Knut R, Karis O, et al. submitted.

[41] Pu NW, Wang CA, Sung Y, Liu YM, Ger MD. Mater Lett 2009;63:1987.
[42] Rangappa D, Sone K, Wang M, Gautam UK, Golberg D, Itoh H, et al. Chem Eur J 2010;16:6488.
[43] Wu JS, Pisula W, Müllen K. Chem Rev 2007;107:718.
[44] Zhi L, Müllen K. J Mater Chem 2008;18:1447.
[45] Clar E, Ironside CT. Proc Chem Soc London 1958;150.
[46] Simpson CD, Brand JD, Berresheim AJ, Przybilla L, Räder HJ, Müllen K. Chem Eur J 2002;8:1424.
[47] (a) Wu J, Gherghel L, Watson MD, Li J, Wang Z, Simpson CD, et al. Macromolecules 2003;36:7082; (b) Yang XY, Dou X, Rouhanipour A, Zhi LJ, Räder HJ, Müllen K. J Am Chem Soc 2008;130:4216.
[48] Liu W, Li H, Xu C, Khatami Y, Banarjee K. Carbon 2011;49:4122.
[49] Bhaviripudi S, Jia X, Dresselhaus MS, Kong J. Nano Lett 2010;10:4128.
[50] Wintterlin J, Bocquet ML. Surf Sci 2009;603:1841.
[51] Klink C, Steensgaard I, Besenbacher F, Laegsgaard E. Surf Sci 1995;342:250.
[52] Generalov AV, Dedkov YS. Carbon 2012;50:183.
[53] Reddy M, Gledhill AD, Chen C-H, Drexler JM, Padture NP. Appl Phys Lett 2011;98:113117.
[54] Gao L, Guest JR, Guisinger NP. Nano Lett 2010;10:3512.
[55] Land TA, Michely T, Behm RJ, Hemminger JC, Comsa G. Surf Sci 1992;264:261.
[56] Coraux J, N'Diaye AT, Busse C, Michely T. Nano Lett 2008;8:564.
[57] Loginova E, Bartelt NC, Feibelman PJ, McCarty KF. New J Phys 2009;11:063046.
[58] Wang B, Caffio M, Bromley C, Fruchtl H, Schaub R, ACS Nano 2010;10:5773.
[59] Marchini S, Gunther S, Wintterlin J. Phys Rev B 2007;76:075429.
[60] Kawanowa H, Ozawa H, Yazaki T, Gotoh Y, Souda R. Jpn J Appl Phys 2002;41:6149.
[61] Hu Z-P, Ogletree DF, Van Hove MA, Somorjai GA. Surf Sci 1987;180:433.
[62] Johansson A-S, Lu J, Carlsson JO. Thin Solid Films 1994;252:19.
[63] Reina A, Jia X, Ho J, Nezich D, Son H, Bulovic V, et al. Nano Lett 2009;9:30.
[64] Kim KS, Zhao Y, Jang H, Lee SY, Kim JM, Kim KS, et al. Nature 2009;457:706.
[65] Huang L, Chang QH, Guo GL, Liu Y, Xie YQ, Wang T, et al. Carbon 2012;50:551.
[66] Li X, Cai W, An J, Kim S, Nah J, Yang D, et al. Science 2009;324:1312.
[67] Wood JD, Schmucker SW, Lyons AS, Pop E, Lyding JW. Nano Lett 2011;11:4547.
[68] Gao T, Xie SB, Gao YB, Liu MX, Chen YB, Chang YF, et al. ACS Nano 2011;5:9194.
[69] Guermoune A, Chari T, Popescu F, Sabri SS, Guillemette J, Skulason HS, et al. Carbon 2011;49:4204.
[70] Ruan G, Sun Z, Peng Z, Tour JM. ACS Nano 2011;5:7601.
[71] Dong X, Wang P, Fang W, Su C-Y, Chen Y-H, Li L-J, et al. Carbon 2011;49:3672.
[72] Ishihara M, Koga Y, Kim J, Tsugawa K, Hasegawa M. Mater Lett 2011;65:2864.
[73] Hu B, Ago H, Ito Y, Kawahara K, Tsuji M, Magome E, et al. Carbon 2012;50:57.
[74] Regmi M, Chrisholm MF, Eres G. Carbon 2012;50:134.
[75] Pirkle A, Chan J, Venugopal A, Hinojos D, Magnuson CW, McDonnell S, et al. Appl Phys Lett 2011;99:122108.
[76] Kim B-J, Lee C, Jung Y, Baik KH, Mastro MA, Hite JK, et al. Appl Phys Lett 2011;99:143101.
[77] Bae S, Kim H, Lee Y, Xu X, Park JS, Zheng Y, et al. Nature Nanotech 2010;5:574.
[78] Chang H, Wang G, Yang A, Tao X, Liu X, Shen Y, et al. Adv Funct Mater 2010;20:2893.
[79] Pham VH, Cuong TV, Hur SH, Shin EW, Kim JS, Chung JS, et al. Carbon 2010;48:1945.
[80] Widenkvist E, Li J, Jansson U, Grennberg H. Carbon 2007;45:2732.
[81] Cote LJ, Kim F, Huang J. J Am Chem Soc 2009;131:1043.
[82] Banhart F, Kotakoski J, Krasheninnikov AV. ACS Nano 2011;5:26.
[83] Boukhvalov DW, Katsnelson MI. J Phys: Condens Matter 2009;21:344205.
[84] Zheng Q, Geng Y, Wang S, Li Z, Kim J-K. Carbon 2010;48:4315.
[85] (a) Compton OC, Nguyen ST. Small 2010;6:711; (b) Swager TM, ACS Macro Lett 2012;1:3; (c) Špitalsky Z, Danko M, Mosnaček J. Curr Org Chem 2011;15:1133.
[86] Ruff O, Bretschneider O, Ebert FZ. Anorg Allgem Chem 1934;217:1.
[87] Robinson JT, Burgess JS, Junkermeier CE, Badescu SC, Reinecke TL, Perkins FK, et al. Nano Lett 2010;10:3001.

[88] (a) Baraket M, Walton SG, Lock EH, Robinson JT, Perkins FK. Appl Phys Lett 2010;96:231501; (b) Khare BN, Wilhite P, Meyyappan M. Nanotechnology 2004;15:1650; (c) Felten A, Bittencourt C, Pireaux JJ, Van Lier G, Charlier JC. J Appl Phys 2005;98:074308.
[89] Zboril R, Karlický F, Bourlinos AB, Steriotis TA, Stubos AK, Georgakilas V, et al. Small 2010;6:2885.
[90] Nair RR, Ren W, Jalil R, Riaz I, Kravets G, Britnell L, et al. Small 2010;6:2877.
[91] Elias DC, Nair RR, Mohiuddin TMG, Morozov SV, Blake P, Halsall MP, et al. Science 2009;323:610.
[92] Boukhvalov DW. Nanotechnology 2011;22:055708.
[93] Sanyal B, Eriksson O, Jansson U, Grennberg H. Phys Rev B 2009;79:113409.
[94] Colman VA, Knut R, Karis O, Grennberg H, Jansson U, Quinlan R, et al. J Phys D 2008;41:062001.
[95] Jafri SHM, Carva K, Widenkvist E, Blom T, Sanyal B, Fransson J, et al. J Phys D 2010;43:045404.
[96] (a) Lim H, Lee JS, Shin HJ, Shin HS, Choi HC. Langmuir 2010;26:12278; (b) Bekyarova E, Itkis ME, Ramesh P, Berger C, Sprinkle M, de Heer WA, et al. J Am Chem Soc 2009;131:1336; (c) Sun S, Kohama S, Zhang Z, Lomeda JR, Tour JM. Nano Res 2010;3:117.
[97] Liu L-H, Lerner MM, Yan M. Nano Lett 2010;10:3754.
[98] (a) Quintana M, Spyrou K, Grzelczak M, Browne WR, Rudolf P, Prato M. ACS Nano 2010;4:3527; (b) Zhang X, Hou L, Cnossen A, Coleman AC, Ivashenko O, Rudolf P, et al. Chem Eur J 2011;17:8957; (c) Georgakilas V, Bourlinos AB, Zboril R, Steriotis TA, Dallas P, Stubos AK, et al. Chem Commun 2010;46:1766.
[99] (a) Subrahmanyam KS, Ghosh A, Gomathi A, Govindaraj A, Rao CNR. Nanosci Nanotech Lett 2009;1:28; (b) Manna AK, Pati SK. Chem Asian J 2009;4:855.
[100] Wang X, Tabakman SM, Dai H. J Am Chem Soc 2008;130:8152.

Chapter 6

Multiferroics: Theory, Mechanisms, and Materials

Paolo Barone*, **Biplab Sanyal**[†] **and Silvia Picozzi***
*Consiglio Nazionale delle Ricerche, Istituto Superconduttori, Materiali Innovativi e Dispositivi (CNR-SPIN), Italy
[†]Division of Materials Theory, Department of Physics and Astronomy, Uppsala University, BOX-516, 75120 Uppsala, Sweden

Chapter Outline Head

I. INTRODUCTION

The research field on "multiferroics" has boomed enormously in the last decade both concerning experiments and theory. The advent of sophisticated experimental techniques for the synthesis and characterization of complex multiferroic oxides as well as the development of accurate theoretical methods have helped the subject grow exponentially. Already a number of excellent review articles [1] exist. In this chapter, we have provided an overview of the state of the art. The interested readers are referred to the review articles for detailed discussions.

According to the conventional notion, a ferromagnetic compound has the magnetization as the order parameter and a ferroelectric compound has the electric polarization as the order parameter. In the former one, magnetism is controlled by an external magnetic field whereas in the latter one, an external electric field controls the polarization. It is known that magnetism occurs due to

Advanced Functional Materials. http://dx.doi.org/10.1016/B978-0-44-453681-5.00006-6

partially filled d or f shells whereas ferroelectricity is caused by d^0-ness, i.e., the off-centering of transition metal ions with empty d orbitals such as Ti^{4+}, that forms covalent bonds with neighboring O ions. These two effects are mutually exclusive. Multiferroics constitute a special class of system, where both these order parameter coexist. Therefore, these systems exhibit both magnetic and ferroelectric order in the same compound. Apart from these two "ferroic" orders, ferroelasticity may also be present. However, the most interesting scenario is that when one is controlled by the other, viz., magnetism is controlled by electric field and ferroelectricity via a magnetic field. In certain multiferroics, this cross coupling exists and it gives rise to the so-called "magnetoelectricity." This effect is extremely important for technology but unfortunately, no such compound has been discovered so far where a strong magnetoelectric coupling is observed at room temperature. Magnetoelectricity can be described by Landau theory where the free energy can be expressed in terms of applied electric and magnetic field. If the free energy expansion is retained up to linear order, a linear magnetoelectric coefficient α_{ij} is defined as

$$P_i = \alpha_{ij} H_j,$$
$$\mu_0 M_i = \alpha_{ji} E_j.$$

In the above equations, $P,\ H,\ M,\ E$ are polarization, applied magnetic field, magnetization, and applied electric field respectively.

The discussion of multiferroics is itself extremely interesting as a rich variety of physical mechanisms play an instrumental role in establishing ferroelectric polarization. Conventionally, multiferroics are broadly classified into two types, (i) proper and (ii) improper. In proper multiferroics, ferroelectricity is caused either by the stereochemically active lone pair of electrons of the cations, such as Bi^{3+} in $BiFeO_3$ or by special geometry related features giving rise to certain structural instabilities. Both these mechanisms are demonstrated below. It is interesting to note that the breaking of inversion and time reversal symmetries is caused by different ions in proper multiferroics. This is certainly a disadvantage in spite of very high magnetic and ferroelectric transition temperatures. This cross coupling is genuinely achieved in the case of improper multiferroics, where the same ion is responsible for both symmetry breaking, therefore carrying an immense potential of strong cross coupling across order parameters and hence, magnetoelectricity. Unfortunately, no improper multiferroic compound has been discovered so far where the coupling occurs at room temperature. Nevertheless, magnetically driven ferroelectricity in improper multiferroics is an extremely rich subject of study and will be discussed in detail below.

II. THEORY OF PROPER MULTIFERROICS

Let us first start the discussion with the most popular proper multiferroic $BiFeO_3$. Bulk $BiFeO_3$ has a ferroelectric Curie temperature of about 1100 K.

The crystal structure of $BiFeO_3$ (BFO) has rhombohedral symmetry (R3c space group), indicated by neutron and X-ray diffraction experiments. The unit cell of this distorted perovskite structure has neighboring O octahedra counter-rotated around the [111] axis along which the ferroelectric polarization develops. It is known that the stereochemically active lone pair of $6s^2$ electrons in Bi^{3+} gives rise to a large displacement of Bi with respect to O. A wide range ($\sim$2–160 $\mu C/cm^2$) of the experimentally measured values of polarization has been reported in the literature. These include experiments on bulk single crystals, thin films, and polycrystalline films of various thicknesses. The interest of the scientific community on the research on BFO has been boosted after the work of Wang et al. on epitaxial thin films of BFO on $SrTiO_3$ J. Wang et al., Science 299, 1719 (2003), where the coexistence of a large spontaneous polarization and magnetization were reported above room temperature.

First principles calculations have played an important role in the understanding of magnetic and ferroelectric properties of BFO. While the calculations of the magnetic structures and corresponding moments are routinely done by DFT calculations using local approximations or beyond, the calculations of spontaneous polarization were not frequently done before the advent of the modern theory of polarization put forward by King-Smith and Vanderbilt [2], where the geometric Berry phase of the wave functions defines the electronic polarization. The ferroelectric polarization is defined through

$$\vec{P}_e^{(\lambda)} = -\frac{ife}{8\pi^3} \sum_{\text{bands}} \int_{\text{BZ}} \mathrm{d}^3\boldsymbol{k} \langle u_{n\boldsymbol{k}}^{\lambda} | \nabla_{\boldsymbol{k}} | u_{n\boldsymbol{k}}^{\lambda} \rangle,$$

where f is the occupation number of the states in the valence band, λ is the parameter to specify the change in Kohn–Sham potential upon atomic displacements, and $|u_{nk}^{\lambda}\rangle$ is the cell-periodic part of the Bloch function. Since there is no specific phase relation between the eigenvectors $|u_{nk}^{\lambda}\rangle$ generated by the numerical subroutines, a direct evaluation of the above equation is not possible. In actual practice, the integration over Brillouin zone is divided into two parts: (1) a two-dimensional integration over an area A performed by sampling k-points over 2D-mesh called $\boldsymbol{k}_\perp$ and (2) a line integral along $\boldsymbol{k}_\parallel$ parallel to a given lattice vector $\vec{G}_i$, $i = 1, 2, 3$ given by

$$\vec{G}_i \cdot \vec{P}_e^{(\lambda)} = -\frac{ife}{8\pi^3} \sum_{\text{bands}} \int_A d\boldsymbol{k}_\perp \int^{|\boldsymbol{G}_i|} d\boldsymbol{k}_\parallel \langle u_{n\boldsymbol{k}}^{\lambda} | \frac{\partial}{\partial \boldsymbol{k}_\parallel} | u_{n\boldsymbol{k}}^{\lambda} \rangle$$

or

$$\boldsymbol{G}_i \cdot \boldsymbol{P}_e^{(\lambda)} = -\frac{fe}{8\pi^3} \int_A \phi^{\lambda}(\boldsymbol{k}_\perp) d\boldsymbol{k}_\perp.$$

The Berry phase $\phi^{\lambda}(\boldsymbol{k}_\perp)$ for a given $\boldsymbol{k}_\perp$ is computed along a string of J k-points defined by

$$\boldsymbol{k}_j = \boldsymbol{k}_\perp + \frac{j\boldsymbol{G}_\parallel}{J},$$

with $j = 0, 1, \ldots, J-1$. The Berry phase is given by

$$\phi^{\lambda}(\boldsymbol{k}_{\perp}) = \mathrm{Im}\left[ln \prod_{j=0}^{J-1} \det(\langle u^{\lambda}_{n\boldsymbol{k}_j} | u^{\lambda}_{mc\boldsymbol{k}_{j+1}}\rangle)\right],$$

where n and m indicate the valence bands.

In polar insulators or semiconductors there are macroscopic electric fields produced by the long-ranged nature of Coulomb forces. These fields are coupled to longitudinal optical phonons (LO) with long wavelength ($\mathbf{q} \to 0$) phonons [3]. Such couplings can be expressed in terms of Born-effective charges (BEC) through

$$Z^{*}_{\kappa\alpha\beta} = \Omega \frac{\partial P_{\beta}}{\partial \tau_{\kappa,\alpha}},$$

$Z^{*}_{\kappa\alpha\beta}$ being the Born-effective charge tensor. Therefore BEC's serve as proportionality constants at linear order between macroscopic polarization caused along the direction β due to the ionic displacement along direction α under the condition of zero external electric field. Another equivalent way of defining the BEC's is

$$Z^{*}_{\kappa\alpha\beta} = \frac{\partial F_{\kappa\alpha}}{\partial E_{\beta}},$$

which expresses BEC's as the proportionality constant between force on the ion κ along direction α due to the macroscopic electric field E_{β} along β in the limit when no ion is displaced. BEC's therefore give the extent of electrostructural coupling in the material.

In bulk crystals, due to the Born–von Karman periodic boundary conditions, the total polarization can only be well defined up to the polarization quantum, expressed as $\mathrm{e}\boldsymbol{R}/\Omega$, where e, $\boldsymbol{R}$, and Ω are electronic charge, lattice vector along the polarization direction, and the volume of the unit cell respectively. One should note though there is an ambiguity in the absolute value of polarization, the difference in polarization between the ferroelectric and suitably chosen centrosymmetric paraelectric structures is a well-defined quantity. Neaton et al. [4] have nicely demonstrated these issues for bulk BFO and have shown that their calculated value of polarization was consistent with the recent experiments. They have also discussed several factors that may be responsible for the difference between theory and earlier experiments in terms of leakage currents or different switching paths between two opposite ferroelectric phases.

Unlike BFO, bismuth manganite (BMO) exhibits ferromagnetism with an ordering temperature of ~100 K and magnetization of ~3.2 μ_B/f.u. In some experiments, polycrystalline and thin films of BMO showed ferroelectric responses indicative of noncetrosymmetric C2 space group whereas a more recent experiment on $BiScO_3$ indicated that BMO with a similar crystal structure should have a non-polar centrosymmetric C2/c space group. A careful DFT + U calculation of Baettig et al. [5] indicated that, although the C2 phase has a spontaneous polarization, the energetically more stable phase is the non-polar C2/c crystal structure with zero polarization. A deeper analysis revealed that, in spite of the local polar distortion around an individual Bi^{3+} lone pair, the adjacent lone pairs orient in such a way to cancel each other's polarity giving rise to an antiferroelectric arrangement.

Not only oxides, but fluoride-based multiferroics are also interesting due to proper geometric ferroelectricity. It should be noted that fluorides are less (more) covalent (ionic) than the oxides and hence, geometry can play an important role in ferroelectricity. A class of isostructural barium fluorides $BaMF_4$, (M = Mn, Fe, Co, Ni, Zn, and Mg) crystallizing in the base-centered orthorhombic structure exist with the polar space group Cmc21. The spontaneous polarization occurs along the *c* direction. From first principles calculations, Ederer and Spaldin [6] showed that the ferroelectricity in this class of compounds occurs from the softening of a single polar phonon mode. Unlike conventional ferroelectrics, the Born effective charges are similar to the nominal charges indicating that the charge transfer between the cations and anions are insignificant. Here, the size effects as well as the special geometry regarding the connectivity between fluorine octahedra induces inversion symmetry breaking and hence, gives rise to ferroelectricity.

Magnetic double perovskites with formula $A_2BB'O_6$ (A is an alkaline rare-earth, B and B' are transition metals) have also attracted a lot of attention in the last decade as potential multiferroics [7]. They have been discussed in terms of polarization, anomalous dielectric properties [8], spin–phonon coupling [9], etc.

One such example is Bi_2CoMnO_6 (BCMO) that crystallizes in the double perovskite structure. It is among a few, rare materials which show existence of both ferroelectric (FE) as well as ferromagnetic (FM) behavior. Due to the presence of $6s^2$ lone pair of Bi^{3+}, there is a non-centrosymmetric (space group C2 (No. 5)) structural distortion which leads to a ferroelectric transition temperature around 600 K [10], above which the centrosymmetric structure appears with space group C2/m (No. 12). Below the magnetic ordering temperature of 95 K, ferromagnetism appears due to the 180° cation–anion–cation ferromagnetic superexchange interaction [11] in the chain Co^+–O^-–Mn^+ where Co and Mn ions are ordered in a rock-salt configuration. A similar system, Bi_2NiMnO_6 has been studied theoretically where the authors have reported electronic structure, magnetism, and ferroelectric polarization. Also, there is a strong interest in the study of magnetically driven electrostructural coupling where different magnetic structures may produce different phonon frequencies arising from exchange-striction effects [8].

The following sections will contain the results obtained from calculations [12] based on density functional theory (DFT) within the Perdew–Burke–Ernzerhof generalized gradient approximation (GGA)[13] on the basis of the projector augmented-wave method [14] implemented in the Vienna *ab initio* simulation package (VASP) [15]. To treat strong electron interaction effect

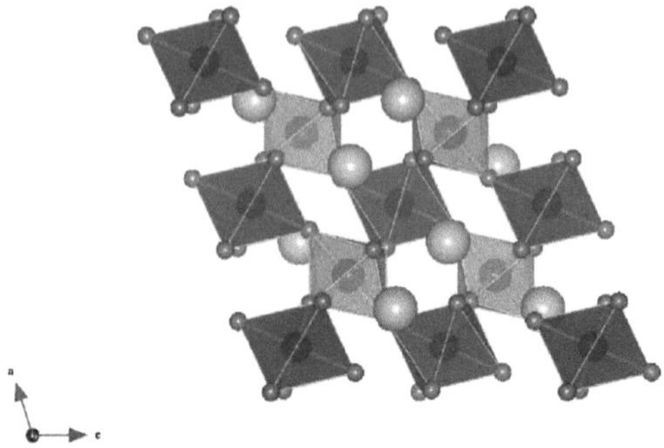

FIGURE 1 Crystal structure of BCMO, as viewed along the short-axis b, showing octahedral tilt patterns of CoO_6 (blue) and MnO_6 (red) octahedra (orange balls are Bi). (For interpretation of the references to color in this figure legend, the reader is referred to the web version of this book.)

in Co and Mn we have carried out GGA + U calculations in the Hubbard model framework with the choice of Coulomb parameter $U=3$ eV for Co-d and $U=4$ eV for Mn-d orbitals. The value of the exchange parameter J used was 1 eV for both the transition metals. We have done a thorough study with different values of U on Co and Mn and have found that the results, e.g., occupation, magnetic moments, ferroelectric polarization, etc. do not depend significantly on the values of U considered. The unit cell consists of four formula units in the C2 phase. In Wyckoff notations, the four Co atoms occupy 2a and 2b sites respectively while Bi, Mn, and O occupy 4c sites. In Glazer's notation, the tilt pattern of the octahedra is given by $a^-b^0c^-$ (Figure 1) which implies that there is no tilting of octahedra along [0 1 0] direction, while the neighboring octahedra tilt in opposite directions to each other along both [1 0 0] and [0 0 1] directions. The dielectric and dynamical properties such as Born effective charge matrices and zone-center phonon frequencies were calculated using density functional perturbation theory (DFPT) [16].

In Figure 2, we show the density of states (DOS) obtained by GGA + U method for the ferromagnetic and ferrimagnetic (FiM) states. The choice of our U values opens up a band gap of around 0.3 eV in the FM state. From the calculations with different values of U, one observes that the band gap increases slightly with the increase in U. The value of the band gap is close to what has been reported (~0.45 eV) for Bi_2NiMnO_6 [17]. In Figure 2, one observes that the top of the valence band is dominated by Co-d and O-p states (not shown) while the Mn states are pushed about 1 eV below the Fermi energy. On the other hand the bottom of the conduction band is due to Mn-d and O-2p states. A comparison of the total energies of iso-structural FM and FiM structures yields that the FM state is 163 meV/f.u. lower in energy compared to the FiM state. The magnetic moment of Co (both in 2a

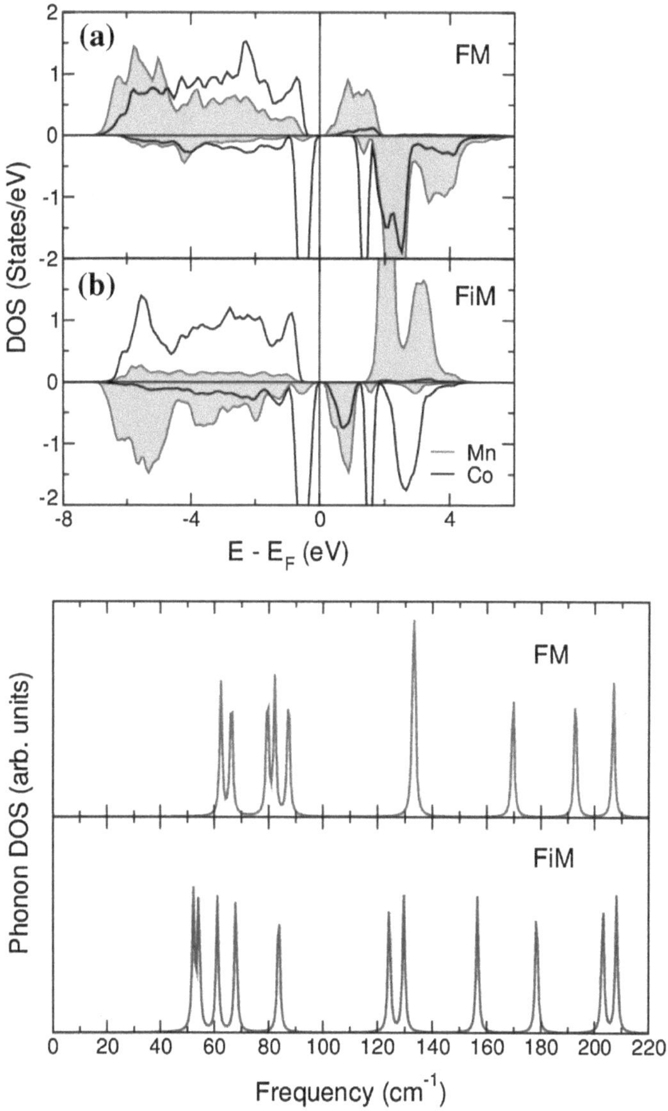

FIGURE 2 (Top) Mn and Co DOSs projected on d-orbitals for (a) FM and (b) FiM configurations. (Bottom) Phonon DOS of IR-active modes for FM and FiM configurations. (For interpretation of the references to color in this figure legend, the reader is referred to the web version of this book.)

and $2b$ sites) is found to be $2.5\,\mu_B$, while Mn carries a magnetic moment of $3.2\,\mu_B$, yielding a magnetic moment of $6\,\mu_B$/f.u. in the FM state. The analysis of orbital-projected charges in both FM and FiM magnetic structures shows that Co is in a high-spin state with around 7d electrons (Co^{2+} state), while the Mn sphere contains around 5d electrons. It seems that Mn is in an oxidation state which is associated with an ionicity of about 2+ rather than 4+, which is the expected ionic state from the counting of oxidation states. This deviation from the expected ionicity may be attributed to the low symmetry structure and small Mn–O bond lengths. Similar ill-defined oxidation states of Mn and O were mentioned by Ciucivara et al. [17] for an iso-structural compound Bi_2NiMnO_6. The structural analysis shows that in the FM state, the angle ∠Mn–O–Co(2a) is equal to 159.45° while the angle ∠Mn–O–Co(2b) is 150.83° giving an average ∠Mn–O–Co angle of 155.14°. In both the Mn–O–Co chains, the Co–O bond is about 0.18 Å longer than the Mn–O bond. This clearly indicates that the Mn–O bond is stronger in comparison with Co–O bond, which is reflected in the DOS where we see that more Mn-d states are pushed down below the Fermi level.

Spontaneous ferroelectric polarization was computed in the FM state by Berry phase method as described before. The ferroelectric polarization is nonzero only along the short-axis **b** in agreement with the C2 symmetry of the crystal. Our estimated value of polarization (5.88 $\mu C/cm^2$) is close to what has been measured for Bi_2NiMnO_6 thin films. The polarization weakly varies with U, similar to an iso-structural compound Bi_2NiMnO_6 [17] with GGA (16.84 $\mu C/cm^2$) and GGA + U (16.63 $\mu C/cm^2$) respectively using $U_{Mn} = 3.5$ eV and $U_{Ni} = 4.0$ eV. Such behavior is due to the fact that most of the contribution to the polarization is due to off-centering of Bi ion, which is unaffected by the changes of electronic states of Mn and Co due to the variation of U. The small variation in the values of the polarization is observed due to the small differences in the optimized volumes of the unit cells for different values of U.

Finally, we analyze the zone center phonon modes for FM and FiM structures. The eight lowest infra-red (IR) active modes are shown in Figure 2. One can observe that the typical softening of IR-active modes could be as large as 10 cm^{-1}. The least shifted mode is at 133 cm^{-1} where the softening is associated with the further lifting of degeneracy of this mode. Subsequently there is an increase in the static dielectric constant in FiM phase. To understand the spin–phonon coupling better, we analyzed the softest mode (at 52 cm^{-1}) in the FiM state more carefully. By adding the atomic displacement due to this mode to the equilibrium atomic positions we find that the atomic displacements are such that the increase in Mn–O–Co angle with respect to FM state is very small. We find an average increase of about 2° of Mn–O–Co angle. This fact may contribute to the observation that the spin–phonon coupling, though present, is small. Moreover, the contribution of Mn to the change in mode effective charges of the IR-active modes of two magnetic states as well as to dielectric constant is smaller compared to Co.

III. THEORY OF IMPROPER MULTIFERROICS

A. Basic Concepts

Despite being prophetized by Pierre Curie at the end of nineteenth century, the first concrete prediction of a coupling between magnetic and electric degrees of freedom is due to Dzyaloshinskii [18], who identified in Cr_2O_3 a potential magnetoelectric crystal, soon after experimentally confirmed by Astrov [19]. His prediction was based on a phenomenological approach that successfully mixes symmetry analysis and thermodynamics, the well-known Landau theory of phase transitions [20]. It also strongly relied on the development of magnetic symmetry theory [21], that soon after would put forward the idea that there may exist systems in which magnetic and ferroelectric ordering can coexist and spontaneously develop without external applied fields [22]. Nonetheless, despite some early attempts during the 70s, the field of magnetoelectric multiferroics started to fully develop only in very recent years.

From a theoretical point of view, two major factors apparently opposed earlier progress in the field. First, the phenomenological approach based on symmetry considerations, despite its predictive character, does not give any clue on the microscopic mechanisms responsible for ferroelectricity, (ferro) magnetism, and their possible coupling: it simply states that these phenomena are allowed by symmetry. One has also to note that a comprehensive microscopic theory of ferroelectricity is indeed still lacking. As opposed to magnetic order, whose microscopic nature is in principle the same in most oxide-based magnets (exchange interactions between localized magnetic moments), there are different types of ferroelectrics and many microscopic mechanisms. As previously pointed out, the definition of polarization itself in crystals has been given only in the 90s, leading to the formulation of what is nowadays known as the modern theory of polarization [2,23,24]. Second, most of the conventional ferroelectrics are transition metal oxides in which transition metal cations have empty d shells [25]; as rationalized only recently [26], empty d shells have a strong tendency to form covalent bonds with surrounding anions and may cause polar shifts of ions. The empirical observation of the so-called d^0-ness apparently rules out magnetism, which on the other hand requires partially filled d shells on transition metal ions, since the electron spins in completely filled shells would add to zero without participating in magnetic ordering. The microscopic mechanism leading to this "exclusion rule" and the general question of why magnetism and ferroelectricity are so scarcely observed together was apparently left untouched until the very end of the 90s [27,28]. Its elaboration significantly contributed to the revival of the multiferroics field, constituting the starting point for the search of different mechanisms that may induce a ferroelectric ordering in magnetic systems.

Before discussing in detail some of the most popular microscopic mechanisms for multiferroicity, let us briefly introduce some basic symmetry considerations that one should always keep in mind when dealing with magnetoelectric multiferroics. They allowed to bypass the aforementioned "exclusion rule," introducing the idea that ferroelectricity can be indeed induced by magnetic order, especially in frustrated magnets [30,29] (as we will see, the opposite is not true, i.e., a spontaneous electric polarization cannot induce magnetic ordering). This finding opened the search for other improper ferroelectrics, where the electric polarization appears as a by-product of some other ordering, such as charge or orbital ordering [31,32].

The order parameters that describe the ferroelectric and magnetic order are governed by different symmetries. The polar vector $\mathbf{P}$ (and the conjugated electric field $\mathbf{E}$) changes sign on the inversion of all coordinates, $\mathbf{r} \rightarrow -\mathbf{r}$, but remains invariant under time reversal $t \rightarrow -t$. On the other hand, the magnetic order parameter $\mathbf{M}$ (and magnetic field $\mathbf{H}$) transforms in precisely the opposite way, being invariant under space inversion and changing sign on time reversal. A linear cross-coupling between these quantities is then possible only when they vary both in time and space. Since the thermodynamic potential has to be invariant under symmetry transformations [20], the coupling between static $\mathbf{P}$ and $\mathbf{M}$ can be only nonlinear, and is expected to be rather weak. However, if an inhomogeneous magnetic order, typical of frustrated magnets, is considered, symmetries allow to include a third-order invariant term $PM\partial M$ [29,30]. This term is linear in $\mathbf{P}$, which enters as a secondary order parameter, implying that electric polarization may arise from arbitrary weak interactions, provided the proper kind of magnetic ordering sets in. In other words, magnetic order in frustrated magnets may break inversion symmetry and induce an electric polarization; on the contrary, a spontaneous ferroelectric polarization does not break time-reversal symmetry, thus it cannot induce any magnetic ordering as a secondary effect.

Frustration in magnetic systems emerges from competing interactions, such that all pairs of magnetic moments cannot simultaneously be in their ground state. A simple but concrete realization of a frustrated system is a classical spin chain with isotropic nearest-neighbor ferromagnetic interaction ($J < 0$) and next-nearest-neighbor antiferromagnetic interaction ($J' > 0$):

$$H = \sum_n \left[J\, \mathbf{S}_n \mathbf{S}_{n+1} + J'\, \mathbf{S}_n \mathbf{S}_{n+2} \right]. \tag{1}$$

For ferromagnetic interaction only, the chain displays a ground state with all parallel spins. Competition between ferro- and antiferromagnetic interactions, however, frustrates this uniform ordering and eventually, for $J' > |J|/4$, induces a spiral: $\mathbf{S}_n = \mathbf{e}_1 \cos Qx_n + \mathbf{e}_2 \sin Qx_n$ where $\mathbf{e}_1,\ \mathbf{e}_2$ are orthogonal unit vectors and the wavevector Q satisfies $\cos Q = |J|/4J'$. The resulting magnetic ordering is therefore highly non-homogeneous. For the chain system, as for cubic crystals, the allowed form of the magnetically induced electric polarization is $\mathbf{P} \propto [\mathbf{M} \cdot \partial\mathbf{M} - \mathbf{M}(\partial \cdot \mathbf{M})]$ [29], meaning that a cycloidal spiral

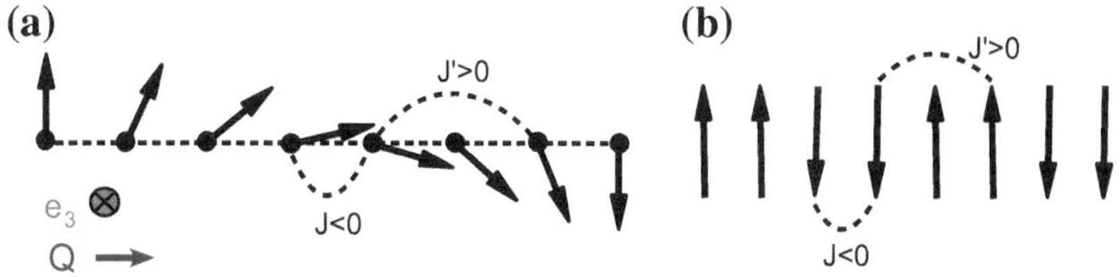

FIGURE 3 Two examples of frustrated spin configurations in spin chains with competing FM ($J < 0$) and AFM ($J' > 0$) interactions. (a) The ground state with cycloidal spiral ordering in the isotropic Heisenberg spin chain when $J' > |J|/4$. (b) The up–up–down–down ground state found in the ANNNI model when $J' > |J|/2$.

state induces a polarization[1]

$$\mathbf{P} \propto \mathbf{e}_3 \times \mathbf{Q}. \tag{2}$$

In the example considered $\mathbf{Q} = (Q, 0, 0)$ is parallel to the chain direction, whereas $\mathbf{e}_3 = \mathbf{e}_1 \times \mathbf{e}_2 \propto \mathbf{S}_n \times \mathbf{S}_{n+1}$ is the spin-rotation axis.

Another simple example of frustrated spin system is obtained by replacing $\mathbf{S}_n$ with Ising spins S_n^z in model (1), that reduces to the so-called ANNNI (antiferromagnetic nearest-next-neighbor Ising) model [34]. The ferromagnetic ordering remains stable as long as $J' < |J|/2$, whereas for $J' > |J|/2$ the ground state displays a $\uparrow\uparrow\downarrow\downarrow$ ordering. The coupling term in the Landau description is again linear in P, being $\propto P(L_1^2 - L_2^2)$, where $L_1 = S_1^z + S_2^z - S_3^z - S_4^z$ and $L_2 = S_1^z - S_2^z - S_3^z + S_4^z$ represent two possible realization of the up–up–down–down configuration. Minimization of the thermodynamic potential predicts an induced polarization $P \propto S_n^z \cdot S_{n+1}^z$ parallel to the chain direction.

The previous examples, sketched in Figure 3, introduce two popular mechanisms for magnetically induced polarization, the so-called spin-current (or inverse Dzyaloshinskii–Moriya) mechanism [35,36] and the symmetric exchange-striction mechanism [37,38], only on the basis of general symmetry considerations. In the next section we will discuss some microscopic models that clarify the origin and the nature of such mechanisms.

B. Microscopic Mechanisms

As anticipated in the previous section, a ferroelectric state can arise from many different microscopic mechanisms, whereas magnetic order emerges substantially from the competition of two-body (exchange) interactions between predominantly localized magnetic moments. When discussing microscopic

---

[1] This formula for the predicted polarization induced by cycloidal spiral order strictly applies only for cubic or tetragonal crystals; indeed Arima recently showed that also a proper-screw spiral, i.e., with $\mathbf{Q} \parallel \mathbf{e}_3$, can induce polarization in monoclinic, triclinic, or trigonal crystals when the modulation vector $\mathbf{Q}$ is not parallel to the threefold rotation axis [33].

mechanisms in multiferroics, then, complications arise when dealing with the origin of electric polarization, that may depend on the specific material under consideration. Historically, macroscopic bulk polarization in crystals has long evaded microscopic understanding mainly because it lacked a proper and adequate definition; it may be useful then to recall some basic concepts as developed in the modern theory of polarization [2,23,24].

The standard picture of macroscopic polarization relies on the Clausius–Mossotti model [39], that assumes the presence of identifiable polarizable units. Upon breaking of inversion symmetry, these dipole "units" order along some preferred axis, and a bulk polarization emerges as the sum of the dipole moments in a given cell divided by the cell volume. This picture, despite being intuitively simple, is however flawed by the fact that in real materials, with the exception of ionic crystals, it is not possible to identify unambiguously localization centers of the electronic charge, that has some periodic but continuous distribution $\rho(\mathbf{r})$. This observation is particularly relevant when considering conventional ferroelectric oxides, that show a mixed ionic/covalent character. On the other hand, the knowledge of $\rho(\mathbf{r})$ cannot be used for a meaningful definition of bulk polarization.[2] The way out from this theoretical impasse came from a change of paradigm, in which one focuses on differences of polarization, that may occur during some process (like turning on an external field or a structural deformation). Then polarization can be defined in terms of adiabatic flow of currents within the material, which in turn are formally expressed in terms of the geometric Berry phase of the electronic wavefunction. This Berry-phase polarization correctly describes the macroscopic **P** coming from mobile electrons in crystals with covalent character, which adds to the ionic contribution from the positive point-charge-like nuclei. Referring interested readers to Refs. [2,23,24] for details, we only stress that the Clausius–Mossotti picture can be eventually restored by mapping the Berry-phase polarization onto Wannier-function centers, that provide a univoquely defined lattice of negative point charges representative of the continuous electronic charge-density, i.e., the localization centers needed to identify (together with the cation positions) the dipole units.

Within this formulation, one can distinguish between two different contributions to bulk polarization: (i) an ionic contribution due to the point-like charge carried by the ions and to their relative displacements, and (ii) an electronic contribution arising from the itinerant electrons and from their inhomogeneous charge distribution. Following this classification, but keeping in mind that it is somewhat artificial, since the two contributions are strongly

[2] For example, one may assume $\mathbf{P} = (1/V)\int \mathrm{td}\mathbf{r}\rho(\mathbf{r})$; since a macroscopic but finite sample must be assumed within this choice, the problem of disentangling surface contributions from bulk ones would invalidate the definition. On the other hand, by carrying out the integration over one unit cell would imply a non-univoque result depending on the shape and location of the unit cell itself.

entangled and not mutually exclusive, we proceed now in describing some of the microscopic mechanisms at play in multiferroic oxides.

(i) Electronic Mechanisms

Most of the microscopic mechanisms of electronic origin for polarization emerge from the interplay of spin, charge, and orbital degrees of freedom, being ultimately connected to the shape and inhomogeneous charge distribution in individual orbital states. The relevant states in many transition metal (TM) oxides are the fivefold degenerate d levels of transition metals; in cubic crystal fields, these levels are split into triply degenerate t_{2g} and doubly degenerate e_g states (further splittings appear as the crystal field symmetry is lowered), which break the rotational symmetry in orbital space. While e_g states have quenched orbital moment, t_{2g} states display nonzero spin–orbit coupling (SOC) $\lambda \mathbf{L} \cdot \mathbf{S}$, with a consequent non-trivial mixing of these states. Finally, hybridization effects with surrounding oxygens must be taken into account, since hopping processes occur mainly through the ligand O, being d states strongly localized.

The easiest way to understand the interplay between spin and electronic polarization[3] consists in defining a cluster model with two TM ions connected by a ligand oxygen [35,40,41]. One assumes then that the electron energy levels on TM ions are determined by the cubic crystal field, hence splitted in high energy e_g and low energy t_{2g} levels; since SOC is not effective in the e_g manifold, one can restrict the analysis within the t_{2g} subspace. The spin configuration on each TM ion is tuned by a spin-polarization (Zeeman) field $H_{\mathrm{TM}} = -U \sum_{a=\mathrm{l,r}} \hat{\mathbf{m}}_a \cdot \mathbf{S}_a$, describing the local electronic spin on left (l) and right (r) transition metal ion subject to the local field along the unit vector $\hat{\mathbf{m}}_a$. The electron transfer through the ligand oxygen, described by the local term $H_{\mathrm{O}} = \Delta \sum_{\beta=x,y,z} \sum_{\sigma} p^{\dagger}_{\beta\sigma} p_{\beta\sigma}$, is given by $H_{\mathrm{dp}} = \sum_{a=\mathrm{l,\,r}} \sum_{\sigma\alpha\beta} V_{a\alpha\beta}(d^{\dagger}_{a,\alpha\sigma} p_{\beta\sigma} + \mathrm{H.c.})$. The hybridization matrix $V_{a\alpha\beta}$ depends on the d and p orbitals involved (labeled by $\alpha = xy, yz, zx$ and $\beta = x, y, z$, respectively), corresponding to σ and π bonding, and to their relative positions; a widely used parametrization of this matrix can be found in Ref. [42]. Within the chosen cluster geometry shown in Figure 4, neither d_{yz} nor p_x orbitals hybridize with any of, respectively, oxygen p and TM d orbitals.

The analysis of the three-atom cluster model starts from the large-U limit, a reasonable assumption since the strength of the spin-polarization field U mimicks the Hund coupling that may well exceed the SOC strength λ. The model can then be solved in second-order perturbation theory in λ, V [35] or within a projection scheme that truncates out the spin orientation antiparallel to the local Zeeman field [40]. In both cases, one finds that the mixing of d orbitals induced by the SOC and hybridization effects mediated by the ligand oxygen triggers

[3] To be precise, this approach allows only for the dipole moment of the cluster to be directly evaluated; the bulk polarization must then be inferred by combination with symmetry analysis and group theoretical treatment.

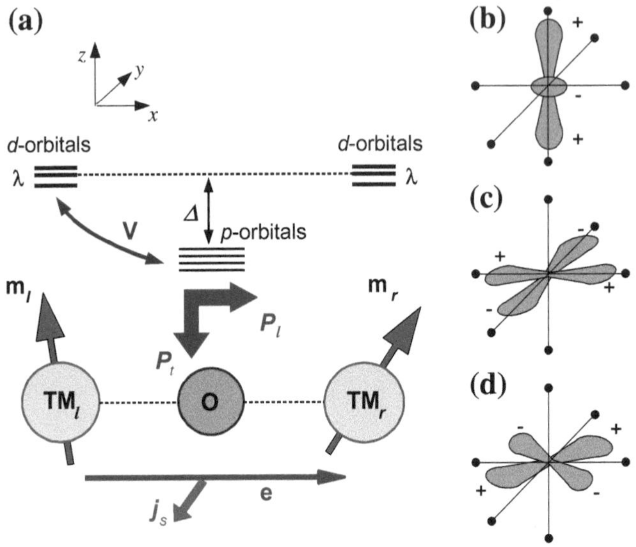

FIGURE 4 (a) The cluster model with two transition metal ions (TM_l and TM_r and the ligand oxygen between them. The noncollinear spin configuration is defined by the local spin directions $\hat{\mathbf{m}}_l$, $\hat{\mathbf{m}}_r$; a longitudinal and a transverse contribution to the local dipole moment are mediated by the SOC (λ) and the p–d hybridization. Schematic shapes of d orbitals are also shown: (b) $3z^2 - r^2$ and (c) $x^2 - y^2$ orbitals with e_g symmetry, (d) xy orbital – the other t_{2g} orbitals have the same shape but they lie on zx and xy planes.

the onset of a local dipole at the ligand site, intimately connected to the shapes of involved d and p orbitals. For example, from the nonzero overlap integral $\langle \mathrm{d}_{a,yz}|\mathbf{y}|\mathrm{p}_z\rangle$, a transverse component of the local electric dipole is found,

$$\mathbf{P} \propto \mathbf{e} \times \mathbf{S}_l \times \mathbf{S}_r, \qquad (3)$$

where **e** is the unit vector parallel to the bond direction [35,41]. This formula corresponds to a microscopic derivation of the mechanism described by Eq. (2). It can be shown that the vector product $\mathbf{S}_\mathrm{l} \times \mathbf{S}_\mathrm{r}$ is proportional to the spin current $\mathbf{j}_\mathrm{s}$, and that the Dzyaloshinskii–Moriya vector **D**, that is related to an antisymmetric relativistic correction to the usual exchange interaction [43,44], acts as its vector potential [35]. In this respect, it is worth noticing that from a symmetry point of view both **P** and $\mathbf{j}_\mathrm{s}$ are invariant under time reversal, since the spin polarization is reversed together with the direction of the spin current, that makes their coupling not surprising.

A longitudinal component of the electric dipole arises from the nonzero overlap integrals $\langle \mathrm{d}_{a,xy}|\mathbf{x}|\,\mathrm{p}_y\rangle$ and $\langle \mathrm{d}_{a,zx}|\mathbf{x}|\mathrm{p}_z\rangle$. This contribution to **P** results from the SOC-mediated variation of the d–p hybridization and it is indeed a single-ion effect rather than a spin-correlation one, as can be seen from the predicted functional form for the three-atom cluster model [40,41]:

$$\mathbf{P} \propto (\mathbf{S}_l \cdot \mathbf{e})\mathbf{S}_l - (\mathbf{S}_r \cdot \mathbf{e})\mathbf{S}_r. \quad (4)$$

In more realistic structures, the consequences of this mechanism can be even enhanced, since the variation in the d–p hybridization may cause an anisotropic charge distribution of p electrons, as suggested, e.g., for $CuFeO_2$ [33] and $Ba_2CoGe_2O_7$ [45,46].

Finally, a third contribution may arise if the inversion symmetry of the cluster is broken, i.e., $\mathbf{P} \propto (\mathbf{S}_l \cdot \mathbf{S}_r)\mathbf{e}$ [41]. This term describes the magnetostriction mechanism and, at odds with the previous two, does not depend on the spin–orbit interaction, but on the local structure and specifically to the absence of an inversion center.

We conclude this discussion on the TM–L–TM model by stressing the fact that ions in the cluster do not contribute to the local dipole moment, being their relative positions held fixed. The spin-current and spin-dependent hybridization mechanisms are indeed driven by the spin–orbit coupling alone, even in the presence of the inversion center at the middle point between the two magnetic TM ions, that rules out the direct Dzyaloshinskii–Moriya interaction (DMI) [43,44] in the cluster. On the other hand, the model does not provide any microscopic explanation of why a noncollinear magnetic configuration sets in; this is assumed to be realized by some other interactions not explicitly included in the model, e.g., competing exchange interactions, as discussed in the previous section, or even the DMI, as we will discuss later.

Despite the clear physical picture it provides, the above-described model only allows to discuss a possible coupling between magnetic (spin) ordering and charge/orbital degrees of freedom. In fact, ordering of the latters can also drive ferroelectricity or, when occurring in magnetic systems, multiferroicity. In order to discuss these types of ordering, however, one needs to go beyond the simple cluster model. The essential mechanism according to which an electric polarization may be induced by charge ordering (CO) can be understood in a single-band tight-binding model for a one-dimensional chain, $H = \sum_{i,\sigma} E_i c^{\dagger}_{i\sigma} c_{i\sigma} - \sum_{\langle ij \rangle} \sum_{\sigma} t_{ij} c^{\dagger}_{i\sigma} c_{j\sigma}$. Assuming a neutral chain, for example at half filling and with each ion providing a unit positive charge, one can then distinguish three cases (for sake of clarity, only paramagnetic solutions are considered). (i) Sites become inequivalent, e.g., $E_i = (-1)^i \Delta$ and $t_{ij} = t$, and a site-centered charge density wave appears, whose strength is modulated by the staggered local potential Δ. Since spatial inversion symmetry is not broken, even if a local dipole develops between neighboring sites, no bulk polarization is expected to develop along the chain. Indeed, by evaluating the Berry-phase polarization and mapping it onto Wannier-function centers, one finds that the localization centers of the electronic negative charge are located on top of the energetically favored site (odd sites in the given example), independently of the strength of Δ; within this framework, no $\mathbf{P}$ appears. (ii) Bonds become inequivalent, e.g., $E_i = 0$ and $t_{ij} = t(1 + (-1)^i \delta)$. This

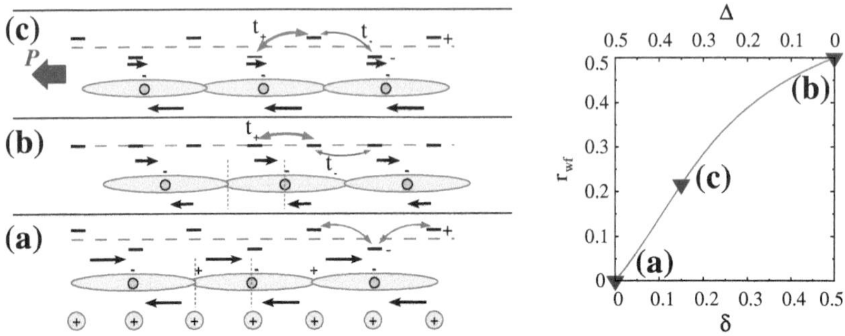

FIGURE 5 Possible orderings in a neutral one-dimensional single-band model: (a) site-centered CO driven by staggered local potential $\pm\Delta$; (b) bond-centered CO, driven by hopping dimerization $t_{\pm} = t(1 \pm \delta)$; (c) a combination of (a) and (b), displaying a finite bulk polarization of electronic origin. Green circles show the position of Wannier function centers specific to each realization of CO, whereas the gray areas schematically show the Wannier functions. Red dashed lines in (a) and (b) show mirror planes of the chain. The right panel shows the displacement of Wannier function centers, in units of the cation–cation distance, as the relevant parameters Δ, δ are changed. (For interpretation of the references to color in this figure legend, the reader is referred to the web version of this book.)

hopping dimerization may happen, e.g., in systems displaying Peierls distortion, and a bond-centered charge density wave develops, with alternating strong and weak bondings. This is reflected in Wannier-function centers located in the middle point of strong bonds, once again without inversion symmetry breaking. Local dipole moments can develop, but they cancel out along the chain with a zero macroscopic polarization. (iii) When both on-site energies and bonds are staggered, $E_i = (-1)^i \Delta$ and $t_{ij} = t(1+(-1)^i \delta)$, a "mixed" CO state emerges. By looking at the electronic localization centers one immediately recognizes that inversion symmetry is broken, local dipoles remain uncompensated and a total $\mathbf{P} \neq 0$ appears (see Figure 5). It is worth noticing that mirror planes of the chain are different for site- and bond-centered CO states, that explains why inversion symmetry disappears when the two types of ordering combine. The mechanism just outlined is purely electronic, since ionic positions are held fixed; however the same mechanism can lead to ferroelectricity also when ions are allowed to move, the essential ingredient being the simultaneous presence of inequivalent sites and bonds [32].

Magnetic ordering can play an important role in triggering bond inequivalency, hence in systems with site-centered CO ferroelectricity may arise as well (see Section C for concrete examples). It may also happen that simultaneous site inequivalency is not a necessary condition for multiferroicity, and an analogous electronic mechanism can lead to a ferroelectric state in the presence of orbital, rather than charge, ordering. A realistic example is found in rare-earth orthorhombic manganites $RMnO_3$ with collinear E-type magnetic configuration (R = Ho, Tm). In these systems manganese ions are all equivalent, showing d[4] (Mn^{3+}) configuration. Due to the large Hund's coupling, only electrons with aligned spins are found on each Mn; a low-energy model suitable for them is the degenerate double-exchange model [47], describing e_g electrons moving on the background of t_{2g} local spins, to which they are strongly coupled:

$$H = -\sum_{\langle i,j\rangle} t_{ij}^{\alpha\beta} \cos\frac{\theta_{ij}}{2} d_{i\alpha}^{\dagger} d_{j\beta} + J \sum_{\langle i,j\rangle} \mathbf{S}_i \cdot \mathbf{S}_j. \tag{5}$$

Here, α, β are orbital index labeling e_g states $x^2 - y^2$, $3z^2 - r^2$, whereas $\mathbf{S}_i$ describe the t_{2g} spins experiencing the superexchange interaction J. Hopping processes depend on the spin configurations through the angle θ_{ij} between neighboring spins. In the limit of infinite Hund's coupling, they are allowed only between sites with parallel local spins, imposing a strong constraint on the electron motion. However, due to the shape of e_g orbitals, hybridization between them is also strongly direction-dependent; for example, $t_{\langle ij\rangle\|x}^{x^2-y^2,z^2} = -t_{\langle ij\rangle\|y}^{x^2-y^2,z^2}$ [42]. When an orbital ordering (OO) develops, therefore, these two ingredients can combine and lead to ferroelectricity, through an "asymmetrization" of hybridization terms. This combination of magnetic and orbital ordering is realized, e.g., in orthorhombic $HoMnO_3$, displaying a staggered OO, with an alternating $d_{3x^2-r^2}/d_{3y^2-r^2}$ pattern along each layer. In the paramagnetic case, there is no inversion symmetry breaking. On the other hand, the E-type magnetic ordering, consisting of zigzag ferromagnetic chains antiferromagnetically coupled, results in an effective dimerization of hopping processes, that differentiate between right and left directions along each chain due to the specific orbitals involved. The other way around, the polarization can be analyzed starting from the E-type configuration and tuning the OO as a function of Jahn–Teller or interorbital Coulomb interaction [48]. Polarization is found to develop as soon as orbital ordering sets in; furthermore, it can be switched by rotating occupied orbitals by 90° in the fixed magnetic configuration, i.e., interchanging $d_{3x^2-r^2}(d_{3y^2-r^2}) \mapsto d_{3y^2-r^2}(d_{3x^2-r^2})$. Correspondingly, Wannier-function centers, that in the insulating phase without orbital ordering sit in the middle point of each bond connecting neighboring Mn sites, move toward the nearest occupied orbital aligned parallel to the bond direction in the OO state. Due to the geometry of the chains, this implies a collective displacement of Wannier-function centers that breaks the inversion symmetry (see Figure 6). Roughly speaking, this mechanism is conceptually similar to the one uncovered by the TM–L–TM model, polarization resulting from the coupling between spin and orbital degrees of freedom: spin configuration determines the way in which d states hybridize and electrons hop in the lattice, and local dipole moments can develop depending on the shape of individual orbitals and their hybridization through oxygen's p orbitals. In this last example, however, the coupling strength is given by Hund's coupling rather than SOC λ, to which e_g

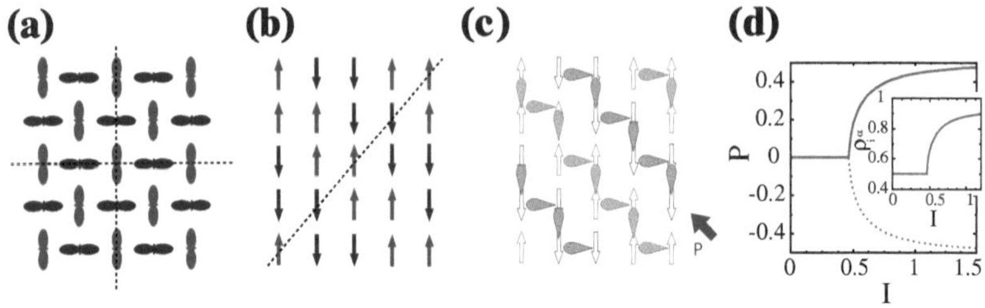

FIGURE 6 (a) Schematic picture of OO pattern in R MnO_3, with alternating $d_{3x^2-r^2}$, $d_{3y^2-r^2}$ orbitals. (b) Schematic picture of E-type antiferromagnetic configuration, with zigzag ferromagnetic chains antiferromagnetically coupled. (c) Combination of E-type and OO state; the asymmetric hopping at each Mn site is pictorially shown, the red arrow indicating the resulting Berry-phase polarization. (d) Evolution of the polarization in the degenerate double-exchange model as a function of the interaction I tuning the OO state. Dashed line is the switched polarization obtained by rotating all occupied orbitals by 90°; the inset shows the average occupancy of $d_{3x^2-r^2}$, ($d_{3y^2-r^2}$) orbital states on even (odd) sites as a function of I, underlying the relationship between P and OO. Mirror planes are indicated by dashed lines in (a) and (b). (For interpretation of the references to color in this figure legend, the reader is referred to the web version of this book.)

orbitals are insensitive, with a consequent expected larger value of polarization (as indeed predicted by *ab initio* DFT calculations [49]).

(ii) Ionic Mechanisms

The microscopic ionic mechanisms are conceptually simpler than those of electronic origin previously sketched, basically relying on the Clausius–Mosotti picture, i.e., they imply the identification of point charges on top of ions whose relative displacements can give rise to local dipole moments. Despite the reliability of this approach has been questioned before, it can provide some reasonable physical insights, even though realistic estimates of polarization values should not be expected (as much as the TM–L–TM model allows to understand the microscopic origin of local dipole moments, rather than bulk polarization). Generally speaking, polar displacements that occur in magnetically ordered systems in order to increase the magnetic energy gain are realizations of the so-called *magnetostriction* (or *exchange-striction*) mechanism. To understand how it works, consider the ANNNI model previously introduced (model (1) with Ising spins): if some energy is gained, e.g., by the shortening of a ferromagnetic bond, in the up–up–down–down configuration the chain will distort in such a way that bonds between parallel spins shorten and those between antiparallel ones stretch (the opposite will happen if one takes $J > 0$, $J' < 0$ with $|J'| > J/2$). If different charges are carried by each site along the chain, a finite polarization emerges. As discussed in relation with charge-ordering-induced ferroelectricity, at the basis of this type of exchange-striction mechanism is the simultaneous presence of inequivalent sites and inequivalent magnetic bonds. Its simplest realization can be found in Ca_3CoMnO_6, where Co^{2+} and Mn^{4+} alternate along one-dimensional chains; at low temperature, magnetic moments order in the up–up–down–down configuration and a polarization appears due to different distortions of ferro- and antiferromagnetic bonds [50,51].

In the case of Ca_3CoMnO_6, exchange striction arises from (and involves) two inequivalent magnetic TM ions with a different valence. A similar mechanism can be also active in transition metal oxides with identical magnetic ions, which involve the ligand oxygens bridging the TM cations. The exchange interaction between magnetic moments is mediated by these bridging oxygens and depends on their positions. Since oxygens carry nominal negative charge (as opposed to the positive one carried by magnetic ions) their collective shift can induce a finite polarization. This type of magnetostriction has been discussed in relation with $RMnO_3$, both in the presence of a noncollinear spiral (R = Tb, Dy) [36] and collinear E-type antiferromagnetic configuration (R = Ho, Tm) [38], even if different microscopic origins of the oxygens displacement were found depending on the magnetic phase. A heuristic explanation of this type of magnetostrictive mechanism goes as follow.

Consider a chain of Mn ions along the $\hat{x}$ direction and assume that bridging oxygens, located on the middle point of each Mn–Mn bond, can move along $\hat{z}$. The spins on each Mn ion experience a superexchange coupling $J\mathbf{S}_n \cdot \mathbf{S}_{n+1}$ ($J > 0$), which is invariant under inversion symmetry and depends only on even powers of z_n, the displacement of each O perpendicular to the bond between $\mathbf{S}_n$ and $\mathbf{S}_{n+1}$, i.e., $J \sim \left[J_0 + \frac{1}{2}J'_\perp z_n^2\right]$, where J_0, $J_\perp$ are constants and higher-order terms are neglected. Assuming $z_n = (-1)^n z_0 + \delta z_n$, which reproduces the oxygen-buckling mode found in the orthorhombic distorted structure of $RMnO_3$, one finds that small O displacements are energetically favorable if $\delta z_n = (-1)^n (J'_\perp z_0/\kappa)\cos\theta_n$, where θ_n is the angle between neighboring (unit) spins and $\frac{\kappa}{2}\sum_n \delta z_n^2$ is the elastic energy associated to the distortion. This semi-empirical result shows that: (i) the symmetric exchange-striction is not effective if the Mn–O–Mn angle is 180° ($z_0 = 0$); (ii) if a spiral spin configuration sets in, then θ_n equals the modulation vector Q and the absolute value of δz_n is the same on each bond, but with opposite directions in even and odd bonds, i.e., $\sum_n z_n = 0$; (iii) in the up–up–down–down spin configuration, $\cos\theta_n = \pm 1$ for ferro- or antiferromagnetic bonds and the factor $(-1)^n$ is compensated, i.e., a collective displacement of ligand O is driven by the symmetric exchange interaction, its direction determined by the sign of the exchange coupling (see Figure 7(a)). Conceptually the same mechanism is at work in the collinear E-type configuration of $HoMnO_3$ or $TmMnO_3$: polarization arises from the simultaneous presence of a distorted perovskite structure and of the frustrated magnetic ordering which mediates inequivalent magnetic bonds [38]. On the other hand, the spiral-induced polarization observed, e.g., in $TbMnO_3$ cannot be driven by the symmetric exchange interaction.

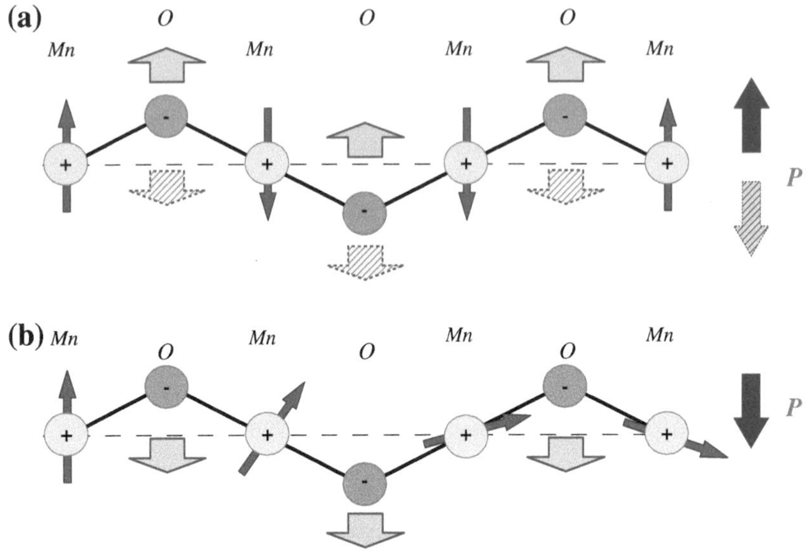

FIGURE 7 Schematic view of allowed magnetostriction phenomena in the distorted Mn–O chain. (a) The symmetric exchange-striction induces different oxygen shifts depending on whether the spins on neighboring Mn ions are parallel or antiparallel. The direction of the collective negatively charged O displacement and of the resulting polarization depends on the sign of the exchange coupling: solid arrows correspond to a ferromagnetic (double-exchange) coupling, whereas dashed ones describe the case of superexchange antiferromagnetic coupling. The Mn–O–Mn angle decreases (increases) for antiparallel (parallel) neighboring spins for $J > 0$ ($J < 0$) and increases (decreases) when spins are parallel (antiparallel). (b) The inverse DMI induces a shift perpendicular to the chain that depends on the angle between spins on neighboring Mn sites; in the cycloidal spiral this angle is given by the modulation vector and is the same for every pair of spins, giving raise to a collective shift of oxygens.

Consider now the antisymmetric Dzyaloshinskii–Moriya interaction [43,44], the relativistic correction to the usual superexchange interaction driven by the spin–orbit coupling, given by $\mathbf{D} \cdot \mathbf{S}_n \times \mathbf{S}_{n+1}$, where the Dzyaloshinskii vector can be expressed as $\mathbf{D} \propto \lambda\,(\mathbf{r_n} \times \mathbf{e_{n,n+1}})$, with $\mathbf{r}_n = (0, 0, z_n)$ the bridging oxygen position and $\mathbf{e}_{n,n+1} = (1, 0, 0)$ a unit vector, parallel to the chain, connecting two Mn ions. Assuming once again the orthorhombic distorted structure, one finds that the corresponding magnetic energy is minimized by $\delta z_n \propto (\lambda/\kappa) \sin\theta_n$. In this case a collective O shift is allowed only by a spiral spin configuration, since the vector product $\mathbf{S}_n \times \mathbf{S}_{n+1}$ has the same sign for all pairs of neighboring spins (see Figure 7(b)). This ionic mechanism, driven by the antisymmetric Dzyaloshinskii–Moriya interaction, is consistent with Eq. (2) and represents a microscopic derivation complementary to the electronic mechanism predicted in the TM–L–TM model. At odds with the cluster model, however, it also provides a possible mechanism for the stabilization of the spiral spin configuration.

(iii) Ionic vs. Electronic Mechanisms

From the previous paragraphs, it is clear that the electronic and ionic microscopic mechanisms can be simultaneously active within the same material – not surprisingly, if one thinks about the mixed ionic/covalent character of many oxides. For example, the observed polarization in perovskite manganites with spiral magnetic order arises both from the electronic and ionic contributions driven by, respectively, the spin-current and the inverse Dzyaloshinskii–Moriya mechanism. Since they both have a relativistic origin, the expected total polarization is small, consistent with that experimentally measured ($P < 0.1\ \mu C/cm^2$) [52], and it is not easy to predict which contribution is larger. One could be tempted to consider the inverse DMI as the microscopic origin of both ionic and electronic contributions, since it provides a natural explanation for the onset of the spiral order, which in turn activates the spin-current mechanism. However, it has been recently shown that a purely electronic model can explain both the spiral and E-type magnetic ordering in $RMnO_3$ [53], hence no conclusions can be univocally drawn but that both contributions may coexist. On the other hand, *ab initio* DFT calculations have shown that ferroelectricity in E-type $HoMnO_3$ has a dual nature, with a purely electronic contribution to P comparable with the ionic one [49]. Therefore, even if the microscopic origin of ferroelectricity can be qualitatively understood, both in terms of the orbital ordering of e_g electrons [48] and of symmetric exchange striction [38], one needs to resort to computational tools for quantitative estimates of polarization.

We conclude this section by describing a possible mechanism where lattice and electronic degrees of freedom must be taken into account together, being their interplay and cooperation at the origin of a predicted ferroelectric state in doped perovskite manganites. Half-doped manganites $R_{1/2}Ca_{1/2}MnO_3$ (R = La, Pr) display different types of magnetic, charge, and orbital orderings in distorted orthorhombic structures, and their physical properties in the low-temperature regime are still debated. It is widely accepted that the local Mn magnetic moments can order in the so-called CE-type antiferromagnetic (AFM-CE) configuration characterized by double-zigzag spin chains, but two possible scenarios have been envisaged concerning the charge and orbital orderings. On the one hand, a checkerboard pattern (CB) of alternating Mn^{4+} and orbital-ordered Mn^{3+} has been proposed (a site-centered CO state – see Figure 8(a)) [54]; on the other hand, a Zener polaron (ZP) picture has been suggested, where equivalent Mn^{3+} ions couple into ferromagnetic dimers, displaying an OO pattern with orbital states aligned along the same direction within each dimer (see Figure 8(b)) [55]. In this picture the CO is bond centered, arising from holes on oxygen atoms that bridge the Mn dimers. According to our previous discussion, any intermediate state between these two types of CO configurations should induce ferroelectricity, as suggested in Ref. [31] for nearly half-doped systems (on the basis of considerations including only exchange interaction between Mn

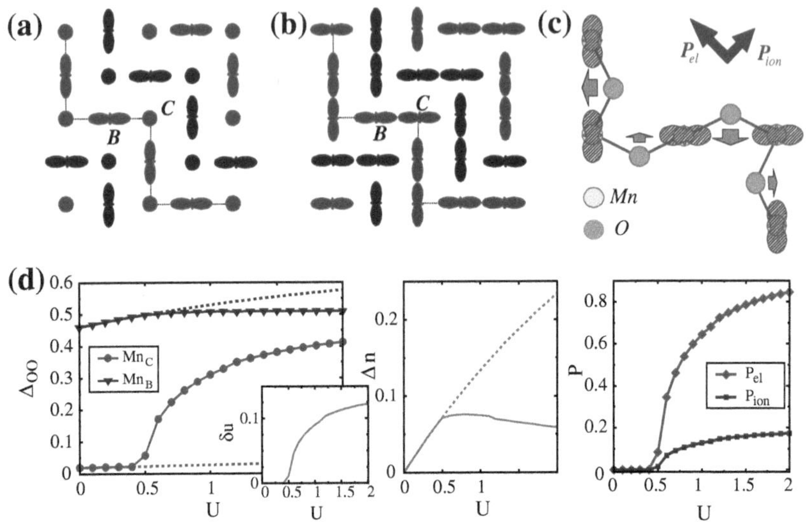

FIGURE 8 (a) Schematic picture of OO pattern in the proposed CB picture of $R_{1/2}Ca_{1/2}MnO_3$, with alternating $d_{3x^2-r^2}$, $d_{3y^2-r^2}$ orbitals on bridge Mn^{3+} sites; no OO develops on corner Mn^{4+} sites. The double-zigzag up (down) spin chains antiferromagnetically coupled are highlighted in red (blue). (b) Schematic view of the proposed ZP picture without CO and with OO developing also on Mn_C sites. (c) Orbital-striction mechanism: the Mn–O–Mn angle between Mn ions with parallel occupied orbitals is increased more than the angle between Mn ions with perpendicular orbitals. The relative shift of O results in a polarization of ionic origin, perpendicular to the chain, whereas the OO pattern causes an electronic polarization parallel to the chain direction. (d) Results from distorted degenerate double-exchange model as a function of Coulomb inter-orbital interaction U. Δ_{OO} measures the orbital ordering as the difference of on-site orbital occupancy between the most ($3x^2-r^2$ or $3y^2-r^2$) and the least (y^2-z^2 or x^2-z^2) occupied orbitals on bridge and corner sites; as bonds dimerize and the O's shifts differentiate between even and odd bonds (shown in the inset is $\delta u = u_n - u_{n+1}$, u_n measuring the displacement of O perpendicular to the Mn–Mn bond) an OO pattern develops as schematically shown in (b). Concomitantly the charge disproportionation Δn between Mn_C and Mn_B is reduced because of the reduced Coulomb energy cost of having all Mn sites with only one orbital state partially filled. Dotted lines show same quantities in the CB picture. The lower right panel shows the evolution of both contributions to P as the correlation-induced dimerized state establishes. (For interpretation of the references to color in this figure legend, the reader is referred to the web version of this book.)

spins). Indeed, by considering a degenerate double-exchange model (as given in Eq. (5)) generalized in order to include the oxygen-buckling mode, that is ubiquitous in perovskite manganites and modulate the hopping processes of e_g electrons, it can be shown that the interplay of lattice and orbital degrees of freedom drives ferroelectricity through a correlation-induced orbital-strictive mechanism [56]. The strong Hund's coupling constraints hopping along the double zigzag spin-chains, characteristic of AFM-CE, changing direction as $\{\ldots x, x, y, y \ldots\}$. One can then distinguish between corner Mn sites (Mn_C), where hopping changes direction, and bridge Mn sites (Mn_B), where it does not. By keeping oxygens fixed, the model reproduces the CB picture, with site-centered CO and with OO developing on bridge Mn ions[4] [57]. However, when oxygen-buckling is included, an OO establishes on corner Mn sites also, in order to reduce the Coulomb energy cost of having both e_g states partially occupied. This is achieved by dimerizing the hopping, which asymmetrizes, through different oxygens displacements on even and odd bonds, between left and right direction along the chain at Mn_C sites. The different shifts of oxygens result in local dipole moments perpendicular to the chain direction, which sum up to give an ionic contribution to the total polarization. On the other hand, the specific OO pattern induces an electronic contribution to P parallel to the chain direction, arising from the asymmetric hopping at corner sites that breaks the inversion symmetry of the AFM-CE configuration, analogously to what happens in E-type manganites. This ferroelectric state shares with the CB picture a CO on Mn sites, even if the charge disproportionation is largely reduced with respect to the pure CB state, whereas the dimerized character of hopping and the OO pattern reminds the ZP picture; it has indeed a "mixed" site- and bond-centered CO character. On the other hand, the orbital-strictive mechanism has neither a purely ionic nor electronic origin, rather both lattice and electronic degrees of freedom cooperate in driving ferroelectricity. The emerging picture is in excellent qualitative agreement with *ab initio* calculations for both La and Pr doped manganites [58,59].

C. Materials and examples

As reported in previous sections, many efforts concentrated on improper multiferroics by using model-Hamiltonian approaches and/or symmetry considerations. Both the latter approaches are extremely helpful to elucidate under which conditions and based on which microscopic mechanism the desired multiferroicity might arise. However, these approaches cannot give any quantitative estimate of the effects, of the resulting polarization, of the magnetoelectric coefficients. To this end, *ab initio* approaches – and density functional theory in particular – were heavily used and contribute significantly to the field. In what follows, we briefly discuss some results concerning (i) an example of CO-induced ferroelectricity, $K_{0.6}Fe^{II}_{0.6}Fe^{III}_{0.4}F_3$, (ii) an example of polarization induced by the cooperation of peculiar charge and spin arrangements in nickelates and (iii) magnetoelectricity induced by spin-dependent p–d hybridization in $Ba_2CoGe_2O_7$. Due to limited space, we here only report the most relevant outcomes and we refer the interested reader to our original papers, respectively Refs. [46,60,61]. All the density-functional

---

[4] As discussed in Ref. [48], this kind of OO cannot drive the magnetic system ferroelectric, as it happens in E-type manganite perovskyte, since it does not break the inversion symmetry of the AFM-CE state.

simulations were performed using the Vienna *Ab Initio* Simulation Package (VASP) [15], the PAW pseudopotentials [14] and the generalized-gradient approximation to the exchange–correlation potential [13]. Whenever needed, in order to improve the treatment of correlation effects for localized d-states, the GGA + U approach was used [62]. The estimate of the polarization was made according to the Berry-phase formalism [2,23,24].

(i) $K_{0.6}Fe^{II}_{0.6}Fe^{III}_{0.4}F_3$ in the Tetragonal Tungsten-Bronze Structure: Fe^{2+}/Fe^{3+} CO as Source for FE

Iron-potassium fluoride, $K_{0.6}Fe^{II}_{0.6}Fe^{III}_{0.4}F_3$, is an interesting fluoride where many coexisting exotic effects such as ferroelectricity, charge-ordering, multiferroicity, and ferroelasticity were put forward (see below). It shares a crystalline arrangement (conventionally referred to as tetragonal tungsten bronze (TTB) structure) with one of the most common ferroelectrics, i.e., barium sodium niobate, $Ba_2NaNb_5O_{15}$ [63]. The latter shows a large ferroelectric polarization along the c axis (about 40 $\mu C/cm^2$), microscopically due to the conventional d^0-ness mechanism [27] (off-centered Nb is here formally 5+) based on cross-gap Nb–O hybridization.

As for $K_{0.6}Fe^{II}_{0.6}Fe^{III}_{0.4}F_3$, early experiments reported coupled ferroelectricity and ferroelasticity to occur below 490 K [64], along with CO [65]. More recently, Mezzadri et al. reported several transitions [66]: a first structural transition at 570 K from tetragonal to orthorhombic, a second transition at 490 K where Fe^{2+}/Fe^{3+} CO arises (with an unknown pattern), and a third transition around 290 K to monoclinic structure, coupled with ferroelasticity.

The orthorhombic $Pba2$ symmetry was experimentally reported for $K_{0.6}Fe^{II}_{0.6}Fe^{III}_{0.4}F_3$, with $a = 12.751$ Å, $b = 12.660$ Å, and $c = 7.975$ Å [66], with the unit cell including two TTB layers along c (see Figure 9(a)). The $Pba2$ group has four symmetry operations: $\{E, C_{2z}, \sigma_x + (\frac{1}{2}\frac{1}{2}0), \sigma_y + (\frac{1}{2}\frac{1}{2}0)\}$; with these constraints, polarization can only occur along the z axis. However, in the experimental structure, we find 10 Fe^{2+}, 6 Fe^{3+}, and 4 mixed-valent $Fe^{2.5+}$ ions (see Figure 9(b) for the charge-density distribution), although a "full" CO would imply the presence of 12 Fe ions to be $Fe^{2+}(d^6)$ and 8 Fe ions to be $Fe^{3+}(d^5)$, consistent with the stoichiometry ($K_{12}Fe^{II}_{12}Fe^{III}_8F_{60}$/cell). By performing DFT calculations [67] for the experimental $Pba2$ structure [60], we find, as expected, a metallic state due to the presence of $Fe^{2.5+}$ ions. Clearly, this is incompatible with reported ferroelectricity and an insulating state must be present. One of the possible routes for an energy-gap to arise is by further reducing the symmetry, so as to obtain a "full" CO (i.e., no mixed-valent $Fe^{2.5+}$ ions).

For computational reasons, we did not consider in our calculations a large monoclinic supercell, previously suggested to explain the twinning satellite observed by electron-diffraction patterns below 290 K [66,68], and we assumed that the CO pattern, determined in the (small) $Pba2$-like unit cell, is not affected

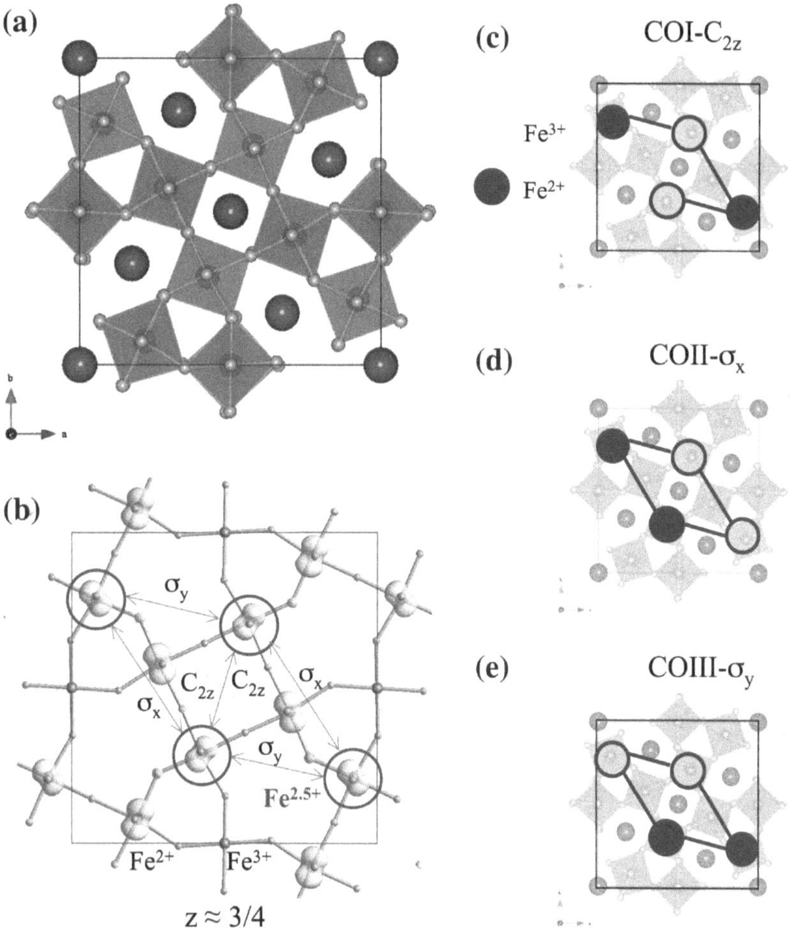

FIGURE 9 (a) Projection in the *ab* plane of the TTB iron-fluoride unit cell; octahedral F_6 cages around Fe ions (light blue spheres) are highlighted. Purple and yellow spheres denote K and F ions, respectively. (b) Down-spin charge of Fe t_{2g} states in the energy range 1 eV below the Fermi level for the $Pba2$ symmetry in the $z \sim 3/4$ plane. Blue circles denote $Fe^{2.5}$ ions. Panels (c), (d), and (e) show how CO can be obtained by turning four $Fe^{2.5}$ ions into two Fe^2, and two Fe^3 ions in different configurations: c) COI, keeping C_{2z} symmetry; (d) COII, keeping σ_x symmetry and (e) COIII, keeping σ_y symmetry. (For interpretation of the references to color in this figure legend, the reader is referred to the web version of this book.)

by expanding the unit cell. Three CO patterns were considered: COI, CO-II, and CO-III, each obtained by keeping one of the three symmetry operations and breaking the other two among C_{2z}, $\sigma_x + (\frac{1}{2}\frac{1}{2}0)$, and $\sigma_y + (\frac{1}{2}\frac{1}{2}0)$, respectively (cf.Figure 9(c)–(e)). Whereas the induced polarization P_z is allowed by the

TABLE 1 **Total energy difference (meV/Fe), energy gap (eV) and induced FE polarization calculated by Berry phase P^{Berry} (μC/cm^2) at partial CO pattern in $K_{0.6}Fe^{II}_{0.6}Fe^{III}_{0.4}F_3$ for the experimental crystal structure (first column), optimized structure keeping experimental symmetry (second column), full CO patterns: COI, COII, COIII (third, fourth and fifth column, respectively, with given symmetry).**

	Exp.	Opt.	COIC_{2z}	COIIσ_x	COIIIσ_y
	Pba2	*Pba2*	*P2*	Pc	Pc
ΔE^{tot}	0	−31.3	−44.3	−57.3	−51.6
E^{gap}	0	0	0.91	1.28	1.08
P^{Berry}	–	–	(0 0 0.09)	(0 − 0.50 − 0.19)	(−5.14 0 0.03)

prototype $Pba2$ crystal symmetry, P_y and P_x are additionally allowed in COII and COIII, respectively. Indeed, when performing a full structural optimization of the internal atomic positions, we found an insulating state and polar structure with sizeable polarization along different directions, as reported in Tables 1 and 2.

In these structures, ferroelectricity is purely induced by CO, with magnetism playing no relevant role; we further remark that the size of polarization is much smaller than in $Ba_2NaNb_5O_{15}$, due to the different microscopic origin of polarization (CO in $K_{0.6}Fe^{II}_{0.6}Fe^{III}_{0.4}F_3$ vs. second-order Jahn–Teller effect in Nb-based oxide). We note from Table 1 that COII is the energy ground state, which should be stabilized when applying an electric field E_y; however, COI and COIII are predicted to be energetically competing states. We here speculate that the competing COIII pattern inducing a large P_x might be realized upon applying an electric field E_x. In the case of a (strong enough) applied electric field which rotates in the xy plane, the induced P along the field would therefore change its saturation value. We define the "ferroelectric anisotropy" (FEA) as the energy required to modify the direction (as well as size) of the polarization by switching the crystal symmetry between different CO phases (in this case COII and COIII); the FEA may find useful applications in future devices, such as multiple-state memories where the information can be stored by exploiting not only the sign of P, but also its direction. The FEA is peculiar for *ferroelectricity induced by electronic degrees of freedom in "improper" multiferroics*. For example, the control over the direction of **P** in spin-spiral manganites by means of a magnetic field was already proven and suggestions toward devices harnessing FEA already came in that framework, so we propose that similar arguments might be valid for CO-induced ferroelectrics [69].

TABLE 2 FE polarization P of $RNiO_3$ (R = Ho, Lu, Pr) in $\mu C/cm^2$ for the experimental structure for Ho, Lu and Ref. [71] for Pr), for either S or T-type magnetic ordering, with $P2_1$ and Pn symmetry, respectively. For $HoNiO_3$, the FE polarization resulting from the N-type ordering is also reported.

Rare earth	T-type			S-type	N-type		
	P_{tot}	P_a	P_c	P_b	P_{tot}	P_a	P_c
Lu	10.31	9.91	2.84	5.21		–	
Ho	8.66	8.05	3.19	3.60	0.10	0.02	−0.10
Pr	14.80	13.23	6.64	1.81		–	

(ii) Nickelates

Despite the relatively simple chemical formula, rare-earth nickelates $RNiO_3$ (R = trivalent rare-earth ion) are rather complex materials. In particular, they show a metal–insulator transition ($T_{MI}(Pr) = 130$ K, $T_{MI}(Nd) = 200$ K, $T_{MI}(Sm) = 400$ K, $T_{MI}(Ho) = 580$ K, and $T_{MI}(Lu) = 600$ K), whose comprehension is still incomplete. Moreover, they also show a charge-ordering transition: nominal Ni^{3+} ions charge-disproportionate in $Ni^{3+\delta}$ and $Ni^{3-\delta}$ arranged in a rock-salt fashion; this is believed to coincide with a simultaneous transition of the crystallographic symmetry from orthorhomic *Pbnm* to monoclinic $P2_1/n$ [70–72]. Experimentally the charge ordering is accompanied by an oxygen breathing distortion of the NiO_6 octahedra and by different magnetic moments on the two inequivalent Ni atoms (1.4/1.4 μ_B and 0.6/0.7 μ_B for Ho/Lu, respectively) [73,74]. As for the magnetic configuration, this has been long debated and several proposed spin orderings were put forward. In our work [61], we considered $RNiO_3$ (with R = Lu, Ho, Pr, and with checkerboard Ni^{2+}/Ni^{4+} CO). We performed calculations [75] for two collinear structures (labeled S and T, see Figure 10(b) and (c)) and for a noncollinear spin structure (labeled N, see Figure 10(d) [76]). In both the S- and T-type magnetic structure, the Ni spins within the *ab* plane form ferromagnetic zigzag chains. Adjacent zigzag chains in the same plane are arranged in AFM fashion. The stacking of the zigzag chains along the *c* axis is different between S- and T-type ordering: in S-type configuration (Figure 10(b)) the zigzag spin chains are all pointing in the same direction, whereas for T-type ordering (Figure 10(c)) the zigzag chains in adjacent planes are pointing in opposite directions. This implies for T-type ordering that the spins in a plane perpendicular to the $[\bar{1}\ 0\ 1]$ direction are aligned in the same direction. Along this axis, the ionic and spin arrangement is as follows $\ldots Ni^{2+} \uparrow Ni^{4+} \uparrow Ni^{2+} \downarrow Ni^{4+} \downarrow Ni^{2+} \uparrow Ni^{4+} \uparrow \ldots$ This clearly breaks inversion symmetry and is expected to give rise to polarization. The noncollinear N structure (Figure 10(d) [17]) is actually a spin-spiral: all spins lie in the *ac* plane, with spins in a plane perpendicular to [1 0 1] pointing

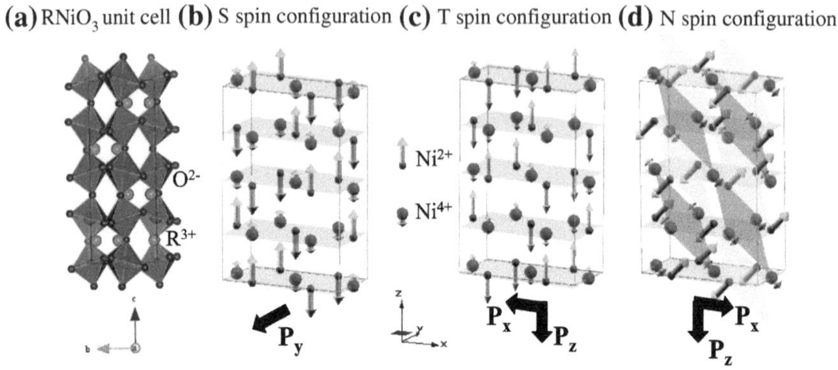

FIGURE 10 (a) Atomic arrangements in the $RNiO_3$ unit cell. Orange, red, purple, and gray spheres denote the rare-earth, Ni^{2+}, Ni^{4+}, and oxygen atoms, respectively. (b) Spin configuration in the collinear S phase; (c) Spin configuration in the collinear T phase; (d) Spin configuration in the noncollinear N phase. The direction of resulting polarization is also shown by black bold arrows. (For interpretation of the references to color in this figure legend, the reader is referred to the web version of this book.)

in the same direction. When moving from plane to plane along [1 0 1], the spins rotate within the *ac* plane.

All the spin configuration were predicted to be ferroelectric, with calculated polarization reported in Table 2.

The Heisenberg-driven polarization in the S case has a microscopic origin which is very similar to what previously discussed for ortho-manganites in the AFM E-like spin arrangement, based on symmetric exchange-striction. In the T case, as previously pointed out, the charge and spin-ordering $Ni^{2+} \uparrow Ni^{4+} \uparrow Ni^{2+} \downarrow Ni^{4+} \downarrow Ni^{2+} \uparrow Ni^{4+} \uparrow \ldots$ are at the origin of the polarization. As previously discussed, in spin-spirals the relativistic spin–orbit coupling causes a FE polarization [1] in the direction **e x Q** (corresponding to [10-1] in our case), where **e** is the vector connecting adjacent spins and **Q** is the spiral wave vector. Indeed, our *ab initio* calculations including spin–orbit coupling predict a polarization in full agreement with these symmetry considerations and with Nagaosa's spin-current model. Remarkably, in the S and T cases the polarization (around 8–10 $\mu C/cm^2$) is found to be much larger than that arising from a relativistic spin–orbit related mechanism (around 0.1 $\mu C/cm^2$), at play in the N-state. According to our findings, an experimental determination of the direction and magnitude of P could therefore solve the long-standing puzzle related to the magnetic ground state in nickelates. Moreover, the fact that we find competing energies of the S-, T-, and N-type magnetic structures but different directions and magnitudes of polarization, suggests that one can stabilize one or the other by cooling in an appropriate electric field, so that an applied external electric field can actually tip the balance in favor of a specific magnetic phase in $RNiO_3$.

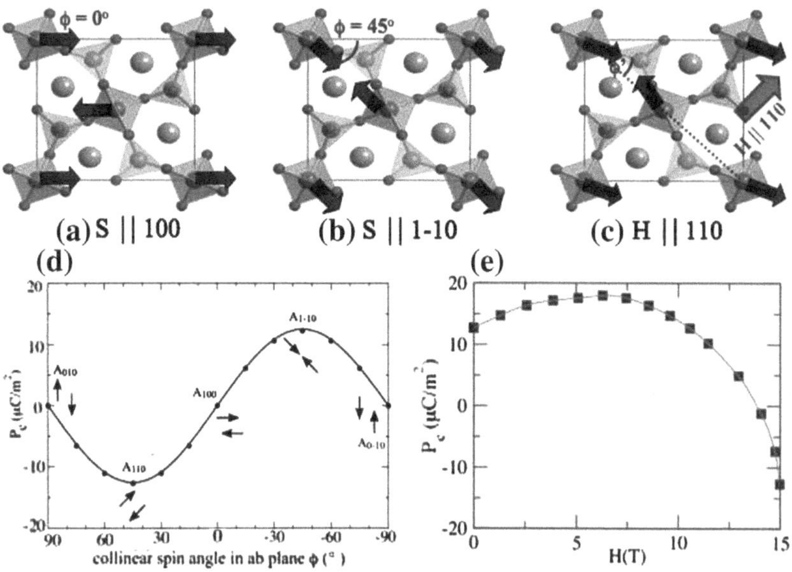

FIGURE 11 Upper row: Crystal structure in the *ab* plane; Co (green spheres) and Ge (orange spheres) ions (both located in O_4 tetrahedron) lie at $c = 0$, whereas Ba ions (gray spheres) lie at $c = 1/2$. Co spin configurations (see black arrows): (a) [1 1 0]-directed, (b) [1 1 0]-directed (collinear) AFM and (c) noncollinear spin-canted under applied H//110. (d) DFT results for P_c as a function of the collinear spin angle ϕ in the *ab* plane, fitted to $f(\phi) = -a\sin(2\phi)$, with $a = 12.7$ (solid line). Spin configurations in the ab plane shown by arrows. (e) DFT results for P_c as a function of the magnetic field *H*, parallel to [1 1 0]. (For interpretation of the references to color in this figure legend, the reader is referred to the web version of this book.)

(iii) $Ba_2CoGe_2O_7$

As mentioned previously when discussing the microscopic mechanisms for magnetoelectricity and multiferroicity, a relatively unexplored phenomenology is based on the spin-dependent p–d hybridization [33], where the SOC affects in an "asymmetric" way the p–d interaction between the central transition metal (TM) and the oxygen ligands (L). Using a perturbative approach to a three-site (TM–L–TM) cluster model, Jia et al. [41] predicted the SOC-induced electric polarization to have the following functional form: $\mathbf{P} \propto \sum_{ij} (\mathbf{S}_i \cdot \mathbf{e}'_j)^2 \, \mathbf{e}'_j$, where $\mathbf{e}'_j$ labels the vectors connecting the TM to the ligands. A recent and paradigmatic example of this interesting mechanism is represented by $Ba_2CoGe_2O_7$, a Co-based melilite.

$Ba_2CoGe_2O_7$ shows tetragonal non-centrosymmetric (non-polar) $P\bar{4}2_1m$ (#113) structure with two inequivalent Co sites (Co1 and Co2 in Figure 11) with a collinear antiferromagnetic (AFM) structure in the *ab* plane being observed below $T_N = 6.7$ K. Experimentally, a ferroelectric P_c along the *c* axis has been reported even without any applied magnetic field [77,78]. Moreover, when

TABLE 3 Predicted P_c for BCGO at different canting angles ϕ', calculated in the fixed non-polar structure (first line) and with optimized internal atomic coordinates (second line). The largest experimental value of P_c is $\sim 120\mu C/m^2$ [78].

P_c ($\mu C/m^2$)	$\phi' = 0°$	30°	90°
fixed structure	12.7	17.6	−12.7
opt. structure	39.8	57.7	−38.9

H is applied, **P** develops finite components along any direction, modulated by the direction and the size of **H**. Here, we will concentrate on an applied magnetic field **H** under two circumstances (as experimentally reported): (i) with fixed magnitude and rotating in the *ab* plane and (ii) applied along the [1 1 0] direction and with increasing magnitude.

DFT calculations using VASP with GGA-PBE potential [79] led to a DFT-optimized structure showing $a = b$=8.28 Å, c = 5.58 Å, and a tilting angle for CoO_4 tetrahedrons given by $\kappa = 23.9°$, consistent with experimental values [78].

First of all, we performed simulations aimed at establishing the magnetic anisotropy energy (MAE) and found that the *ab* plane is an easy plane (in which spins easily follow an applied field by arranging in a direction roughly perpendicular to **H**, with possible small canting) and the *c* axis is the hard axis. This is consistent with experiments and with our expectations: In CoO_4 tetrahedra, the Co^{2+} ion shows an orbital-quenched $e_2^{\uparrow}\ t_3^{\uparrow}\ e_2^{\downarrow}\ t_0^{\downarrow}$ electronic configuration, which is expected to lead to a very small MAE. We therefore evaluated the ME effect as the change of P (calculated via Berry phases) induced when Co spins are kept collinear but rotated in-plane with respect to the crystalline axes in the fixed non-polar crystal structure, thus mimicking the effect of an applied **H**. In Figure 11(a) we show P_c as a function of the angle ϕ (labeling the spin direction with respect to the [1 0 0] axis): the trend of $P_c \propto \sin 2\phi$ is consistent with experiments, with polarization deriving from a purely electronic contribution via SOC. Moreover, the polarization is further enhanced when atomic internal coordinates are optimized in the A_{1-10} configuration (cf Table 3).

We look then at the spin-canting effect induced by an applied field H_{110}. In Figure 11(b) we show the change in P_c induced by artificially canting the spins by an angle ϕ', starting from the A_{1-10} AFM configuration. P_c evolves as $\cos 2(\phi' - 22.7°) + \text{const.}$, with a peak at $\phi' \sim \kappa$. The **P–H** curve shown in the inset in Figure 11(b) – first increasing with **H** and then decreasing and even changing its sign – denotes a peculiar nonlinear ME trend, that emerges from the symmetry properties of $Ba_2CoGe_2O_7$ rather than from high-order terms in the free energy expansion in series of the electric and magnetic fields, as one might in first place naively think.

Due to lack of space, we here only briefly mention two additional results and we refer to Yamauchi et al. [46] where all the details of the derivation are reported.

(1) On the basis of Landau theory, one can indeed derive the trends of polarization both as a function of ϕ and of ϕ', showing that many of the $Ba_2CoGe_2O_7$ relevant properties come from its peculiar symmetry.
(2) On the basis of a simple cluster model (complemented by DFT simulations) that takes into account crystal-field effects, one can show that the microscopic origin of magnetoelectricity is based on two relevant ingredients (i.e., the anisotropic p–d hybridization between Co and O states and the on-site spin–orbit coupling).

REFERENCES

[1] Cheong SW, Mostovoy M. Nat Mater 2007;6:13; Ramesh R, Spaldin NA. Nat Mater 2007;6:21; Eerenstein W, Mathur ND, Scott JF. Nature 2006;442:759; Hill NA. J Phys Chem B 2000;104:6694; Fiebig M. J Phys D 2005;38:R123; Rao CNR, Serrao CR. J Mater Chem 2007;17:4931; Khomskii D. Physics 2009;2:20; Picozzi S, Ederer C. J Phys: Condens Matter 2009;21:303201.
[2] King-Smith RD, Vanderbilt D. Phys Rev B 1993;47:1651.
[3] Baroni S, de Gironcoli S, Dal Corso A. Rev Mod Phys 2001;73:515.
[4] Neaton JB, Ederer C, Waghmare UV, Spaldin NA, Rabe KM. Phys Rev B 2005;71:014113.
[5] Baettig P, Seshadri R, Spaldin NA. J Am Chem Soc 2007;129:9854.
[6] Ederer C, Spaldin NA. Phys Rev B 2006;74:024102.
[7] Shimakawa Y, Azuma M, Ichikawa N. Materials 2011;4:153; Azuma M et al. J Am Chem Soc 2005;127:8889; Singh MP, Truong KD, Jandl S, Fournier P. J Appl Phys 2010;107:09D917; Dass RI, Goodenough JB. Phys Rev B 2003;67:014401; Ynez-Vilar S, et al. Phys Rev B 2011;84:134427.
[8] Das H, Waghmare UV, Saha-Dasgupta T, Sarma DD. Phys Rev Lett 2008;100:186402; Das H, Waghmare UV, Saha-Dasgupta T, Sarma DD. Phys Rev B 2009;79:144403.
[9] Hong J, Stroppa A, Iniguez J, Picozzi S, Vanderbilt D. Phys Rev B 2012;85:054417.
[10] Takata K, Azuma M, Shimakawa Y, Takano M. J Jpn Soc Powder, Powder Metal 2005;52:913.
[11] Goodenough JB. Magnetism and chemical bond. NY: Inter Science Pub; 1976.
[12] Bhattacharjee S, Eriksson O, Sanyal B. J. Phys. Condens. Mat. (in press).
[13] Perdew JP, Burke K, Ernzerhof M. Phys Rev Lett 1996;77:3865.
[14] Blöchl EP. Phys Rev B 1994;50:17953; Kresse G, Furthmüller J. Phys Rev B 1999;59:1758.
[15] Kresse G, Furthmuller J. Phys Rev B 1996;54:11169.
[16] Baroni S, Resta R. Phys Rev B 1986;33:7017.
[17] Ciucivara A, Sahu B, Kleinmann L. Phys Rev B 2007;76:064412.
[18] Dzyaloshinskii IE. Sov Phys JETP 1959;10:628.
[19] Astrov DN. Sov Phys JETP 1960;11:708.
[20] Landau LD, Lifshitz EM. Statistical physics, Part I. Oxford: Pergamon Press; 1980.
[21] Landau LD, Lifshitz EM. Electrodynamics of continuous media. Oxford: Pergamon Press; 1984.
[22] Freeman AJ, Schmid H, editors. Magnetoelectric interaction phenomena in crystals. London: Gordon and Breach; 1975.
[23] Resta R. Rev Mod Phys 1994;66:899.
[24] Ahn CH, Rabe KM, Triscone JM, editors. Physics of ferroelectrics: a modern perspective. Berlin: Springer-Verlag; 2007.

[25] Matthias BT. Phys Rev 1949;75:1771.
[26] Cohen RE. Nature 1992;358:136.
[27] Hill NA. J Phys Chem B 2000;104:6694.
[28] Khomskii DI. Bull Am Phys Soc C 2001;21:002.
[29] Mostovoy M. Phys Rev Lett 2006;96:067601.
[30] Cheong S-W, Mostovoy M. Nat Mater 2007;6:13.
[31] Efremov DV, van den Brink J, Khomskii DI. Nat. Mater 2004;3:853.
[32] van den Brink J, Khomskii DI. J Phys: Condens Matter 2008;20:434217.
[33] Arima T. J Phys Soc Jpn 2007;76:073702.
[34] Fischer ME, Selke W. Phys Rev Lett 1980;44:1502.
[35] Katsura H, Nagaosa N, Balatsky AV. Phys Rev Lett 2005;95:057205.
[36] Sergienko IA, Dagotto E. Phys Rev B 2006;73:094434.
[37] Arima T, Tokunaga A, Goto T, Kimura H, Noda Y, Tokura Y. Phys Rev Lett 2006;96:097202.
[38] Sergienko IA,Sen C, Dagotto E. Phys Rev Lett 2006;97:227204.
[39] Mossotti OF. Azioni e deformazioni nei dielettrici, Memorie di Matematica e di Fisica della Società Italiana delle Scienze Residente in Modena, 1850;24:49; Clausius R. Die mechanische behandlung der electrica. Berlin: Vieweg; 1879.
[40] Jia C, Onoda S, Nagaosa N, Han JH. Phys Rev B 2006;74:224444.
[41] Jia C, Onoda S, Nagaosa N, Han JH. Phys Rev B 2007;76:144424.
[42] Slater JC, Koster GF. Phys Rev 1954;94:1498.
[43] Dzyaloshinskii IE. J Phys Chem Solids 1958;4:241.
[44] Moriya T. Phys Rev 1960;120:91.
[45] Murakawa H, Onose Y, Miyahara S, Furukawa N, Tokura Y. Phys Rev Lett 2010;105:137202.
[46] Yamauchi K, Barone P, Picozzi S. Phys Rev B 2011;84:165137.
[47] van den Brink J, Khomskii D. Phys Rev Lett 1999;82:1016.
[48] Barone P, Yamauchi K, Picozzi S. Phys Rev Lett 2011;106:077201.
[49] Picozzi S, Sergienko IA, Yamauchi K, Sanyal B, Dagotto E. Phys Rev Lett 2007;99:227201.
[50] Choi YJ, Yi HT, Lee S, Huang Q, Kiryukhin V, Cheong S-W. Phys Rev Lett 2008;100:047601.
[51] Wu H, Burnus T, Hu Z, Martin C, Maignan A, Cezar JC, et al. Phys Rev Lett 2009;102:026404.
[52] Kimura T, Goto T, Shintani H, Ishizaka K, Arima T, Tokura Y. Nature (London) 2003;426:55.
[53] Kumar S, van den Brink J, Kampf A. Phys Rev Lett 2010;104:017201.
[54] Radaelli PG, Cox DE, Marezio M, Cheong S-W. Phys Rev B 1997;55:3015.
[55] Daoud-Aladine A, Rodríguez-Carvajal J, Pinsard-Gaudart L, Fernández-Díaz MT, Revcolevschi A. Phys Rev Lett 2002;89:097205.
[56] Barone P, Picozzi S, van den Brink J. Phys Rev B 2011;83:233103.
[57] van den Brink J, Khaliullin G, Khomskii D. Phys Rev Lett 1999;83:5118.
[58] Giovannetti G, Kumar S, van den Brink J, Picozzi S. Phys Rev Lett 2009;103:037601.
[59] Colizzi G, Filippetti A, Fiorentini V. Phys Rev B 2010;82:140101(R).
[60] Yamauchi K, Picozzi S. Phys Rev Lett 2010;105:107202.
[61] Giovannetti G, Kumar S, Khomskii D, Picozzi S, van den Brink J. Phys Rev Lett 2009;103:156401.
[62] Anisimov VI, Aryasetiawan F, Lichtenstein AI. J Phys: Condens Mater 1997;9:767.
[63] Scott JF. J Phys: Condens Matter 2008;20:021001.
[64] Ravez J, Abrahams SC, de Pape R. J Appl Phys 1989;65:3987.
[65] Calage Y, et al. J Appl Phys 1990;67:430.
[66] Mezzadri F, Fabbrici S, Montanari E, Righi L, Calestani G, Gilioli E, et al. Phys Rev B 2008;78:064111.
[67] DFT simulations were performed using the VASP code [15] and the PAW pseudopotentials [14] within the GGA + U approach [62] ($U = 5$ eV and $J = 0$ eV for Fe d-states). The cut-off energy for the plane-wave expansion of the wave-functions was set to 400 eV and k-point shells of (2, 2, 4) for $K_{0.6}Fe^{II}_{0.6}Fe^{III}_{0.4}F_3$ were used for the Brillouin zone integration The internal atomic coordinates were fully optimized until the atomic forces were less than 0.01 eV/Å while the lattice parameters were taken from experiments [66]. The FE polarization P was calculated using the Berry phase method, by comparing the FE and paraelectric (PE) state; the latter is constructed by imposing the x, y, z reflection in the atomic structure. For simplicity, $K_{0.6}Fe^{II}_{0.6}Fe^{III}_{0.4}F_3$ was treated as a ferromagnet.
[68] Fabbrici S, Montanari E, Righi L, Calestani G, Migliori A. Chem Matter 2004;16:3007.

[69] Schierle E, et al. <http://arXiv:0910.5663>.
[70] Alonso JA, et al. Phys Rev B 2000;61:1756; Alonso JA et al. Phys Rev Lett 1999;82:3871.
[71] Medarde M, et al. Phys Rev B 2008;78:212101.
[72] Girardot C, et al. Phys Rev B 2008;78:104101.
[73] Fernandez-Diaz MT, et al. Phys Rev B 2001;64:144417.
[74] Mazin II, et al. Phys Rev Lett 2007;98:176406.
[75] In our VASP simulations, we included strong Coulomb interactions between the Ni 3d electrons, within SGGA + U calculations for $U = 8$ eV and a Hunds rule exchange of $J_H = 0.88$ eV.
[76] Scagnoli V, et al. Phys Rev B 2006;73:100409(R).
[77] Yi HT, et al. Appl Phys Lett 2008;92:212904.
[78] Murakawa H, et al. Phys Rev Lett 2010;105:137202.
[79] We checked our results by using also GGA + U potential with $U = 3$ or 5 eV for Co-d states.

Chapter 7

Multiferroic Complex Metal Oxides: Main Features of Preparation, Structure, and Properties

Sergey Ivanov
X-ray laboratory, Department of Inorganic Materials, Karpov Institute of Physical Chemistry Moscow K-64 Russia

Chapter Outline Head

Advanced Functional Materials. http://dx.doi.org/10.1016/B978-0-44-453681-5.00007-8

I. MULTIFERROIC MATERIALS

It is necessary first to define the expression "multiferroic". The original definition of ferroics was aimed to categorize materials and effects that allow the formation of switchable domains. Since domains can be formed in the case of ferroelectric, ferromagnetic, ferrotoroidic or ferroelastic ordering, primary multiferroics are compounds exhibiting forms of ordering from this group. The term "ferroic" has been suggested by Aizu to describe the crystals in which several order parameters (spontaneous magnetization, spontaneous polarization, and spontaneous deformation) are coexisting [1,2]. Unfortunately, this useful concept was not a serious influence during 25 years on the progress in the field of compounds in which two or all three of "ferroic" properties occur simultaneously. Five years later the thermodynamic classification of ferroics was proposed by Newnham in [3,4] and the ferroelectrics, ferromagnetics, and ferroelastics were considered as *primary ferroics*, whereas six kinds of *secondary ferroics*(including *ferromagnetoelectrics*) have been defined. In 1958 Smolenski predicted the possibility of simultaneous ordering of electric dipoles and magnetic moments (*ferroelectromagnets*) in compounds with the structure of perovskite ABO_3 [5,6]. In such multifunctional compounds the presence of cations with a lone pair of 6s electrons in the A sublattice and transition-metal elements with an inert-gas shell after removal of the s and d electrons in the B sublattice create favorable conditions for ferroelectric (FE) ordering. For this reason, to satisfy the necessary conditions for the appearance of FE and magnetism, the following cations were introduced in octahedral positions: ferroelectrically active ions on the one hand and magnetic ions on the other. Thus, in 1961 the first ferroelectromagnetic compounds combining FE and antiferromagnetic (AFM) properties were obtained in the form of polycrystalline solid solutions $(1 - x)(PbFe_{2/3}W_{1/3}O_3 – xPbMg_{1/2}W_{1/2}O_3$ [7] and then single-crystal FE–AFM $PbFe_{1/2}Nb_{1/2}O_3$ [8]. In the first case, Mg^{2+} and W^{6+} cations were diamagnetic and caused ferroelectricity with Pb cations whereas d^5Fe^{3+} cations are responsible for the magnetic ordering. Other examples include B-site ordered Pb_2CoWO_6 [9], which is ferroelectric and ferromagnetic, B-site disordered $PbFe_{1/2}Nb_{1/2}O_3$ [6,10] and $PbFe_{1/2}Ta_{1/2}O_3$ [11], which are both ferroelectric and antiferromagnetic with weak ferromagnetism around 10 K. As a result of the dilution of the magnetic ions, these materials all have rather low Curie or Neel temperatures. Following this approach about 50 ferroelectromagnets and several tens of solid solutions were discovered over the next two decades [12–15], including first ferromagnetic ferroelectric boracites ($Ni_3B_7O_{13}I$, $Cr_3B_7O_{13}Cl$) [16,17], fluorides ($BaMF_4$, M = Mn, Fe, Co, Ni) [18,19], magnetite Fe_3O_4 [20], $(Y/Yb)MnO_3$ [21] and $BiFeO_3$ [22]. However, such a combination in these ferroelectromagnets has been proven to be unexpectedly tough. Moreover, a successful combination of the two orders does not necessarily guarantee a strong magnetoelectric coupling and convenient mutual control between them [23]. Finally, a single phase material with large and robust magnetization and polarization has not been identified and a very insular and limited progress has been made in this direction [24,25]; known ferromagnetic ferroelectrics tend to have low magnetic Curie temperatures are often weak ferromagnets or are not strong enough insulators to sustain a ferroelectric polarization at room temperature. During initial search of new multiferroics the particular interest was related with materials which exhibit a combined ferroelectric and magnetic phase transition. It was found that $Ni_3B_7O_{13}I$ boracite shows coupled ferroelectric, ferromagnetic, and ferroelastic properties at the same phase transition around 62 K [16]. The multiferroic behavior in this compound arises from the fact that the magnetic transition is connected to a lattice distortion, which in turn allows the appearance of ferroelectric order [17].

The term “multiferroic” was introduced by Schmid in 1993 to define materials in which two or three types of ferroic order (ferroelectricity, ferromagnetism, and ferroelasticity) occur simultaneously in the same phase [26] (the overlap required for ferroelectromagnets is shown schematically in Figure 1). This drastic definition of multiferroics is often relaxed to include the compounds which exhibit combinations of any type of long ferroelectric, magnetic, and ferroelastic order.

Magnetic order is conventionally driven by exchange interactions between magnetic dipoles, originating from unfilled shells of electron orbitals. Electric order is the result of ordering of local electric dipoles. Elastic order is the result of ordering of atomic displacements due to strain. Today, this classification of multiferroics has been extended to include materials that exhibit any type of long-range magnetic ordering (including a antiferromagnetic ordering) together with spontaneous polarization [27]. Ferro ordering occurs when all dipoles are oriented in the same direction, whereas antiferro-ordering involves a ferro-ordering of two or more sublattices in opposite directions. In the recent years the term "multiferroic" has also been used to denote a ferroelectromagnet. The redesignation of "ferroelectromagnet" by "multiferroic" is unfortunate, since the latter term is uninformative. It requires clarification as to which subsystems are ordered. In addition, practically all research publications using the term "multiferroics" concern precisely a ferroelectromagnet [28]. Another important group of materials are the "linear magnetoelectrics" (ME), very often known as "magnetoelectrics", which possess long-range magnetic ordering but no spontaneous polarization.

A considerable development of MF materials has been started with successful synthesis of multiferroics' thin films [29] or even more, when a novel class of multifunctional materials has been discovered, where a multiferroic state is not only mutual arrangement of particular subnets but ferroelectricity is exclusively determined by magnetically ordered state [30,31]. More recently the definition of MF has also been extended to include so called ferrotoroidic order [32].

Until recently, the studies were focused on ferroelectricity in crystals with classical ferromagnetic or antiferromagnetic order. In these systems ferroelectric properties were established at high temperature and only below lower temperature phase transition some kind of magnetic ordering takes place. One does not expect remarkable interaction between dielectrical and magnetic properties because the two phenomena are essentially independent of one another.

However, limited number of investigations was connected with materials which have more complex magnetic structures and where ferroelectricity appears simultaneously with incommensurate-modulated magnetic order. In these crystals with a strong coupling between ferroelectric and magnetic order the ferroelectric order appears only in one specific magnetic phase whose appearance strongly depends in the value of the magnetic field.

Interest in ferroelectromagnetic compounds has increased hardly in recent years (see Figure 1) and the number of publications on this subject has increased dramatically. The study of MF materials has largely been confined to oxides. As shown in Figure 1, there are various subclasses of dielectric crystals having magnetic and electrical ordering. However, practically all the well-studied examples are oxides and many are perovskites (see Figure 2).

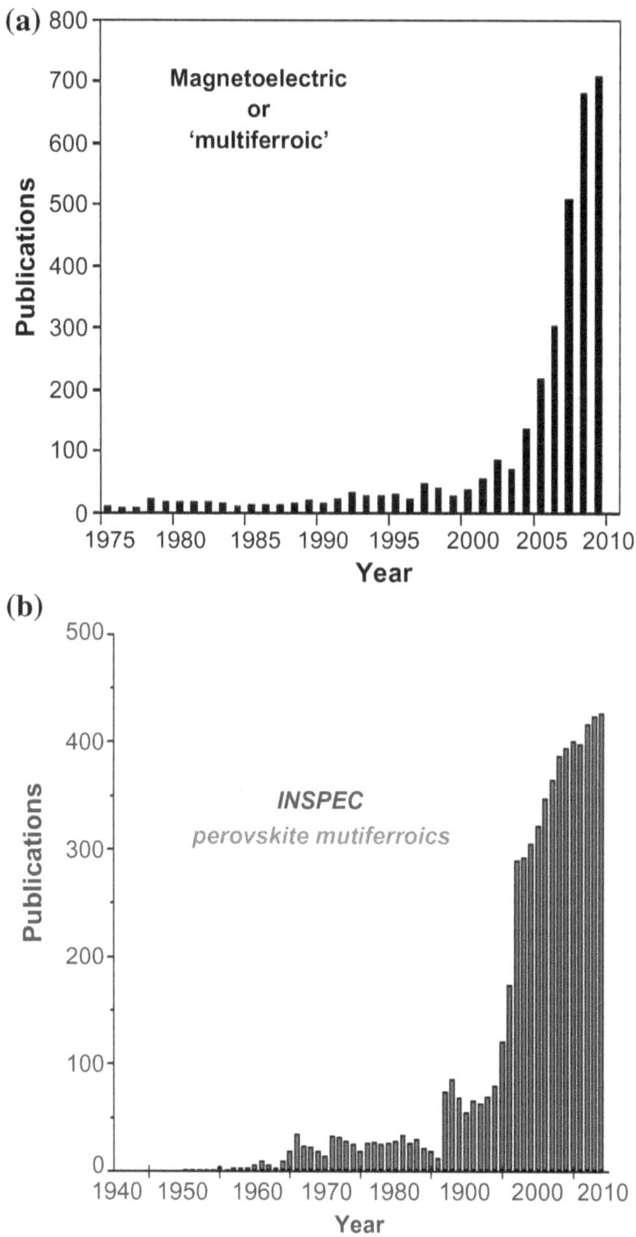

FIGURE 1 a) Evolution of the number of publications per year containing the keyword "magnetoelectric"or "multiferroic", b) Number of papers containing keyword "perovskite multiferroic" (as cited by INSPEC).

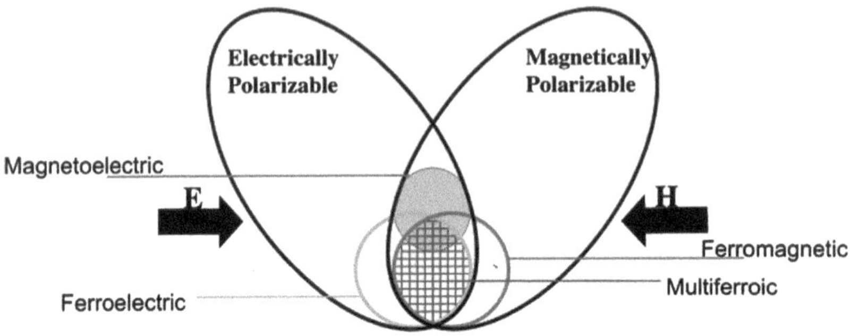

FIGURE 2 Families of magnetically and electrically ordered MF crystals: coupling between magnetic and electric degrees of freedom and the relationship between multiferroics and magnetoelectrics.

In Figure 1, the diagram illustrates a wide range of overlapping materials and shows the relationship between MF and ME materials. Ferromagnets (ferroelectrics) form a subset of magnetically (electrically) polarizable materials such as paramagnets and antiferromagnets (paraelectrics and antiferroelectrics). The intersection (red hatching) represents materials that are MF materials. ME coupling (blue hatching) is an independent phenomenon that can, but need not, arise in any of the materials that are both magnetically and electrically polarizable. In practice, it is likely to arise in most of these materials, either directly or via strain.

A host of factors have driven the current spate of interest in MF including the detail studies of structure–property–composition relationships. Preparation of a material in which ferroelectricity and ferromagnetism coexist would be a milestone for modern electronics and functionalized materials. The most appealing applications of such a material are new types of storage media using both magnetic and electric polarization, and the possibility of electrically reading/writing magnetic memory devices (and vice versa). In developing MF materials researchers seek to enhance and design the weak properties while retaining the dominant property.

The coexistence of various electronic order parameters that can interact in MF system opens new opportunities to design unconventional spintronic devices and also poses new challenges for physics, chemistry, and technology [33]. However, the inapplicability, depends on the nature of the actual coupling between the electronic, magnetic, and lattice order parameters. In this context, there are various stumbling blocks. The most challenging ones are how we can couple directly their molecular magnetic and electric fields responsible for ferromagnetism and ferroelectricity, and control eventually this coupling? Moreover, if this intimate coupling is achieved, how important is the material's response to an external applied field? Answers to these questions are hence

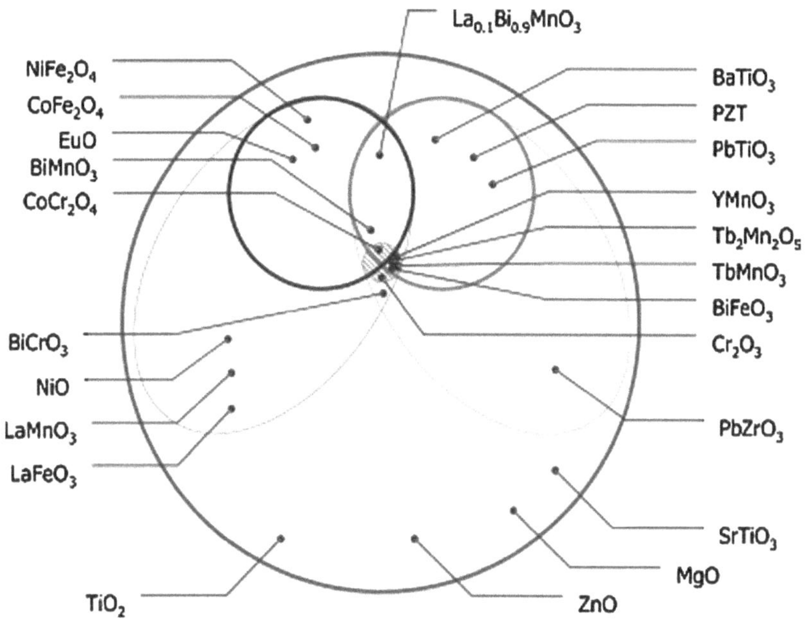

FIGURE 3 As in Figure 2 but with examples of oxides. (mostly perovskites)

crucial for understanding the microscopic mechanism behind the ferroic coupling and for their applications.

In this context, double perovskites $A_2B'B''O_6$ (where B′ has partially filled e_g-orbitals while B″ has empty e_g-orbitals or vice-versa) provide a unique opportunity to promote strong multiferroic behaviors, because they are ferromagnetic insulators governed by the 180° superexchange process [34] with a potential to generate a polar character. In fact, the polar nature can be obtained if one can promote different chemical valencies for the B′ and B″ cations in a self-ordered phase with B′ and B″ alternating along specific crystallographic directions (i.e., with a B′–O–B″–O–B′ sequence). Of course, such effect is absent in a disordered phase.

There has been a scarcity of magnetic ferroelectrics due to the so-called "d^0 vs d^n problem" [35]. That is, while the ferroelectric materials have transition metal ions with empty d shells (e.g., Ti^{4+} ions in $BaTiO_3$), ferro-, ferri- and antiferro-magnetic materials require the presence of (3d) transition metals or ions with partially filled d shells [36] (see Figure 3). Against this well-known scarcity of the magnetic ferroelectrics, in a ABO_3 perovskite oxide, the stereochemically active $6s^2$ lone pair electrons of Bi^{3+} (and Pb^{2+}) ions in the A-site offers an alternative way to get the FE ordering irrespective of the kind of (magnetic) ions present in the B-site [37].

Table 1 gives some of the perovskite ferroelectromagnets and indicates the type of electric and magnetic ordering and corresponding temperatures of the electric and magnetic transitions. The complete lists of the ferroelectromagnets and their physical characteristics can be obtained in early reviews [12–15,23] (See Table 2).

II. INCOMPATIBILITY BETWEEN FERROELECTRICITY AND MAGNETISM

A modern period of multiferroic's history began with new investigation of the problem using a density functional-based approach when the provocative question "Why are there so few magnetic ferroelectrics?" was formulated by Hill in [35]. The important prompt for correct answer concerning this scarcity, at least for the perovskite oxides with the chemical formula ABO_3, might be found after a comparison of a detailed list of magnetic and ferroelectric perovskites in the tables compiled in [38,39], respectively. In spite of hundreds of magnetic oxides and ferroelectric oxides, there is practically no overlap between them [40].This leads to an unfortunate but clear argument that magnetism and ferroelectricity tend to exclude each other. Key insights into this scarcity of MF phases can be understood by investigating a number of factors including symmetry, electronic properties, and chemistry. The new activity attracted serious interest in order to have a correct answer on above question and gave the discovery of MF properties of $TbMnO_3$ [41] and $TbMn_2O_5$ [42]. These materials were named as "improper"MF [43].

Given a definition of MF materials, the incompatibility between ferroelectricity and magnetism comes out as the first issue to address. From a point of view of symmetry consideration, ferroelectricity needs the broken spatial inverse symmetry while the time reverse symmetry can be invariant. A spontaneous polarization would not appear unless a structure distortion of the high-symmetry paraelectric (PE) phase breaks the inversion symmetry. The polarization orientation must be different from those crystallographic directions which constrains the symmetry of the point group. In contrast, the broken time-reversal symmetry is the prerequisite for magnetism and spin order, while invariant spatial inverse symmetry applies for most conventional magnetic materials in use but not a prerequisite. Among all of the 233 Shubnikov magnetic point groups, only 13 point groups, i.e., 1, 2, $2'$, m, m', 3, $3m'$, 4, $4m'm'$, $m'm2'$, $m'm'2'$, 6, and $6m'm$, allow the simultaneous appearance of spontaneous polarization and magnetization. This is quite restrictive, but in fact, the symmetry can be reduced and allow polarization and magnetization if there is an incommensurate modulation. In this case, many possibilities exist. Above mentioned restriction in the crystallographic symmetry results in the fact that multiferroics in nature are rare. Even so, it is known that some compounds

TABLE 1 **Ferroelectric and magnetic properties of single phase perovskite compounds (from our data and results published in [12–15]).**

Compounds	Structure at room temperature	(Anti)Ferroelectric transition temperature (K)	Magnetic transition temperature (K)	Properties
$PbFe_{2/3}W_{1/3}O_3$	Cubic	178	363	RFE-AFM
$SrFe_{2/3}W_{1/3}O_3$	Tetragonal	470	370	AFE-FIM
$BaFe_{2/3}W_{1/3}O_3$	Hexagonal	570	210	AFE-FIM
$BaFe_{2/3}Te_{1/3}O_3$	Hexagonal	573	228	AFE-FIM
$BaFe_{2/3}Re_{1/3}O_3$	Hexagonal	443	243	AFE-FIM
$BaFe_{2/3}Mo_{1/3}O_3$	Hexagonal	470	160	AFE-AFM
$PbMn_{2/3}W_1/3O_3$	Monoclinic	473	<77	FIE-FIM
$PbFe_2/U_{1/3}O_3$	Cubic	170	315	AFE-FIM
$PbFe_{2/3}Te_{1/3}O_3$	Cubic	170	370	FE-AFM
$PbMn_{2/3}Te_{1/3}O_3$	Cubic	540	<77	AFE-AFM
$PbMn_{2/3}Mo_{1/3}O_3$	Cubic	530	90	AFE-AFM
$PbFe_{1/2}Nb_{1/2}O_3$	Cubic	387	143, 12	RFE-AFM
$PbFe_{1/2}Ta_{1/2}O_3$	Cubic	233	180	RFE-FIM
$PbMn_{1/2}Nb_{1/2}O_3$	Monoclinic	415	11	AFE-AFM
$PbFe_{1/2}Mn_{1/4}W_{1/4}O_3$	Cubic	265	143	FE-FIM

Continued.

TABLE 1 Continued.

Compounds	Structure at room temperature	(Anti)Ferroelectric transition temperature (K)	Magnetic transition temperature (K)	Properties
$PbCo_{1/2}Nb_{1/2}O_3$	Cubic	284	213	FE-AFM
$PbCo_{1/2}Ta_{1/2}O_3$	Cubic	203	153	FE-AFM
$CdFe_{1/2}Nb_{1/2}O_3$	Monoclinic	753	48	AFE-AFM
$PbCo_{1/2}W_{1/2}O_3$	Monoclinic	305	9	AFE-FIM
$PbCo_{1/2}Te_{1/2}O_3$	Hexagonal	210, 370	16	AFE-AFM
$PbMn_{1/2}W_{1/2}O_3$	Monoclinic	445	45	AFE-AFM
$PbNi_{1/2}W_{1/2}O_3$	Cubic	290	56	AFE-AFM
$PbMn_{1/2}Re_{1/2}O_3$	Monoclinic	393	103	AFE-FIM
$PbCo_{1/2}Re_{1/2}O_3$	Cubic	403	<77	AFE-FIM
$PbNi_{1/2}Re_{1/2}O_3$	Monoclinic	343	<77	AFE-FIM
$SrMn_{1/2}Re_{1/2}O_3$	Monoclinic	473	115	AFE-FIM
$SrCo_{1/2}Re_{1/2}O_3$	Tetragonal	623	60	FE-AFM
$SrNi_{1/2}Re_{1/2}O_3$	Tetragonal	573	17	FE-AFM
$BaMn_{1/2}Re_{1/2}O_3$	Cubic	373	105	FE-AFM

Continued.

TABLE 1 Continued.

Compounds	Structure at room temperature	(Anti)Ferroelectric transition temperature (K)	Magnetic transition temperature (K)	Properties
$BaFe_{1/2}Re_{1/2}O_3$	Cubic	343	300	FIE-FIM
$BaCo_{1/2}Re_{1/2}O_3$	Cubic	253	40	AFE-AFM
$BaCo_{1/2}Te_{1/2}O_3$	Hexagonal	280	19	AFE-AFM
$BaFe_{1/2}Sb_{1/2}O_3$	Hexagonal	573	12	AFE-AFM
$Ba_{1/2}Eu_{1/2}TiO_3$	Cubic	165	4	AFE-AFM
$SrFe_{1/2}Sb_{1/2}O_3$	Monoclinic	613	36	AFE-AFM
$SrCu_{1/2}W_{1/2}O_3$	Tetragonal	920	43	FE-AFM
$Pb(Fe_{0.6}W_{0.2}Nb_{0.2})O_3$	Cubic	250	300	RFE-AFM
$SrMn_{1/2}Sb_{1/2}O_3$	Tetragonal	723	10	FIE-AFM
$SrCo_{1/2}Mo_{1/2}O_3$	Tetragonal	560	36	FIE-AFM
$LaFeO_3$	Ortorhombic	>300	740	FIE-AFM
$DyFeO_3$	Ortorhombic	4	4 (Dy), 645(Fe)	FE-AFM
$YCrO_3$	Monoclinic	~475	140	FIE-AFM
$LaCr_{1/2}Fe_{1/2}O_3$	Ortorhombic	>300	264	FIE-AFM

(Note: FE-ferroelectric, AFE-antiferroelectric, RFE-relaxor ferroelectric, FIE-ferrielectric, FM-ferromagnetic, AFM-antiferromagnetic, FIM-ferrimagnetic).

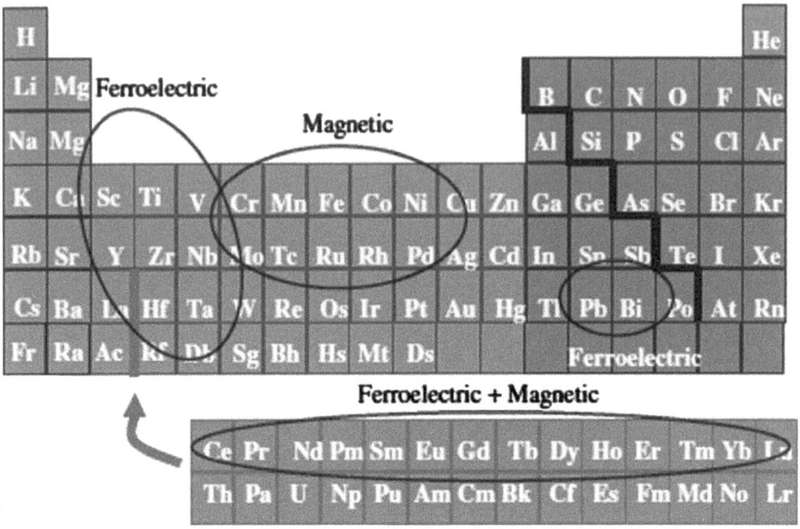

FIGURE 4 Periodic table of the elements highlighting the elements generating magnetic and ferroelectric behavior.

belonging to the above 13 point groups do not show any multiferroicity. Therefore, approaches different from simple symmetry considerations are needed [44].

The coexistence of ferromagnetism and ferroelectricity is difficult to achieve for many reasons. First, and most fundamentally, that all conventional ferroelectric perovskite oxides contain transition metal cations with a formal configuration d^0, such as T^{i4+}, Ta^{5+}, W^{6+}, at B-sites (i.e., the cations with an empty d-shell). The empty d-shell seems to be a prerequisite for ferroelectricity generation, while it does not mean that all perovskite oxides with empty d-shell cations must exhibit ferroelectricity. Magnetism, on the contrary, requires the cations at B-site with partially filled shells (always d- or f-shells), such as Cr^{3+}, Mn^{3+}, Fe^{3+}, because the spins of electrons occupying completely the filled shell add to zero and do not participate in magnetic ordering (Figure 4). Therefore the coexistence of the two phenomena, although not prohibited by any physical law or symmetry consideration, is discouraged by the local chemistry that favors one or the other but not both. In practice, alternative mechanisms for introducing off-centering on ions other than the magnetic transition metal ions, such as stereochemically active lone pairs or geometrically driven distortions can be used to circumvent this restriction. Second, and important from a practical point of view, is the fact that ferroelectrics must be robust insulators, so that an applied field can reorient the spontaneous polarization without causing conductivity. In practice, many ferromagnetic materials are metallic, and magnetic insulators tend to show antiferromagnetic coupling between the magnetic ions. Finally, magnetic transition metal ions tend to be fairly easy to oxidize or reduce, and are associated with multiple valence states accompanied by anion nonstoichiometry. Those few ferromagnetic insulators that do exist often show hopping conductivity at room temperature.

Magnetoelectric effect is large in compounds where the electrical and magnetic energy scales are comparable. In the most proper MF one expects the electrical energy scale to be in general large compared to magnetic energy scale, this appears as $T_E \gg T_M$ giving a weak magnetoelectric effect. In improper MF, this inequality ceases to be valid and in many such MF magnetoelectric effect is large. Therefore, in order to maximize the value of magnetoelectric coupling, one should attempt to increase the magnitudes of electric and magnetic susceptibilities of the crystal. Since susceptibilities are largest at the ordering transition, crystals with magnetic and ferroelectric phase transition at the same temperature should show maximum large magnetoelectric coupling that has been confirmed for $Sr_{3/2}Ba_{1/2}Zn_2Fe_{12}O_{22}$ multiferroic [45].

III. THE CLASSIFICATION OF MULTIFERROIC COMPOUNDS

The microscopic origin of magnetism is basically the same in all magnets: it is the presence of localized electrons, mostly in the partially filled d or f shells of transition-metal or rare-earth cations, which have a corresponding magnetic moment. Exchange interactions between the localized moments lead to magnetic order. The situation with ferroelectrics is quite different. There are several different microscopic sources of ferroelectricity, and accordingly one can have different types of multiferroics. Generally speaking, the analysis of various pathways to multiferroism is aided by classification of all multiferroic materials into two main groups. Let us briefly review few multiferroics from these groups (for more information, see [46]).

Type I multiferroics are materials in which ferroelectricity and magnetism have different sources and appear largely independent of one another, though there is some coupling between them. This group of multiferroics is the largest and incorporates the complex metal oxides from different structural families (namely, perovskites). In these materials, ferroelectricity typically appears at higher temperatures than magnetism, and the spontaneous polarization P_s is often rather large (of order 10–100 $\mu C/cm^2$). It is possible to create a Type I multiferroics in the perovskites (ABO_3) where one can make use of the stereochemical activity of an A-site cation with a lone pair (i.e., 6s electrons in Bi or Pb) to induce a structural distortion and ferroelectricity while inducing magnetism with the B-site cation. Examples are $BiFeO_3$ ($T_{FE} \sim 1100$ K, $T_N = 643$ K, $P_s \sim 100$ $\mu C/cm^2$) and $YMnO_3$ ($T_{FE} \sim 800$ K, $T_N = 76$ K, $P_s \sim 6$ $\mu C/cm^2$) but the coupling between magnetism and ferroelectricity is usually rather weak. The materials, challenge is to keep all their positive features and try to enhance the coupling. First example ($BiFeO_3$) is referred to as *proper* ferroelectrics, where a structural instability towards a polar state is the main driving force for the transition. Highly polarizable lone pairs of (Bi or Pb) cations are electrons not participating in bonds. If these lone pairs are localized and ordered in one direction, inversion symmetry is broken and an electric polarization results. If polarization is only a part of a more complex lattice distortion or if it appears as a result of some other ordering, the ferroelectricity is called *improper*. In the case of the second example (hexagonal manganite $YMnO_3$ with smaller ionic radius of heavy rare earth) improper ferroelectricity can be realized through geometrically driven effects where long-range dipole–dipole interactions and anion rotations drive the system towards a stable state with a spontaneous polarization.

Another new possible pathway by which one can achieve improper ferroelectricity in a Type I multiferroic is *via* charge ordering where non-centrosymmetric charge ordering arrangements (coexistence of inequivalent sites with different charges and inequivalent bonds) result in ferroelectricity in magnetic materials as is found in $LuFe_2O_4$. More recently it has been suggested that the coexistence of charge ordering in $Pr_{1-x}Ca_xMnO_3$ leads to a non-centrosymmetric charge distribution and a non-zero net electric polarization. As a conclusion, in this type of multiferroics magnetism and ferroelectricity result from two unrelated and very different mechanisms and it is very difficult to expect a strong magnetoelectric coupling in these materials.

The second group, which we can call Type II multiferroics, are the relatively recently discovered materials [41,42], in which a particular type of magnetic spiral or even for collinear magnetic structures causes ferroelectricity, implying a strong coupling between the two order parameters. However, the polarization in these materials is usually much smaller (10^{-2} $\mu C/cm^2$).

Most of the Type II multiferroics belong to this subgroup with a spiral Type II multiferroics and it is usually found in frustrated systems, where ferroelectricity and antiferromagnetism coexist with gigantic coupling. The well-known examples of this sort of behavior are multiferroics $TbMnO_3$, $TbMn_2O_5$, and $MnWO_4$ in which ferroelectricity is induced by the formation of a symmetry-lowering magnetic ground state that lacks inversion symmetry. For instance, in $TbMnO_3$, the appearance of ferroelectricity is directly correlated with the onset of spiral magnetic order around ~28 K. The intimate connection between magnetic and ferroelectric order results in extraordinary coupling – including the ability to change the direction of electric polarization with an applied magnetic field in $TbMnO_3$ and switching from positive to negative polarization in $TbMn_2O_5$ with a magnetic field.

The true nature of the mechanism for ferroelectric ordering in these materials is still under debate. Current theories have noted that for most of these materials the ferroelectric state is observed only when there is a spiral or helicoidal magnetic structure. Similar effects have also been observed in $Ni_3V_2O_8$. The key questions for this group of multiferroics are how is it possible that magnetic ordering can induce ferroelectricity and what is the role of frustration.

In the second case of Type II multiferroics with collinear magnetic structures all magnetic moments are aligned along a particular axis without the necessary involvement of the spin–orbit interaction. Polarization can appear in these materials as a consequence of exchange striction because the magnetic coupling varies with the atomic positions. The simplest example is in Ca_3CoMnO_6 [47] consisting of one dimensional chains of alternating Co^{2+} and Mn^{4+} cations. At high temperature the distances between the ions along the chain are identical, the chain has inversion symmetry, and polarization is absent. Magnetic ordering, however, breaks inversion symmetry: the spins form a $\uparrow\downarrow\downarrow$ type magnetic structure. Due to an exchange striction the distortion of ferro and antiferro bonds ($\uparrow\uparrow$ and $\uparrow\downarrow$) is different and Ca_3CoMnO_6 becomes ferroelectric.

IV. PEROVSKITES: INTRODUCTION AND SCOPE

The understanding of ABO_3 perovskite remains among the most challenging topics at the interface between solid state chemistry and physics, both from academic and application point of view. Among the studied multiferroics, the perovskites are particularly important (around 90% of known multiferroics are perovskites) since they exhibit an extraordinary range of structures, chemical bonding and physical properties. Although ferroelectric perovskites with magnetic ordering have been known for a long time, new intriguing physical effects are periodically discovered and their understanding remains an internationally highly competitive area. One of the key questions for the future development and understanding of these multifunctional materials concerns the mutual coupling between the properties, its underlying mechanism, and whether it can be used in applications.

The structure of the ideal simple ABO_3 perovskite is cubic, *Pm*-3*m* (see Figure 5). In the complex perovskite structure two limiting cases can be distinguished. First, the distribution of cations among octahedral sites is random and the structure can still be described by the *Pm*-3*m* space group. Second, the cations are ordered. As a result the lattice parameter doubles, in comparison with the simple perovskite, and the structure is called as a double perovskite and described by the *Fm*-3*m* space group (see Figure 6). In addition to these two cases, the symmetry lowering of the cubic structure may occur. Moreover, crystals may exhibit different degrees of short-range order (see Figure 7) forming ordered nanodomains spread inside a disordered matrix [48].

Data analysis of more than 300 perovskites showed that the random distribution of cations among octahedral sites can be realized only when the ionic radii difference is less than 0.2 Å and the charge difference is 2 [49]. Ionisation potentials, cation coordination geometry, and the A-cation/B-cation size ratio have been also mentioned as factors that influence the degree of ordering [50]. According to Galasso [51], the ratio r_{B1}-r_{B2}/r_{B1} determines the

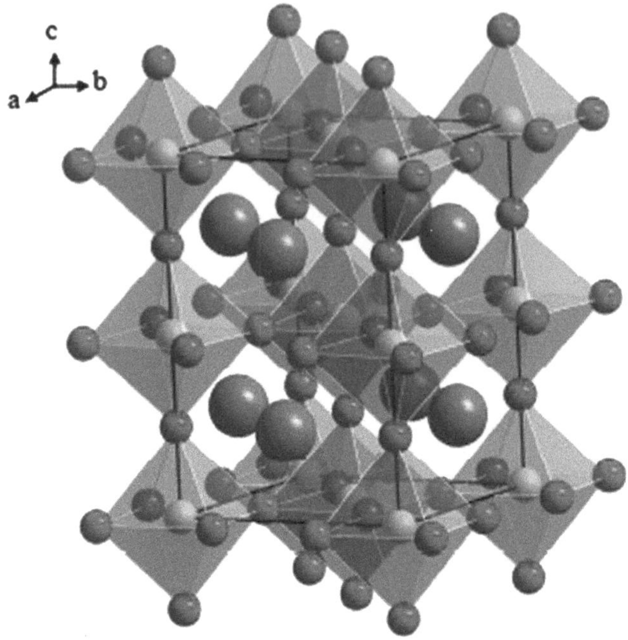

FIGURE 5 Perovskite structure ABO_3 using a polyhedral representation.

ordering in complex perovskite: if s is greater than 0.09, the cationic distribution should be ordered.

Obviously, the criteria mentioned above, though generally fulfilled, have exceptions; some complex perovskites have been found which apparently do not follow them [52].

In this case, the size and charge differences criteria seem not to hold. Finally, it is important to stress that, although the perovskite oxides are believed to be well understood from a structural point of view, there are still many points remaining unclear. The geometrical and electrostatic picture of ionic compounds has been useful for decades, but is clearly unsatisfactory for the complex perovskites. Nowadays, some other approaches to inorganic solids are being developed [53].

A. Perovskites: crystallochemistry and size effects

Perovskite lattice instabilities are often understood using the tolerance factor t.

The cell axis, a, in the ideal cubic unit cell depends on the ionic radii (r_A, r_B, and r_O) as described by equation:

$$a = \sqrt{2}(r_A + r_O) = 2(r_B + r_O).$$

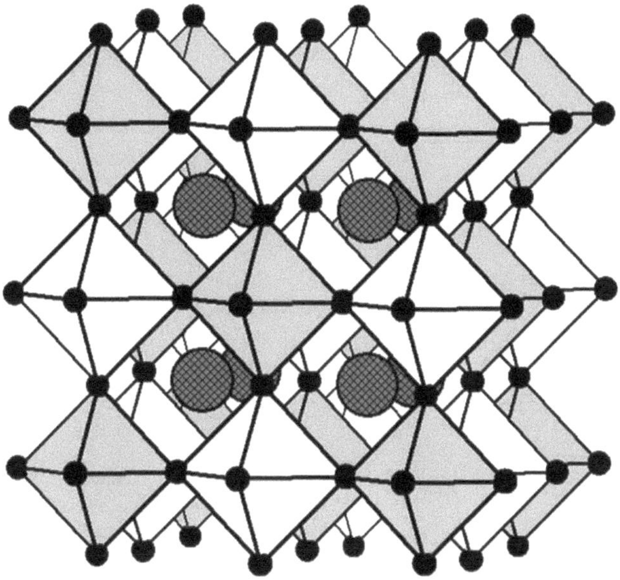

FIGURE 6 Perovskite structure with 1:1 ordering of two different B-site cations (space group *Fm*3*m*) : conventional view showing corner-sharing $B'O_6$ (light grey octahedra) and $B''O_6$ (empty octahedral) with the A cations (crosshatched dark grey spheres) residing in the cavities formed by the octahedral network.

The ratio of the two expressions for the cell length is called the Goldschmidt *tolerance factor t*and allows us to estimate the degree of distortion as described in equation:

$$t = (r_A + r_O)/\sqrt{2}(r_B + r_O).$$

The cubic perovskite $SrTiO_3$ has $t = 1$. If the size of the A cation decrease then t becomes smaller than 1. As a result the $[BO_6]$ octahedra will tilt in order to fill space. Ferroelectrics, such as $BaTiO_3$, have $t > 1$, indicating that the B-site ion is too small for its site in the ideal cubic structure (so-called B-site driven materials, where the ferroelectricity is from off-centering of B-site cations). There is another important class of ferroelectric perovskites, so-called A-site driven materials. In these, t is normally less than unity, and the ferroelectricity is from off-centering of A-site cations. This family includes important Pb and Bi-based ferroelectromagnets. The essential physics is lone pair stereochemistry, specifically hybridization of Pb and Bi 6p states with O p states. This class includes the few known magnetic ferroelectrics with strong ferroelectric properties, e.g., $BiFeO_3$, $BiMnO_3$, and $PbVO_3$ [54]. Without Pb or Bi, $t < 1$ perovskite structures generally derive from BO_6 octahedral tilts and not A-site off-centering. Ions with d electrons are generally larger than d^0 ions. The majority of magnetic perovskites have $t < 1$, with lattice structures based on tilts of the BO_6 octahedra and not ferroelectricity. Lower values of t will lower the symmetry of $GdFeO_3$ with $t = 0.81$ is orthorhombic (see figure 8). With values less than 0.8 finally the ilmenite structure is more stable. On the other hand if t is larger than 1 due to a large A or a small B ion then hexagonal variants of the perovskite structure are stable, e.g., $BaNiO_3$ type structures. In this case the close-packed layers are stacked in a hexagonal manner in contrast to the cubic one found for $SrTiO_3$, leading to face sharing of the $[NiO_6]$ octahedra (see Figure 8). The t value for$BaNiO_3$ is 1.13. The tolerance factor is only a rough estimate and also the nature of the A and B ions influences the structure. For example $LnFeO_3$ (Ln = lanthanide ion) have all-perovskite type structures, this is true for $LnMn^{3+}O_3$ only if Ln is La or Ce–Dy. The compounds in which Ln = Y or Ho–Lu adopt a different hexagonal structure, with 5- and 7-coordination of Mn and A, respectively [55].

B. Distorted perovskites – Jahn–Teller effects

In some perovskites the distortion of the structure can be assigned to Jahn–Teller active cations at the B position. For example in $LnMnO_3$ (Ln = La, Pr or Nd) with Mn^{3+} ions the $3d^4$ electrons divide up into 3 t_g electrons and 1 e_g electron. The odd number of electrons in the e_g orbital causes an elongation of the $[MnO_6]$ octahedron (see Figure 8).

C. Superstructures

Doubling the unit cell edges of the ideal cubic perovskite structure makes it possible to occupy equivalent positions with atoms of different elements (see Figure 7(a)). If the B position is partly replaced by a second metal a double perovskite of the general formula $A_2B'B''O_6$ may form where the $B'O_6$ and the $B''O_6$ octahedra are alternatively arranged (see Figure 7(b)). The double perovskites frequently exhibit distortions, e.g., when small cations occupy the A-site and the octahedra are then tilted resulting in a lowering of the symmetry.

The ideal cubic $Fm3m$ symmetry is usually expected for double perovskites when the tolerance factor t is close to unity. For the majority of these perovskites, however, the size of the A-cation is too small to fit in the cavity formed by the 12 anions. In such case, the tilting of octahedra generally occurs leading to a lower symmetry (see Figure 9). Woodward [56] has, based on the Glazer's [57] classification of the tilt systems, derived 13 possible space groups for double perovskites by considering the cationic ordering and the rotation of octahedral occurring simultaneously. Howard and Stokes [58] undertook a group theoretical analysis, identifying 12 space groups under the same conditions. The structures of double perovskites are usually determined from powder diffraction data using profile analysis by the Rietveld method [59]. Although the different tilt systems can be recognized by inspecting diagnostic reflections, this is, in practice, not always straightforward as the sample quality,

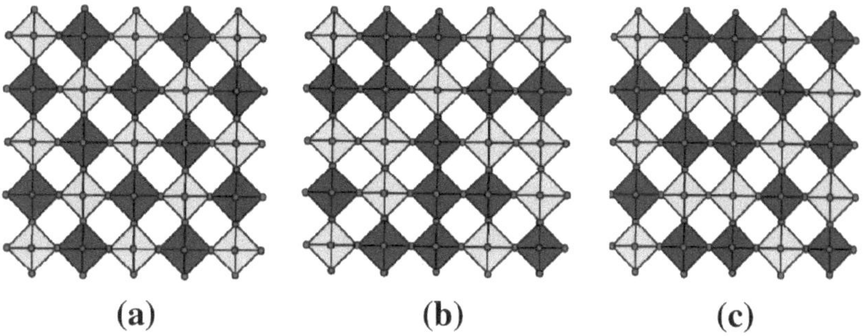

FIGURE 7 B-cation ordering in double perovskites: a) full order, b) partial order, c) full disorder.

instrument resolution, and the presence of pseudo-symmetry may hinder proper observation.

Several factors lead to deviations from ideal composition and symmetry in double perovskites [60]. Those include:

1. Structural vacancies (e.g., anion deficiencies in $A_2B'B''X_{6-\delta}$).
2. Incomplete B'/B''-site cation ordering.
3. Electronic factors, such as the presence of a Jahn–Teller cations (e.g., Mn^{3+}) or a cation with a stereoactive lone pair of electrons (e.g., Bi^{3+}) on the A or B'/B''-site.
4. Octahedral tilting distortions.

The rich physics of double perovskites is related to the fact that they are very flexible with respect to variations of the magnetic and/or non-magnetic B-site as well as the A-site cations. In this way they offer a wide range of possibilities to influence the structure and/or the magnetic interactions and, hence, to tailor the magnetic properties:

1. The magnetic ion on the B-site can be varied using either 3d or 4d transition metals (e.g., Fe, Cr, Mn, Ni , Co, Ru) with localized states or 4f lanthanides (e.g., Gd, Dy, Ho, Er) with highly localized states.
2. The non-magnetic ion on the B''-site can be varied. Here, we have to distinguish between cations with delocalized electrons (e.g., Mo^{5+}, W^{5+}, Re^{5+}, Re^{6+}) and those with a noble gas electronic configuration (e.g., Nb^{5+}, W^{6+}) or a completely filled 3d- or 4d-shell (e.g., Sb^{5+}, Te^{6+}).
3. Real possibility for doping of electrons into the system can be realized by partially replacing the divalent alkaline earth ions on the A-site by a trivalent rare-earth ion such as La^{3+}.
4. The crystal structure can be changed by using alkaline earth ions on the A-site with different ionic radii. This results in considerable steric effects,

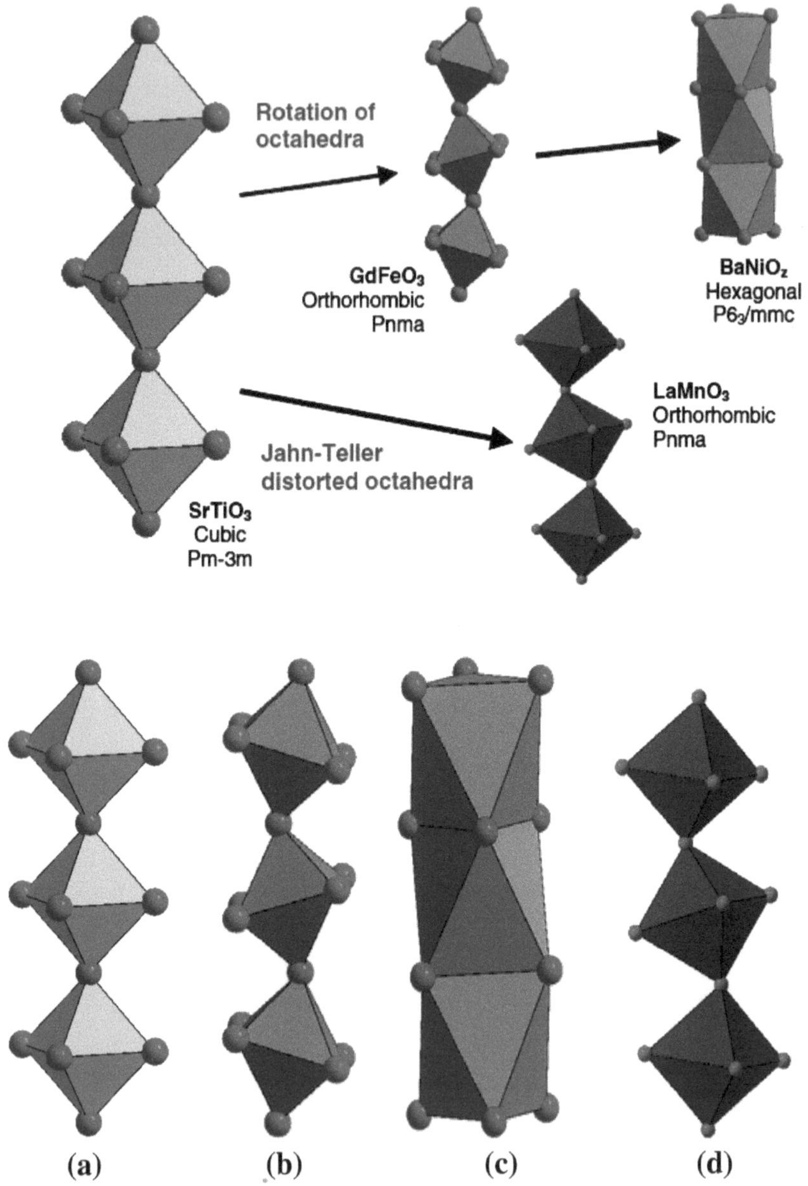

FIGURE 8 Ideal cubic perovskite structure with $t = 1$ (a), tilted $[FeO_6]$ octahedra in the orthorhombic $GdFeO_3$ with $t < 1$ (b), hexagonal variant of perovskite structure for$BaNiO_3$ with $t > 1$ (c), Jahn–Teller distorted Mn octahedra in $Ln$$MnO_3$ (d)

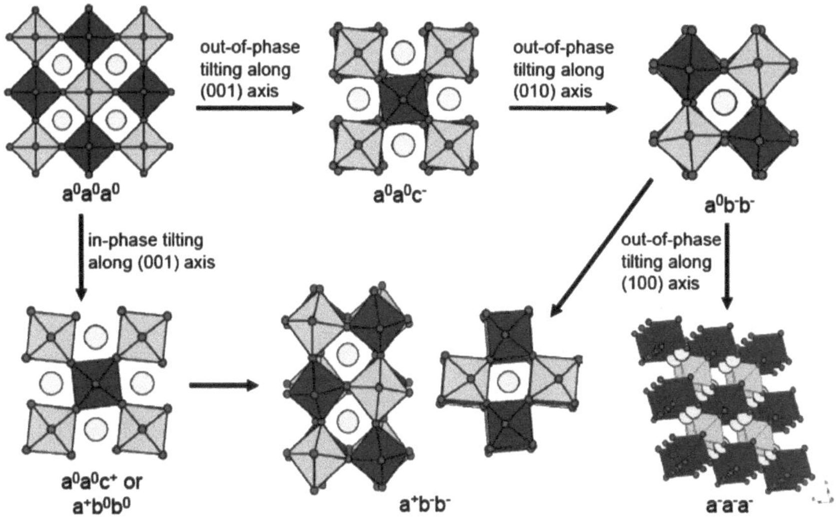

FIGURE 9 Combinations of in-and out-of phase tilting in double perovskites [56].

which in turn change the bond angles thereby modifying the magnetic interactions.

5. It is possible also to use magnetic ions on both the B′- and B″-sites as, e.g., La_2CrFeO_6. This class of double perovskites may be interesting for the realization of half-metallic antiferromagnets.

V. COEXISTENCE OF FERROELECTRICITY AND MAGNETISM IN PB-BASED PEROVSKITES

As mentioned above, the conceptually simplest approach is to synthesize multiferroics with two structural units functioning separately the ferroelectricity and magnetism.

The first route towards perovskite multiferroics was taken by Russian researchers (see reviews [12,13,15,61]. They proposed to mix both magnetic cations with *d* electrons and ferroelectrically active cations with d^0 configurations at the B-sites (i.e., substituting partially the d^0-shell cations by magnetically active 3d ions while keeping the perovskite structure stabilized). It is hoped that the magnetic ions and d^0-shell cations favor separately a magnetic order and a ferroelectric order, although this may be difficult if the magnetic doping is over-concentrated. Two sorts of B-cations at the octahedral sites may be ordered, disordered, or partially ordered, and this has a tremendous influence on the multiferroic properties of the perovskites. The typical (and one of the most studied) compounds are Pb-based complex metal oxides. The coupling between magnetic order and ferroelectric order in this kind of multiferroics is, in most cases, very weak because these two orders originate from different kinds of ions.

VI. TERNARY FE-CONTAINING PEROVSKITE OXIDES

Complex ferroelectric and antiferroelectric perovskite-family oxides $Pb(B'_{1-m}B''_m)O_3$ ($m = 1/2$, $1/3$) have been studied intensively in recent years [62–67]. Of special interest, is the possibility of varying the degree of ordering of B′ and B″ ions over equivalent crystallographic positions. Such ordering is usually called compositional to differentiate it from the ferroelectric and antiferroelectric order. For the characterization of such ordering, a long-range order parameter S is usually introduced, which is equal to unity for a fully ordered state, and to zero, for a disordered state [63]. The perovskite oxides with B^{3+} = Fe are of particular interest. The weak superstructural reflections on X-ray patterns associated with a partial ordering of cations have been observed only in $Pb(Fe_{2/3}W_{1/3})O_3$ (PFW) [65]; numerous attempts to fabricate ordered ceramics $PbFe_{1/2}Ta_{1/2}O_3$ (PFT) and $PbFe_{1/2}Nb_{1/2}O_3$ (PFN) by long annealing at high temperatures have not met with success [62–67]. At the same time, the experimental values of the antiferromagnetic Néel temperature T_N for PFW, PFN, and PFT are located approximately halfway between calculated values of this temperature for $S \sim 1$ and $S \sim 0$, which is usually interpreted as an evidence of a partial ordering of B cations (T_N depends strongly on S), because the ordering changes the number of magnetic ions in neighboring unit cells [64]. The ordering of cations in these oxides seems to occur locally; i.e., mesoscopic regions of the size of several nanometers with different values of S arise in a crystal. Small sizes of ordered regions are responsible for the difficulties in observing superstructural reflections on X-ray diffraction patterns.

A. $Pb(Fe_{1/2}B_{1/2})O_3$ perovskites

Lead ferroniobate $PbFe_{1/2}Nb_{1/2}O_3$ (PFN) also belongs to the family of ferroelectromagnets with perovskite structure and exhibits ferroelectric and antiferromagnetic properties [64,65]. The phase transition from rhombohedral (ferroelectric) to cubic (paraelectric) phase takes place at ~385 K and the transition to the antiferromagnetic phase occurs at the Néel temperature T_N, which, depending on the sample preparation conditions, lies in the range from 140 to 170 K. The magnetic properties of PFN are related to the presence of Fe^{3+} ions in the octahedral lattice positions. In the antiferromagnet phase, the magnetic moment of each Fe^{3+} ion is directed antiparallel to the magnetic moments of the nearest neighbors. Below T_N, in the region of coexistence of the antiferromagnetic and ferroelectric phases, magnetoelectric effect can be observed in PFN [68].

Among the Pb perovskites, $PbFe_{1/2}Ta_{1/2}O_3$ (PFT) is unique in the sense that it can only be obtained in the completely long range disordered form. PFT crystals undergo two successive ferroelectric phase transitions that take place at about 270 and 220 K [65,66] each one occurring in a quite narrow temperature range. The sequence of the ferroelectric symmetry changes was defined as cubic (Pm3m) $\rightarrow$ tetragonal (P4mm) $\rightarrow$ monoclinic (Cm), with Fe and Ta cations randomly distributed over the octahedral sites in all the three phases although the possible presence of nanoscale short-range ordered regions is still under discussion. The dielectric behavior shows, in fact, pronounced relaxor properties, with the two maxima of the dielectric constant, corresponding to the two structural transitions, displaced towards higher temperatures for increasing frequencies. The magnetic susceptibility shows an antiferromagneticlike anomaly at a temperature which is strongly sample dependent, and equal to 133, 160, and 180 K according to different authors [65,66]. The magnetic structure is G-type antiferromagnetic, with the lattice parameter doubled with respect to the crystallographic Pm3m one, disregarding the weak ferroelectric pseudo-rhombohedral distortion. In the undistorted antiferromagnetic G-type structure, each magnetic 12 Fe^{3+} cation interacts with its six nearest neighbors via oxygen-mediated 180° antiferromagnetic superexchange. There is experimental evidence of a second magnetic transition at 9 K which may be connected with the clusters of ordered perovskite phase.

In Pb_2CoWO_6, the B-site is occupied by two different species of cations: a complex perovskite structure is obtained. Owing to the stoichiometry, and the charge and cationic size differences of Co and W, this structure exhibits cationic ordering. The physical properties are closely related to an interesting original phase transition sequence involving normal to incommensurate phase transitions: the monoclinic incommensurate phase is stable in a finite range of temperatures (230–298 K). The antiferromagnetic phase transition takes place at $\sim$9 K [67].

The possible leading role of Pb cations in both magnetic and ferroelectric coupling in $Pb(B'_{1-m}B''_m)O_3$ perovskites was discussed in [68]. The ferroelectric coupling in these perovskites is usually enhanced owing to the strong ferroelectric activity of the Pb cation through its lone-electron pair. It was found that T_N values of $PbFe_{1/2}B_{1/2}O_3$ (B = Nb, Ta) perovskites ($\sim$150 K) [69,70] differ noticeably from T_N of $AFe_{1/2}B_{1/2}O_3$ (A = Ba, Sr, Ca; B = Nb, Ta) ($\sim$25 K) [71,72]. Along with it, Pb perovskites have AFM G-type magnetic structure whereas a spin-glass state is typical for lead-free compounds. This is in contrast with a hypothesis concerning a determining role of Fe–O–Fe interactions for the fundamental understanding of magnetic interactions. It was assumed that the magnetic coupling can be enhanced either because of the involvement of the Pb cations into the superexchange or because of the influence of the Pb-related polarization onto the magnetic properties. These factors do not exist in lead-free perovskites. An important role of empty 6p-orbitals of non-magnetic Pb^{2+} and Bi^{3+} cations can be used as an additional instrument for design of new multriferroic complex metal oxides [72].

B. $A(B_{2/3}B_{1/3})O_3$ perovskites

Since it is well known that properties of complex perovskites featuring two different ions sharing the B-sites may be strongly affected by order/disorder phenomena involving the B ions, it is important to study the degree of order in perovskites. It is not surprising that complex perovskites featuring two different ions sharing the B-sites, such as $A(B'_{2/3}B''_{1/3})O_3$, $A(B'_{1/2}B''_{1/2})O_3$, or $A(B'_{1/3}B''_{2/3})O_3$, very often display interesting electrical and magnetic properties [49,51,56–62]. The properties of this type of compounds may be strongly affected by the order–disorder phenomena, mainly involving the B ions. Although many compounds of these types have been investigated in recent years and their structural characteristics determined, information about the multiferroic properties remains scarce.

Lead iron-tungstate $PbFe_{2/3}W_{1/3}O_3$ (PFWO) is known as an important candidate for multilayer ceramic capacitors and for possible piezoelectric applications. Typical relaxor properties with a broad maximum of the low-field permittivity at $T = 180$ K and a frequency dispersion at low frequencies were found [73]. Relaxor ferroelectrics are well known in the literature since with 1950s. and different groups have different opinions on the origins of this behavior but confirming that the characteristic signature of relaxors is the existence of polar nano-scale ordered ferroelectric regions in a paraelectric disordered matrix [74]. No structural change was detected by X-ray diffraction experiments, the material remaining in a pseudocubic phase down to 10 K. The relaxors do not possess a non-zero macroscopic polarization, but local polarized nanoregions (with short range ordering) [75]. The nanoregions tend to grow and join by decreasing the temperature. However, as they are inhibited by the quenched compositional fluctuations and by the induced random fields, they will not establish adequate interaction to result in a long range ordering as in normal ferroelectric.

A very interesting peculiarity of PFW system by comparison with other relaxors is related to the presence of magnetic ions Fe^{3+} with an occupancy of 67% of the B-octahedral sites. Two magnetic transitions at $T_{N1} \sim 20$ K from weak ferromagnetism to antiferromagnetism and from antiferromagnetism to paramagnetism at $T_{N2} \sim 350$ K were found. Due to the anomalies found both on the dielectric constant and on the dissipation factor at ~20 K, a multiferroic behavior and a possible magnetoelectric coupling was considered for the present system. The magnetoelectric coupling could be explained through microstructural characteristics of PFW ceramics, which obtain both disordered and partially ordered polar clusters between the off-valence cations on the

B-site [75]. The magnetic interactions that give rise to the observed behavior might be ascribed as follows: (1) in the Fe/W ordered polar domains, a weak superexchange of Fe^{3+}–O–W–O–Fe^{3+} type of interaction responsible for the magnetic anomaly at low-temperature T_{N1} and (2) in the Fe/W disordered regions, a stronger superexchange of Fe^{3+}–O–Fe^{3+} pathway interaction is responsible for the magnetic ordering at higher temperatures T_{N2}.

It can be noticed that other complex oxides from the lead series of compounds, namely $PbFe_{2/3}M_{1/3}O_3$ (M = W, Te, Mo, U), also adopting the cubic perovskite structure, show similar magnetic properties (Table 4) [76]. However, the difference between the values of T_N is quite striking. It is important to understand the possible influence of the M cations causing this difference. The difference in size of the M^{6+} cations is not large enough to explain the different values of T_N (except for the U^{6+} cation). It should, however, be particularly emphasized that this analysis involves difficulties arising from the fact that the available structural data are very restricted. Consequently, it is difficult to find a simple explanation for the different values of T_N based on only crystallochemical considerations using the tolerance factor (t), the lattice parameter (a), and the volume (V) of the unit cell. It is, however, clear that the decrease of the lattice parameter a in this type of cubic perovskite structure (which is equivalent of decreasing Fe–Fe distance) could result in the increase of T_N. This simple rule holds for the compared Pb-containing perovskite compounds. It is also evident that the off-center polar distortion which is created by the Pb and O atoms is responsible for the ferroelectricity. At the same time the magnetic properties strongly depend on the B-site ordering and on the relative concentration of the magnetic cation. With a decrease of the Fe content in the B-perovskite sublattice will follow a progressive reduction of the magnitude of T_N, which is vividly illustrated in Figure 10.

It is interesting to note that another oxides $A_3Fe_2WO_9$ (A = Ba, Pb, Sr, Ca) with perovskite structure also show the magnetic properties [77] (see Table 3). It is not possible to find a simple explanation for different values of T_{CM} for these perovskites based on only crystallochemical considerations using the tolerance factor t, lattice distortions, and volume of the unit cell. It is seen that the magnetic phase transition temperature of all the samples is nearly the same showing that the temperature of magnetic ordering is insensitive to the changing in A-site. This is in agreement with the band structure calculations in the double perovskites that contain B" cations such as W^{6+} where the substitution in the A-site does not change the band gap significantly [78].

C. Influence of oxygen deficiency

Although the perovskite family crystallographic structures have been studied for a long time, their basic physical properties are still leading to surprising discoveries. It has been shown that when oxygen vacancies are introduced

TABLE 2 New potential multiferroics among $BiMO_3$compounds.

M	Space group	Ferroelectricity	Magnetism	Synthesis conditions	Reference
Fe	*R*3*c*	$T_C \sim 1100$ K $P_S = 100$ μC/cm^2	$T_N = 643$ K AFM G-type	900 C Ambient *P*	[89,90]
Cr	*C*2/c	Centrosym.	$T_N = 382$ K AFM G-type	1380 C 6 GPa	[103]
Mn	*C*2/*c*	Centrosym.	$T_C = 373$ K FM	500-1110 C 4-6 GPa	[104]
Co	*P*4*mm*	$T_C > 520$ K $P_S = 175$ μC/cm^2	$T_N = 470$ K AFM C-type	970 C 6 GPa	[105]
Ni	*P*-1	Centrosym.	$T_N = 300$ K AFM	1000 C 6 GPa	[106]
$Cr_{0.5}Fe_{0.5}$	*R*3*c*	$Tc > 300$ K $P_S = 63$ μC/cm$_2$	$T_N = 400$ K AFM	1000 C 6 GPa	[107]
$Ni_{0.5}Mn_{0.5}$	*C*2	$Tc = ?$ $P_S = 20$ μC/cm$_2$	$T_C \sim 680$ C FM	800 C 6 GPa	[108]
$Mn_{0.67}Ni_{0.33}$	*Pn*21*m*	$Tc = ?$ $P_S = 60$ μC/cm^2	Short range order.	800 C Ambient *P*	[109]

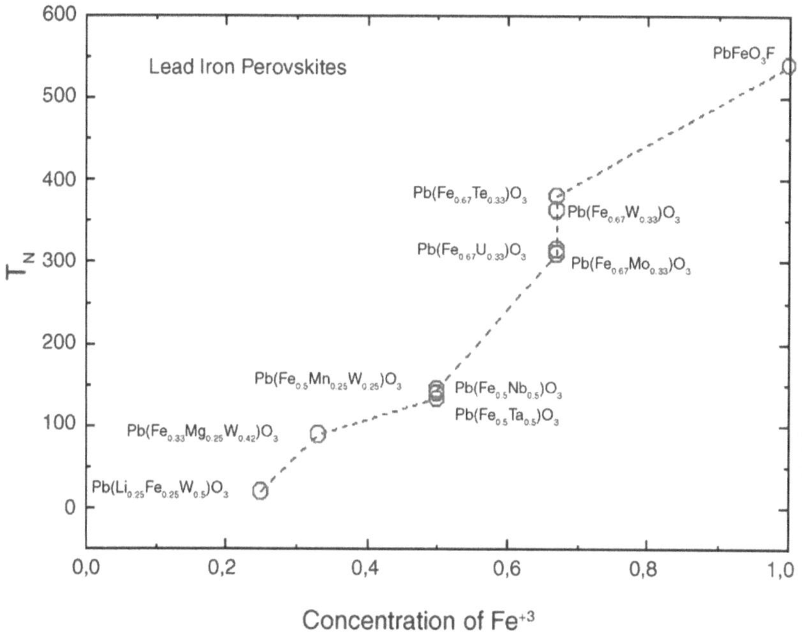

FIGURE 10 The dependence of T_N on the concentration of Fe^{3+} cations in Pb-based perovskites.

TABLE 3 Crystallographic and magnetic properties of different $AFe_{2/3}W_{1/3}O_3$(A=Ba, Pb, Sr, Ca) with perovskite structure (from our data and results published in [15,16]).

Cation A^{2+}	r_{ion},	s.g. at 295K	T_{CM}(K)	T_{CE}(K)	Properties
Ba	1.61	P63/mmc	330	573	AFE-WFM (order)
Pb	1.49	Pm3m	340	180	FE-AFM (disorder)
Sr	1.44	I4/m	373	473	AFE-WFM (partial order)
Ca	1.34	$P2_1/n$	340	>400	AFE? –WFM (partial order)

into the lattice, hexagonal perovskites such as $SrMnO_3$ transform to cubic perovskites [79,80].

The magnetic and structural properties of oxygen deficient perovskites with composition $Ba_3Fe_2WO_{9-\delta}$ have been also systematically studied for two different oxygen contents corresponding to $\delta = 0.00$ and $\delta = 0.55$ in the chemical formula in order to determine and correlate their chemical composition, structural, and magnetic properties [81]. The evolution of nuclear and magnetic structures with temperature has been investigated by neutron powder diffraction. It was shown that at room temperature the stoichiometric

TABLE 4 Crystallographic and magnetic properties of different $PbFe_{2/3}B_{1/3}O_3$ (A=Te, Mo, W, U) with perovskite structure (from our data and results published in [15,16]).

B_{6+}	r_{ion}(A)	t	a(A)	T_N	T_C
Te	0.56	1.013	3.981	380 AFM	510 FE
Mo	0.59	1.008	3.994	310 AFM	500 FE
W	0.60	1.007	3.978	340 AFM	180 FE
U	0.73	0.986	4.071	315 FIM	170 AFE

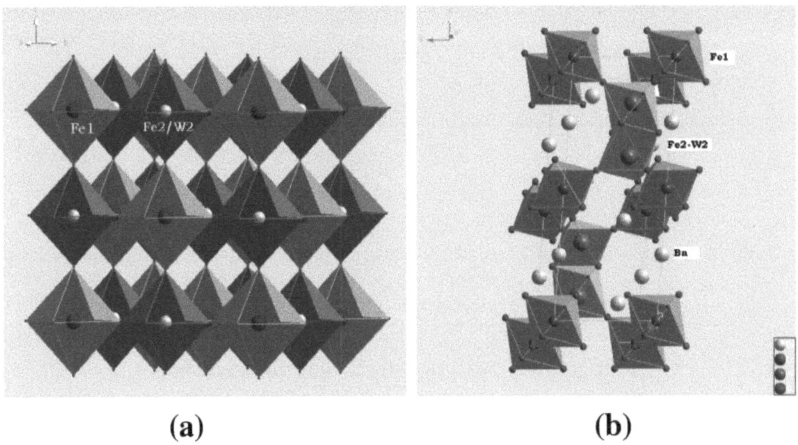

FIGURE 11 Crystal structure of $Ba_3Fe_2WO_{9-\delta}$ ($\delta = 0.00$ (a), $\delta = 0.55$ (b)).

compound ($\delta = 0.00$) adopts a hexagonal 6H-perovskite structure (space group $P6_3/mmc$) (see Figure 11). This phase, when heated at high temperature under a stream of Ar gas, transforms to an oxygen-deficient phase with $\delta = 0.55$, which is an ordered cubic perovskite structure (space group $Fm-3m$) (see Figure 11). It was shown that a relatively small concentration of oxygen vacancies is effective for controlling the crystal and magnetic structure (Figure 12) and fundamental physical properties of this perovskite. The oxygen deficiency implies a partial reduction of the Fe sublattice and therefore a change of magnetic properties (Figure 13). An oxygen nonstoichiometry tends to stabilize the long-range magnetic order and to increase the temperature of magnetic phase transition.

On introducing oxygen vacancies the $Ba_2FeMn_{0.5}W_{0.5}O_6$ compound also undergoes a structural transformation from hexagonal to cubic

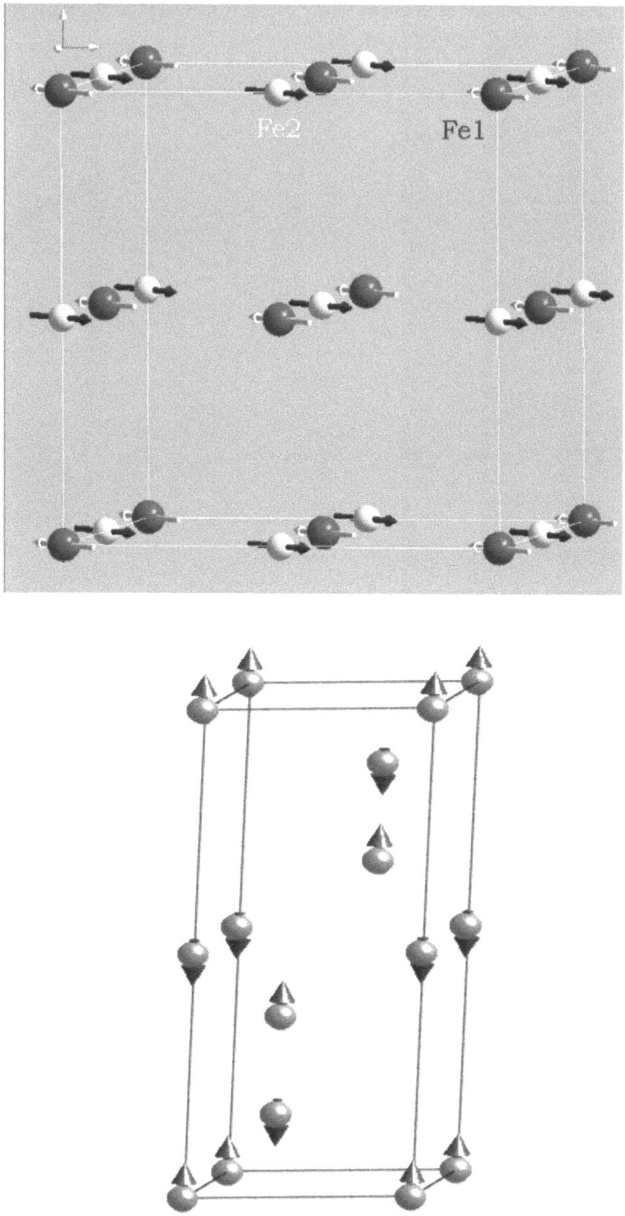

FIGURE 12 Magnetic structure of $Ba_3Fe_2WO_{9-\delta}$ ($\delta = 0.00$ (a), $\delta = 0.55$ (b)).

and this transformation is reversible. The oxygen deficient member $Ba_2FeMn_{0.5}W_{0.5}O_{5.72}$ exhibits a significant enhancement in its magnetic

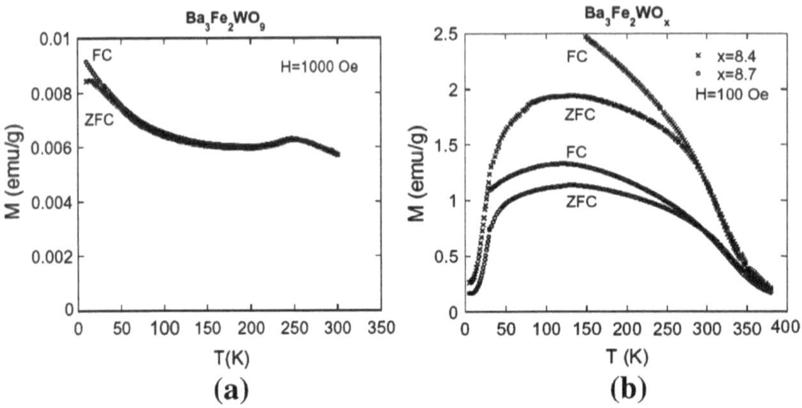

FIGURE 13 Magnetic properties of $Ba_3Fe_2WO_{9-\delta}$ ($\delta = 0.00$ (a), $\delta = 0.55$ (b)).

behavior compared to that of the oxygen stoichiometric parent composition [82].

Finally, these studies principally underline the importance of slight oxygen nonstoichiometry as an effective parameter for the preparation of new complex metal oxides with a perovskite structure having the specified structural and magnetic properties.

VII. ROOM-TEMPERATURE MULTIFERROIC $BIFEO_3$

$BiFeO_3$ is another well-known multiferroic material because it is one of the few multiferroics with both ferroelectricity and magnetism above room temperature. This perovskite is a model multiferroic material because it presents a number of properties making the compound more tractable including a simple chemical formula, exchange interactions only in the Fe^{3+} sites and the Bi sublattice with a stereo-chemically active $6s^2$ lone electron pair as a main origin of ferroelectric ordering. $BiFeO_3$ receives also noticeable attention due to its potential applications (spintronics, data storage microelectronics, etc.) as well as due to the fascinating physics behind its properties.

Since dozens of papers have been published on this compound we were unable to cite all of them and had to omit some references saving only the principal results and detailed reviews [83].

The rhombohedrally distorted perovskite structure can be indexed with $a = b = c = 5.633$ Å, $\alpha = \beta = \gamma = 59.4°$ and space group *R3c* at room temperature (see Figure 14), owing to the shift of Bi ions along the [111] direction and distortion of FeO6 octahedra surrounding the [111] axis [84, 85]. The displacement of Bi is larger than the displacements of Fe. These

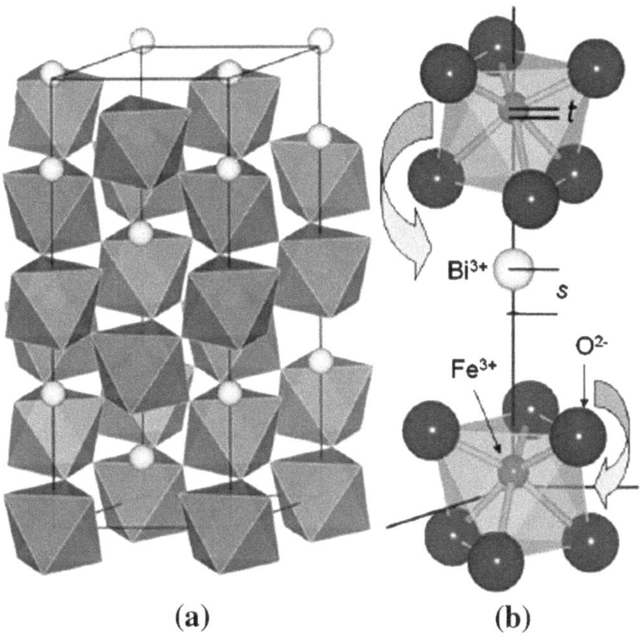

FIGURE 14 (a) Hexagonal unit cell of BiFeO3 displayed with Bi3+ cations and FeO6 octahedra. (b) Part of the hexagonal unit cell where the FeO6 octahedra rotate anitparallel about the [001] hexagonal axis.

displacements are cooperative, in the same direction, as parallel displacements are energetically strongly favored compared to antiparallel due to electrostatic repulsion between the cations. $BiFeO_3$ at ambient conditions possesses an antiferrodistortive rotation angle of ~12° and octahedral strain equal to 0.997. The polyhedral volumes of $BiFeO_3$ are $V_A = 51.587$ Å^3 for the BiO_{12} dodecahedron and $V_B = 10.913$ Å^3 ratio for FeO_6 octahedron, thus $V_A/V_B = 4.73$. The Fe–O–Fe angle is smaller than 180° due to the distortions of the unit cell compared to the ideal cubic structure. This angle affects the overlap of Fe^{3+} 3d orbitals and O 2p orbitals, and is thus important for the magnetic and electronic properties of $BiFeO_3$. The spontaneous polarization P_S is || [0 0 1] hex in $BiFeO_3$, and is due to the cooperative displacements of Bi and Fe cations along this axis. Recent measurements on high quality thin films, single crystals, and ceramics have shown that P_S is 90–100 μC/cm^2 along the polar axis [86]. Its ferroelectric order ($TC \sim 1103$ K) originates from the sterochemical activity of the Bi^{3+} lone electron pair [87]. The magnetic structure of $BiFeO_3$ is of *G*-type antiferromagnetism below $T_N \sim 643$ K [88], with a modulated cycloidal spin structure having a long periodicity of 62 nm [89] (see Figure 15). As for $BiFeO_3$ bulk ceramics, the synthesis of single-phase perovskite is quite complicated, on one hand, the kinetics of phase formation most frequently

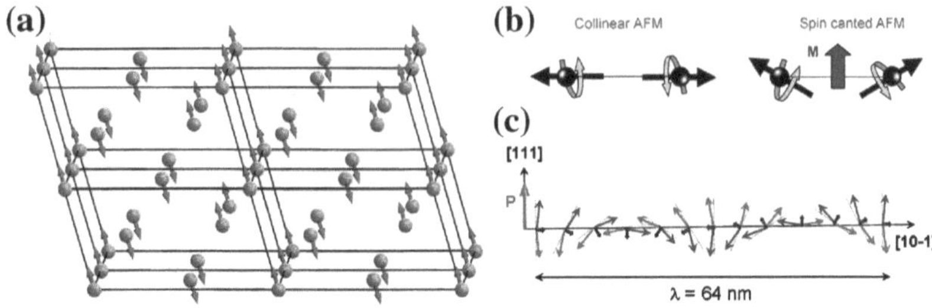

FIGURE 15 (a) G-type antiferromagnetic ordering. (b) Weak ferromagnetic moment induced by canted spin sublattices caused by the Dzyaloshinskii-Moriya interaction. (c) Cycloidal modulation of the spin structure, directions are given with pseudocubic indices. Adapted from references 110, 126 and 168 respectively.

leads to impurity phases, such as $Bi_2Fe_4O_9$, Bi_2O_3, and $Bi_{25}FeO_{39}$ at the grain boundaries of polycrystalline ceramics and, on the other hand, by the strong dependence of multiferroic properties on the oxygen stoichiometry and crystal perfection. Nitric acid leaching is applied to eliminate the impurity phases after the calcination of mixed bismuth and iron oxides. In bulk samples, leakage problems make the observation of intrinsic saturated ferroelectric hysteresis loops difficult. Leakage in $BiFeO_3$ ceramics is likely induced by the existence of Fe^{2+} and oxygen vacancies, and both are also detrimental for the intrinsic magnetic properties. The space group of the paraelectric phase has however been under debate. Structural transition sequence with temperature *R3c* → *bnm* → *Pm-3m* at 830 ± 5, 925 ± 5, and 940–970 °C was proposed in [90].

$BiFeO_3$ has a complicated magnetic configuration with antiferromagnetic spin order of G-type ordering (each Fe3+ with spin up is surrounded by six nearest neighbors with spin down) is not spatially homogenous but rather a spatially modulated structure [89], manifested by an incommensurate cycloid structure of a wavelength of $\lambda \sim 62$ nm, as shown in Figure 15. The spiral spin propagation wavevector q is along the [110] directions and the polarization is along the [111] directions. These two directions define the (1 2 1) cycloidal plane where the spin rotation occurs. Due to this feature, the antiferromagnetic vector is locked within the cycloid, averaged to zero over a scale $\sim\lambda$, and responsible for the very weak magnetization of bulk BiFeO3. Tilting of the FeO_6 octahedra reduces the Fe–O–Fe angle from 180°, reducing the overlap of Fe d and O 2p orbitals. At ambient temperature and pressure the Fe–O–Fe angle is 154–155°. Different ways of the cycloid degradation are known [91]. For example, nanodomain films can induce destruction of such incommensurate structure when the sizes of the epitaxial domains are smaller than the modulation length.

VIII. DOPING STRATEGIES FOR INCREASED PERFORMANCE IN $BIFEO_3$

In the last few years, attention has been given to study doped $BiFeO_3$ (both A-site and B-site doping) in an attempt to improve the current problems with high leakage currents, small remnant polarizations, high coercive fields, ferroelectric reliability, and inhomogeneous magnetic spin structures. The introduction of doping cations into $BiFeO_3$ favors separation of the perovskite phase from impurity phases. There are different types of doping including isovalent doping with magnetic and non-magnetic cations in the magnetic Fe sublattice, isovalent doping in the Bi sublattice and the so-called self-doping $Bi_{1-x}FeO_{3-1.5x}$. In all indicated cases, the oxidation state of iron is kept at +3.

Spontaneous magnetization was revealed both for the $Bi_{1-x}RE_xFeO_3$ (RE = Nd, Sm, Tb) solid solutions doped by magnetically active rare-earth cations [92–94] and for the diamagnetically substituted $Bi_{1-x}A_xFeO_3$ (A = La, Ba, Pb) compounds [95–98]. It was found that the value of net magnetization depends strongly on the kind of the diamagnetic substituting element. For instance, the spontaneous magnetization of the Ba-doped samples is at least an order of magnitude greater than that observed for the La-doped solid solutions at the same doping concentration [95,96]. While the dependency between a kind of the substituting element and the magnetization value might be natural for the rare-earth substitution, which can induce additional magnetic contribution at sufficiently high concentrations (at low temperature), the nature of correlation between the kind of a diamagnetic dopant and magnetic properties of samples is not quite clear.

The diamagnetic A-site substitution with the biggest ionic radius ions leads to the effective suppression of the spiral spin configuration, thus resulting in the appearance of a net magnetization. Although there are a number of studies on B-site alloyed $BiFeO_3$, greater attention has been given to A-site alloyed phases. The most widely studied dopants are materials from the lanthanide series – especially La, Dy, Gd, etc. The studies probed the effect of La-alloying on the magnetic structure of $BiFeO_3$ and showed that the spin-modulated structure disappeared in single crystals with only 20% addition of La. Other studies have also investigated Ba, Sr , Ca, and Pb-doping, and many others. It should be noted that there are numerous studies of A-site alloying, too many to be covered thoroughly here.

In order to enhance the ferromagnetism and ferroelectricity in $BiFeO_3$, researchers focused on a theoretical prediction [99,100] that Bi_2FeCrO_6 would exhibit huge macroscopic magnetization and polarization, due to the ferromagnetic superexchange interaction between Fe and Cr ions which induces the ferromagnetic state in La_2FeCrO_6 [101,102]. However, it is challenging to synthesize materials with ordered Fe and Cr ions. Meanwhile, compared with pure $BiFeO_3$, samples with disordered Fe/Cr configuration showed no significant improvement of the multiferroicity.

IX. BI PEROVSKITES ($BIMO_3$)

Although rare, there are several compounds that are known to exhibit magnetism and ferroelectricity and some even exhibit a coupling between the two. This coupling is often manifested as a change in the dielectric constant at the magnetic ordering temperature or a change in magnetization at the ferroelectric Curie temperature. To illustrate the role of Fe^{3+} in $BiFeO_3$, the structures and properties of closely related Bi-based non-ferrite perovskites are presented here. Selected properties and structures of Bi-based perovskites are given in Table 2 [103–109]. Understanding of each compound requires a special individual approach. It should be noted that all the other simple $BiMO_3$ materials are metastable at ambient conditions, hence they must be prepared by high-pressure synthesis. $BiFeO_3$ is also the only simple compound exhibiting multiferroism at ambient temperature.

For example, $BiMnO_3$ is both ferromagnetic and ferroelectric [104]. This compound is still the most controversial among $BiMO_3$ materials despite a remarkable number of research activities devoted to it. It is a high pressure phase that is synthesized at 4 GPa. At room temperature it has a monoclinic structure (space group C2/c), which changes to orthorhombic (e.g. Pbnm) at the ferroelectric transition temperature of ~760 K. The magnetic properties of $BiMnO_3$ are well established and it is a ferromagnet where a ferromagnetic ordering occurs below 102 K and the material has a saturation magnetization of 3.6 μ_B per formula unit. Multiferroic behavior had been observed at ~80 K with a magnetization of ~1 μ_B per Mn atom and a polarization of <0.15 $\mu C/cm^2$. A change of ~0.6% in the dielectric constant was recorded at the ferromagnetic temperature under the application of an applied magnetic field of 8 T. It is postulated that the Bi ions play a role is stabilizing the ferromagnetism and inducing the ferroelectric distortion.

However, there are still unresolved and puzzling problems concerning ferroelectric properties of $BiMnO_3$. Experimental results [110] showed that a pure stoichiometric phase of $BiMnO_3$ crystallizes in the centrosymmetric C2/c structure but some thin-film samples demonstrate ferroelectric hysteresis loops and have different magnetic properties (reduced T_C and saturated magnetization) from those of the bulk material. This difference may be related with a dissimilar chemical composition of these compounds and needs a detailed characterization of stoichiometry.

X. $AFEO_3$, A = Y, LA, PR, … , YB, LU

To illustrate the significance of Bi^{3+} for $BiFeO_3$, the structure and properties of closely related ferrite perovskites are shortly summarised. These perovskites are known as orthoferrites with the centrosymmetric orthorhombic space group

Pbnm at ambient conditions, illustrating the importance of the $6s^2$ lone pair of Bi^{3+} for stabilizing $BiFeO_3$ in the polar structure. The orthoferrites are also G-type antiferromagnets with high T_N of 350–467 °C. Rare earth Ln cations with unpaired *f* electrons have the possibility of magnetic exchange striction between Ln 4f orbitals and Fe 3d orbitals, which can induce a small polarization. However, the magnetic ordering temperature of Ln is very low, 3.5 K in Dy in $DyFeO_3$, where a magnetic field can induce polarization below 3.5 K. A similar effect has been observed in GdFeO3, but here the small polarization of 0.12 $\mu C/cm^2$ at 2 K is stable even in zero magnetic field.

XI. SOLID SOLUTIONS OF $BIFEO_3$–ABO_3

A few solid solution systems of $BiFeO_3$ with ferroelectric materials have been extensively studied [111–116], the earliest being $BiFeO_3$–$PbTiO_3$ [13,15]. From a fundamental point of view, simultaneous substitution on both the A-site and B-site can hinder interpretations of the significance of Bi^{3+} and Fe^{3+}, thus substitution only on either the A- or B-site is a better proxy for this purpose. Optimizations of functional properties has however been demonstrated, e.g., in $(BiFeO_3)_{1-x}(Pb_{1-y}La_yZr_{1-z}Ti_zO_3)_x$ where simultaneous weak ferromagnetism and low ferroelectric polarization was reported to occur at room temperature. Coexistence of ferroelectricity of ferromagnetism at room temperature, albeit weak, has been reported for $(BiFeO_3)_{0.3}(BaTiO_3)_{0.7}$ and $(BiFeO_3)_{0.8}(BaTiO_3)_{0.2}$ ceramic materials with monoclinic *Cm* structure exhibiting spontaneous polarization and magnetization at room temperature. The crystal structure, morphotropic phase boundary and multiferroic properties of $(BiFeO_3)_{1-x}(PbTiO_3)_x$ and $(BiFeO_3)_{1-x}(PbZrO_3)_x$ solid solutions have also recently been studied [116–118].

XII. EXPERIMENTAL TECHNIQUES

The following section presents a brief survey of the major preparation methods and important experimental techniques which are generally applicable to the study of multiferroics – generally applicable in the sense that meaningful results can be obtained from almost any ferroelectromagnets. Some of the main advantages and disadvantages of each technique, distinctive merits, headaches, and limitations are outlined, although no attempt is made to provide a detailed discussion of the various techniques. This section is presented only to summarize these methods, and the reader should consult the individual reviews specifically addressing the various experimental methods for details. References are cited in order to provide a starting point in the pertinent literature. They are not exhaustive and are not intended as a complete bibliography.

In this part of chapter, some chemical methods which were used for the preparation of $BiFeO_3$ are described. Sample preparation part is not a general review but rather a recipe how to prepare the $BiFeO_3$ as a ceramic material. Although especially the details of sample preparation techniques are different from material to material, the methods described as follows are rather fundamental and quite generally applicable to a variety of multiferroic complex oxides.

XIII. SAMPLE PREPARATION

While the majority of today's investigations focus on the physics of multiferroics, their study and exploitation would not be possible without the vital contribution of ceramic science and sample preparation engineering, including phase equilibria, sintering mechanisms, microstructure control, and defect chemistry. Great strides have been made in the preparation of high quality complex metal oxide multiferroics in the last few decades. Much of the performed work fought the struggle against impurity and parasitic phases and careful control of the molar ratios of the starting materials, heating and cooling rates and gaseous environment. Here we will shortly describe the key features and advances in each method of synthesis [40,119–121].

The preparation of oxide multiferroics can be divided mainly into two categories of methods: the solid state method and wet-chemistry methods including sol–gel and citrate gel methods (see Figures 16 and 17). In the first, starting ingredients are in a powder form and these powders are thoroughly ground and mixed before heat treatments. The ideal starting point for solid state reaction synthesis is a homogenous mixture of small binary oxide particles. Several approaches to mix the binary oxides prior to firing were explored. The most successful with respect to time consumption, control of stoichiometry and reproducibility in terms of phase purity of the fired product was to mix the binary oxides in ethanol using an Agate mortar and a pestle. Ball milling in ethanol with yttria stabilised zirconia balls was less successful with respect to obtaining phase pure samples after firing. Ball milling inevitably introduces impurities to the system, which destabilise the perovskite phase. Complete separation of the milled powder from the zirconia balls also proved difficult, hindering control of the nominal stoichiometry.

Wet chemistry methods differ from the solid state methods in that the starting materials are mixed to form a solution, which is then dried before the heat treatments. The recipes can vary in both methods and especially the heat treatment temperature, duration, and gas flow are parameters that vary greatly. The purity and the quality of obtained multiferroic strongly depend on those variables. Usually used ingredients in the solid state method are oxides or carconates, but other starting compounds have also been used. For wet-chemistry methods, the nitric-acid-based method and citric-acid-based method

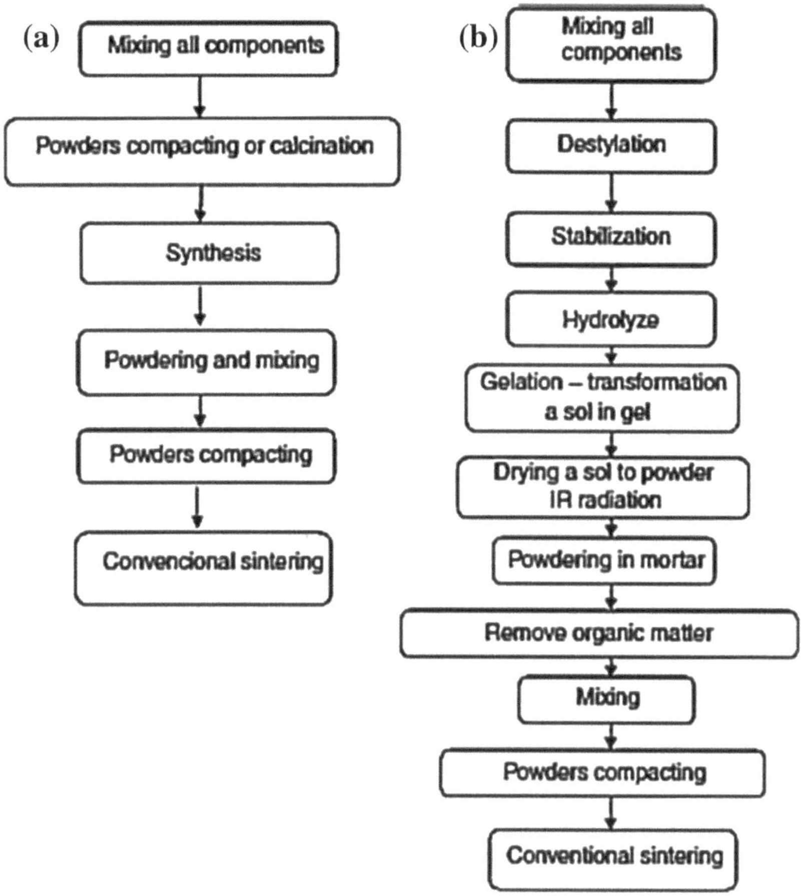

FIGURE 16 A general diagram of the mutliferroic ceramic production: a) the conventional method, b) the sol–gel method.

are used. Stoichiometric amounts of starting ingredients are mixed with either of the acids forming a homogeneous solution. There are several differences between these two types of preparation methods. The sol–gel method is faster than the solid state method, although in principle solid state method seems simpler. The thorough grounding at the beginning can already take several days. The sol–gel method on the other hand is more complex with mixing all the ingredients. The advantages of the wet chemistry methods are that the solution is homogenously mixed on the atomic level and the multiferroic composition is already preformed and cation order is developed, even before the calcination and sintering takes place (comparison to the solid state method, when disorder is much more likely to occur). The solid state method requires higher

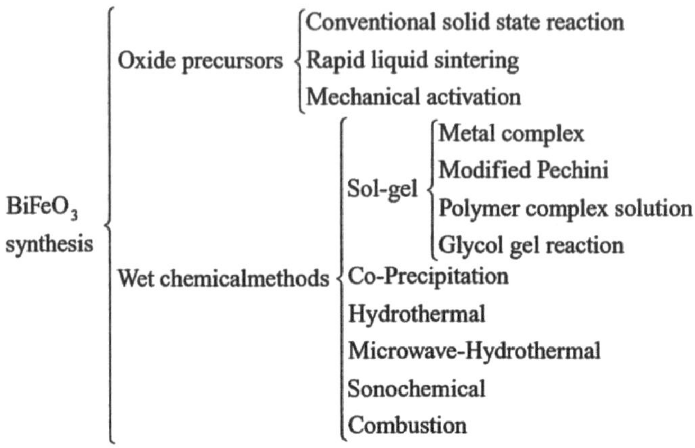

FIGURE 17 A diagram showing the preparation techniques used for multiferroic $BiFeO_3$.

calcination and sintering temperatures to allow interdiffusion of the cations, as well as a longer sintering time than wet-chemistry methods. Solid state reaction synthesis of phase pure perovskite-based multiferroics is relatively difficult, and very often with poor reproducibility, due to the sensitivity towards impurities, local and nominal off-stoichiometry and the intermediate temperature region of metastability. May be the most important difference is in the resulting multiferroic powder itself. With the solid state method it is impossible to reach such a small grain size which is formed inevitably with the sol–gel method. From a solid state reaction the typical obtained grain size is up to several micrometers. With the more controlled wet-chemistry methods the grain size stays on the nanometer scale, which becomes an important factor when considering for example thin film ablation.

In the literature, we can find useful reviews about the preparation methods of multiferroic perovskites [40,119–123] with a special attention to obtaining and sintering of bulk ceramic and thin films.

Most of multiferroics have been prepared in bulk form (such as single crystal, ceramic, or powder) and thin film. However, for the bulk samples, single perovskite phase is hard to obtain and their leakage problems lead to low resistivity, presumably due to defect and nonstoichiometry related issues. Although great strides have been made in the production of high-quality ceramics or powders, much of the research on multiferroics has been driven by progress in understanding the structure, properties, and growth of thin films. No discussion of multiferroics would be complete without reference to the topic of thin films, which today represents approximately 40% of all publications on multiferroics. Current research in multiferroic films focuses on nanoscale effects and heterostructures in which multilayers of alternating compositions are built-up layer by layer on the scale of a few unit cells in order to create novel materials through coupling of ferroelectric and ferromagnetic orderings. Recently, high-quality thin-film growth techniques have opened the door to the design of practical devices based on magnetoelectric coupling. The thin-film growth have been prepared, in some cases through high-pressure route, by pulsed laser deposition (PLD), liquid-phase epitaxy, sol–gel methodology and chemical solution deposition. [40,121]. These syntheses for thin film multiferroic materials all involve multi-steps in various high temperatures or specific environment. As epitaxy is of prime importance in heterostructure studies, sol–gel techniques are not suited and vacuum techniques are employed almost universally. However, for technical applications requiring thick layers, sol–gel methods (which can be realized in almost any laboratory equipped with glassware and a fume cupboard) can offer a significantly more rapid process rate than the vacuum techniques. Additional details concerning these techniques can be found in [40,119–123].

To optimize conditions of a new multiferroic production, a thermal analysis method is used. Recorded DTA, TG, and DTG curves were subjected to a detailed analysis and interpretation. By combining the information obtained in such a way with material test results (XRPD, EDS, and SEM), the optimum temperatures of synthesis, sintering, crystallization, and temperature of occurrence of other effects, accompanying the production process of the ceramic multiferroics were determined. During powder calcination and sintering of compacts of a complex chemical composition, mass changes in the specimen often occur. An analysis of the TG curves recorded enables to separate the mass losses improving properties of the ceramic materials (e.g., water loss, burning redundant organic parts, etc.) from those losses which are responsible for worsening of physical properties of those materials (e.g., formation of cation and/or anion vacancies). While testing the multiferroic materials, the thermal analysis methods play an important role, provided they are supplemented by results obtained by other methods.

Perovskite-type multiferroics are traditionally prepared through the solid-state reaction of single oxides at high temperature or by mechanical activation at lower temperatures.

Recently, wet-chemical methods have received abundant attention. Typical techniques in this family are: (a) sol gel, which includes metal complex, glycol reaction, Pechini method and polymeric precursor; (b) hydrothermal; (c) sonochemical; (d) co-precipitation, and (e) combustion synthesis. Figure gives a bird's eye view of the mentioned preparation techniques.

A. Conventional Solid State Reaction

The traditional solid-state reaction route for the synthesis of multiferroics consists of mixing by milling the appropriate metal oxides or carbonates and then calcinating at high temperature to allow interdiffusion of the cations. The chemical reaction occurs by solid-state diffusion of the ions, which is characterized by a slow kinetic rate. The degrees of mixing and powder particle sizes are of particular importance. The reaction starts at the points of contact between the initial components and continues successively by ionic interdiffusion through the product phase $BiFeO_3$. The initial reaction is rapid because of short diffusion paths, but further reaction proceeds slower. In many cases, a number of unwanted phases appear. This traditional method is widely used: it is relatively inexpensive, simple, and it leads to acceptable results. Some drawbacks that may be identified are poor sintering behavior, non-uniformity of particle size and shape, lack of reproducibility, multiphase character, and loss of stoichiometry.

B. Rapid liquid sintering

In rapid liquid sintering, the calcination stage is skipped. Initial steps of the process are the same as in the solid state route: high-purity initial powders are thoroughly mixed, dried, and pressed. The samples are sintered at the selected high temperature above the melting point, which is considered a high-heating rate and a short-sintering period. The appearance of the liquid phase seems to play an important role during the synthesizing reaction. The liquid phase accelerates the synthesis, facilitates the sintering, and probably prevents the formation of a second phase.

C. Mechanical activation

The reactivity of the starting materials can be significantly improved by mechanical activation.

This way, calcination can be completed at relatively low temperatures.The initial oxides are mixed and subsequently milled up to 100 h in a high-energy planetary ball mill. Milling leads to atomic scale mixing of the ingredients, thus forming a homogenous amorphous mixture. With this method, calcination temperature is significantly lower.

D. Sol–gel

(i) Metal complex

In the metal complex method, a homogeneous solution is sought as well as a reduction in the motion of metals. This is achieved by metal-chelating. The most commonly used chelating agents are citric, tartaric, oxalic, and malic acids. Once the solvents evaporate, the resulting powder has an amorphous phase. The calcination step allows the release of organics and formation of the desired phase.

(ii) Modified pechini method

The fundamental idea behind the modified Pechini methods to reduce the mobility of different metal ions, which can be achieved by encircling stable metal-chelate complexes steadily by a growing polymer net. Immobilization of metalchelate complexes in such a rigid organic polymer net can reduce segregations of particular metals during the decomposition process of the polymer at high temperatures. The basic chemistry of this process is the dehydration reaction of a carboxylic acid and an alcohol; i.e., esterification. The resulting product of the ester contains one alcoholic hydroxyl group in its left end and one carboxylic acid group in its right end, so that the former can react with another di-carboxylic acid and the latter with another di-alcohol to form a tetramer. This type of reaction occurs in sequence, leading to a polymer molecule. In fact, the original patent for the Pechini process describes the formation of a polymeric resin produced through polyesterification between metal chelate complexes using hydroxycarboxylic acids (citric acid and malic acid) and a polyhydroxy alcohol such as ethylene glycol. Citric acid and ethylene glycol are most widely employed in the Pechini process.

This method leads to a more homogeneous mixing and to more reactive forms of the starting reagents, minimising phase segregation comparing to the traditional solid-state reaction.

One of the disadvantages of the Pechini method is the large weight loss produced on firing the precursors, due, basically, to the decomposition and combustion of the organic reagents used to prepare the polymeric resins.

(iii) Polymer complex solution (PCS)

Unlike the Pechini method, this method does not involve complex polymeric polyesterification reactions. Polyvinyl alcohol (PVA) is one of the most water-soluble polymers used in this method, and it acts as a coordinating organic polymer to cations. This action strongly modifies the rheological characteristics of the initial aqueous precursor solution. Metal ions are capable of playing a role as cross-linking agents between polymers. Random cross-linking across polymer chains entrap the water, allowing the growth of three-dimensional networks, so the system can turn into a gel.

E. Co-precipitation method

The co-precipitation process involves the separation of a solid containing various ionic species from a solution phase. An ideal situation is that a quantitative and simultaneous precipitation of all the cations occurs (without segregation of constituents in the precipitates) to form a completely mixed-metal precursor. However, this is a very rare situation in most of the solutions containing more than one metal ion. Homogeneous co-precipitation with respect to metal ions at an atomic level is almost impossible in solution precursor oxides with two or three metal ions. This problem can be reduced by introducing precipitating agents that render the cations insoluble.

F. Hydrothermal synthesis

Hydrothermal chemistry has been extensively developed for the synthesis of multiferroics which are difficult to obtain by high temperature solid state reactions. In particular, hydrated materials are mainly prepared through processes involving such chemistry, since the reactivity increases under high-pressure conditions and at moderate temperatures. In using a solvent well above its boiling point, the solvent is heated in a sealed vessel (autoclave, bomb, etc.), so that the autogenous pressure far exceeds the ambient pressure. This automatically raises the effective boiling point of the solvent. In the special case of the solvent being water, the technique is often called hydrothermal.

G. Sonochemical

Sonochemical rates for homogeneous reactions depend on a variety of experimental parameters such as vapor pressure of precursors, solvent vapor pressure, and ambient gas used. To achieve high sonochemical yields, the precursors should be relatively volatile, because the primary sonochemical reaction site is the vapor inside the cavitating bubbles. In addition, the solvent vapor pressure should be low at the sonication temperature, because significant solvent vapor inside the bubble reduces the bubble collapse efficiency. Powders are prepared in solution under ultrasonic vibration. A mixed aqueous solutions of metal nitrates is sonicated until the precipitation is complete. A small amount of decalin is used for proper power transfer and sodium dodecyl sulphate surfactant is also used for preventing agglomeration. The precipitated powder is collected, washed in alcohol, and dried in a vacuum oven at 40 °C.

H. Combustion method

The term combustion covers smouldering reactions (heterogeneous), flaming reactions (involving a homogeneous gas phase), and explosive reactions. The rate of combustion determines the nature of reaction, i.e., decomposition, deflagration, or detonation. Synthesis of ceramic oxides is achieved by smouldering and flaming reactions.

XIV. SAMPLE CHARACTERIZATION

A. X-ray diffraction measurements

X-ray powder diffraction (XRPD) is the most common convenient method for characterizing the crystallinity and the phase purity of sintered samples. Formation of single phase are monitoring by recording XPXD patterns of the polycrystalline samples at each stage of the chemical reaction.

The crystal structure of prepared samples may be studied by the Rietveld refinement method of XRPD patterns.

B. Thermal analysis

Thermogravimetry analysis (TGA) is usually used in order to determine the oxygen nonstoichiometry, The oxygen off-stoichiometry can be determined by using TG analyzer from the weight gain on heating the sample. The stoichiometry of cation composition can be confirmed by the inductively coupled plasma atomic emission spectrometer.

Differential thermal analysis (DTA) is usually used for the monitoring of chemical reactions during sample preparation. First order phase transitions can be determined as the onsets of the thermal events. Specific heat data may be used for a detail study of AFM transitions. Néel temperatures are determined as the temperatures of the maximum value of the calorimetric peaks.

C. Dielectric measurements

The dielectric properties of ceramic samples are investigated over the frequency range of 1×10^0–1×10^6 Hz using an impedance analyzer with some applied voltage and Ag, Pt, or Au sputtered electrodes.

D. Magnetic measurements

The temperature and field dependence of the magnetization are usually studied with a Quantum Design SQUID magnetometer. Temperature dependent magnetization data are recorded at various applied magnetic fields (H) in the temperature range $2 < T < 400$ K after cooling the sample in zero magnetic field (zero field cooled, ZFC) and also while cooling in the presence of the field (field-cooled, FC). The deviation of ZFC and FC data from each other may be an indicator of a magnetic transition.

E. Magnetic structures and their determination

Neutron scattering is a powerful and ideal technique to study both changes in the crystallographic and magnetic structures as well as elementary excitations in magnetic materials. Neutrons have zero charge and non-zero magnetic moment, and are scattered by the magnetic moments of atoms in a sample. The cross section for magnetic scattering is sensitive to the relative orientation of the neutron magnetic moment, the atomic magnetic moment, and the scattering vector. The most straightforward application of this technique is in the investigation of spin configurations and the measurement of the magnetizations of different sublattices within a given material. The most significant questions often concern the orientations of the individual moments, i.e., the spin configuration. Neutron diffraction gives the most definitive answers to such questions, and can usually do so from powdered samples. The complete determination of some magnetic structure involves determination of the magnetic phase transition, the propagation vector, the magnetic moment directions and the magnitudes of the moments. T_N is almost always known from bulk measurement. The determination of the propagation vector could be in fact the most difficult part of the magnetic structure determination. The principal difficulty, apart from the availability of reactors, is the amount of sample required, from 3 to 5 g being desirable for best results with powders. Smaller quantities can be studied, but with lower signal-to-noise discrimination.

There are two basic types of magnetic structures: antiferromagnetic (AFM, for each magnetic moment there exists exactly one that has the same magnitude but opposite direction) and ferromagnetic (FM, having non-zero total magnetization). However, there are many different geometrical ways to arrange individual magnetic moments leading to a large variety of different magnetic structures [124]. Some simple examples are shown in Figure 18. As many of the recently identified multiferroics exhibiting simultaneous magnetic and ferroelectric order are incommensurate magnets, a correct determination of incommensurate magnetic structure is crucial for determining whether or not ferroelectric order is allowed [125]. Detail description of modern powerful formalism in terms of irreducible representations for determination of the incommensurate magnetic structure symmetry in multiferroics with non-trivial unit cells was proposed in [126].

XV. CRYSTAL CHEMISTRY SEARCH OF MULTIFERROICS

Multiferroics comprise a rather sparse class of materials where there exists a limitation on the coexistence of electric and magnetic orderings explained mainly by the conclusion that d electrons of transition metals, which serve as a source of magnetism, reduce the possibility of the existence of ferroelectric distortions resulting into the loss of the symmetry center. The following phenomena are considered as an additional electronic or structural support of a simultaneous realization of ferromagnetic and ferroelectric transitions in a single phase: charge ordering in magnetic oxides, stereochemical activity of electron lone pairs, and the Jahn–Teller distortion. In search of new potential

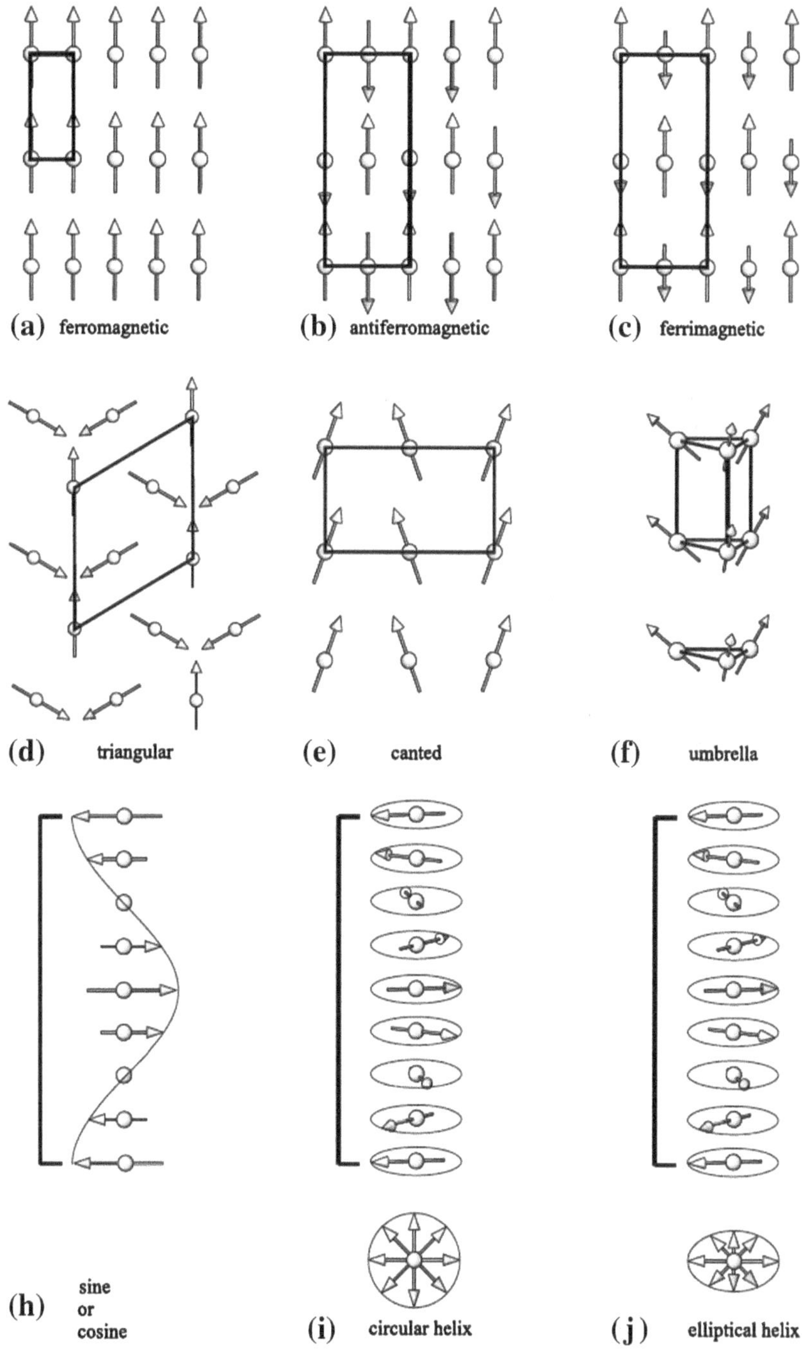

FIGURE 18 Some different types of magnetic structures.

multiferroic compositions, it is important to know how and why both the spin and the dipole ordering emerge in a single phase. A correct solution of the problem of emerging of magnetoelectric ordering can be attained with using quantum chemistry methods based on the "first-principles". The accuracy and correctness of quantum chemistry calculations of the crystal structure, its stability, and properties increase continuously. However, there are still serious difficulties in prediction of the structure and properties from "first-principles". As magnetoelectric coupling is determined by the structure and magnetic symmetry of a crystal, small modifications might alter, eliminate, or allow magnetoelectric effects.

The crystal chemistry does not have answers to the above questions, however, it is capable to determine under which crystal chemical condition this might happen and find the compounds with these conditions fulfilled. Important research in this direction was made in [127] where formulation of the criteria and approaches to the search of novel multiferroics was proposed and search of novel potential multiferroics in the Inorganic Crystal Structure Database (ICSD) among the stoichiometric compounds containing a magnetic cation and a cation with the stereochemically active lone pair of electrons Tl^{1+}, Sn^{2+}, Pb^{2+}, Sb^{3+}, Bi^{3+}, Se^{4+}, or Te^{4+} was performed using the calculation of magnetic interaction parameters by the crystal chemistry method [128]. As a result, new potential multiferroics with a lone pair were proposed.

XVI. CRYSTAL CHEMISTRY CRITERIA FOR THE SEARCH OF MULTIFERROIC COMPLEX METAL OXIDES

The development of the theory of multiferroics allowed understanding and in some cases successful prediction of the ferroelectric and magnetic properties of single-phase compounds, which significantly facilitates the search and design of the new materials with spin and dipole ordering. Approaching the problem of a new promising multiferroics, different researchers can use available information about these materials in different ways: human imagination does not have limits. But a crystallochemical approach for a productive design of new multiferroic compositions, however, is more or less obvious. The main problem of this approach is that one should know what type of crystal structures to select. The presence of special structural building fragments permitting polar ionic shifts is mandatory. But the total number of non-centrosymmetrical structures among a complex metal oxides is strongly restricted and comprises less than 15%. In order to avoid magnetic dilution, paramagnetic cations responsible for both cation displacements and magnetic coupling are very desirable. The structure of MF compounds must provide magnetic interaction pathways between the magnetic cations, with direct exchange or superexchange interaction, as found more frequently in complex metal oxides. As concerns the perovskite family of magnetically ordered ferroelectrics, antiferromagnetic ferroelectrics seem to be more numerous than ferromagnetic ones.

From the crystal chemistry point of view, the common feature of the emergence of both the electrical polarization and the magnetic ordering is the ionic displacements. The ferroelectrics are classified into proper and improper ones depending on the nature of the initially symmetrical crystal structure. In the former case a phase transition into polar phase of the crystal occurs either due to small ionic shifts with a formation of partially covalent bonds (as in $BaTiO_3$) or upon the loss of centrosymmetrical position of lone electron pair (as in $BiFeO_3$). Phase transitions in improper ferroelectrics are related with more complicated lattice distortions when the spontaneous polarization is exhibited as a secondary effect due to the specific magnetic ordering, spontaneous lattice distortion or electron correlations. Thus, the magnetically dependent (the polarization directly results from a definite spin structure) and magnetically independent (the magnetic sublattice exists independently of ferroelectric ordering) types of dipole ordering are distinguished in multiferroic materials. By establishing the relation between the crystal chemistry parameters and the emerging of the magnetically induced ferroelectricity in known multiferroics $TbMnO_3$, $TbMn_2O_5$, and $BiMn_2O_5$, three necessary conditions whose structural availability can produce magnetoelectric ordering under the effect of an external magnetic field were obtained. First, the existence of a competition of magnetic interactions on the specific geometric fragments in the paramagnetic phase. Second, the presence of intermediate ions in critical positions displacements from which might cause reorientation of the magnetic interactions between spins resulting in magnetic ordering. Third, the displacements accompanying the magnetic ordering should be a polar (like in $TbMnO_3$) or create necessary conditions for emerging of polar displacements (for example, the charge disordering, like in $TbMn_2O_5$ and $BiMn_2O_5$) with preservation of the magnetic ordering. The fulfillment of the three above conditions substantially reduces the probability of the multiferroics existence and explains, from the crystal chemistry point, "why are there so few magnetic ferroelectrics" [35].

The crystal chemistry studies of the room-temperature multiferroic $BiFeO_3$ enabled us to correct the criteria of search of novel multiferroics having high temperatures of electrical and magnetic ordering. The main search criteria are the presence of the electrical polarization and the structurally conditioned magnetic ordering in the compounds of interest at room temperature. The existence of the competition of magnetic interactions is acceptable, if it can be eliminated by reorientation of the spins of the competing interactions (AFM–FM transition) through displacements of intermediate ions in critical positions. In the latter case, under the magnetic field effect, there is the possibility of the phase transition from the state with competing magnetic interactions to a uniform magnetic structure with preservation of the electrical ordering and increase of the electrical polarization due to emerging of the linear magnetoelectric effect.

The search should be performed among the compounds with a stereochemically active lone pair or a magnetic ion with a directed distortion of the coordination polyhedron which can be induced by the Jahn–Teller effect or other reasons. Besides, it is possible to use the compounds containing the magnetic ion with an unstable (non-rigid) coordination surrounding.

XVII. SINGLE PHASE MULTIFERROIC MATERIALS

This subchapter provides a short overview of the vast and rapidly increasing number of potential multiferroic among single-phase complex metal oxides. As many questions still remain open, it is safe to claim that the field is immature. The peculiarities of crystal and magnetic structure are considering, the consideration is based on the classification in terms of chemical composition, crystal structure type, and mechanism of ferroelectric ordering together with its connection to the magnetic subsystem (magnetically dependent or independent types of electric ordering) [91].

The current key question in materials science of multiferroics is what kinds of other materials we can also expect in spin and dipole ordering. With a few exceptions, the multiferroics materials that have been investigated are transition metal oxides. It will be interesting to see how different types of magnetic and electric interactions found in new kinds of inorganic compounds affect the magnetic and ferroelectric properties and the magnetoelectric coupling.

A. Multiferroics with charge ordering

Several multiferroics with an electronic origin of polarization and strong correlation effects are ferrites RFe_2O_4 (R = lantanides, In) [129] with a rhombohedral (*R3m*) structure which consists of Fe_2O_2 and RO_2 layers alternating along the c axis with the disposition of cations in the vertices of the triangles [130] (see Figure 19). Iron cations with different valence state (Fe^{2+} and Fe^{3+}) in Fe_2O_2 occupy positions in the trigonal planar structure with the same probabilities that lead to charge frustration. This charge ordering induces polarization.

Another example of multiferroic with the polarization caused by charge ordering is magnetite Fe_3O_4 with an inverse spinel structure and a different valence state of iron [131] but there is still no definite opinion about the origin of spontaneous polarization in this compound.

Doping of lanthanide manganites by divalent cations (Ca, Sr, Ba, Pb) leads to an appearance of different valence state of Mn cations (Mn^{3+} and Mn^{4+}) and the possibility of emergence of different types of ordering among which the charge ordering plays the main role [132]. The appearance of polarization in these crystals was predicted theoretically [133] but additional studies are requested for confirmation that the ferroelectric ordering exists.

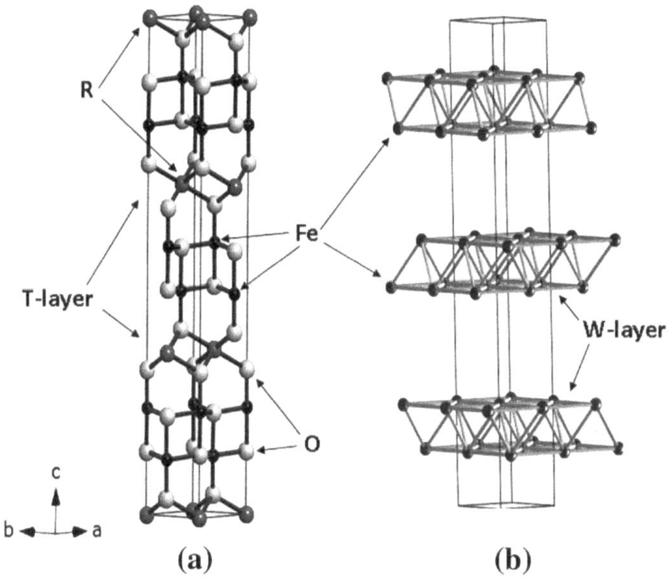

FIGURE 19 Crystal structure of RFe2O4 showing the R, O and Fe triangular layers stacked along the c-axis. (b) The W layer: Fe2+/Fe3+ honeycomb.

B. Spin-frustrated multiferroics

The vanadates $A_3V_2O_8$ (A = Ni, Co, Mn, Cu) have orthorhombic crystal structure (space group *Cmca*) which consists of layers of $A^{2+}O_6$ octahedra perpendicular to the *b* axis and separated by non-magnetic $V^{5+}O_4$ tetrahedra (Figure 20). Magnetic cations are arranged in two crystallograpically non-equivalent positions, either at the common vertices that connect the triangles occurring in one plane (A1) or at the bases of the triangles that lie in different planes (A2) [134]. $Ni_3V_2O_8$ is an well-known example of spin frustrated magnet which undergoes a series of competing magnetic ordering at low temperature [135]. Magnetic lattice is based on an anisotropic staircase Kagome net with magnetic cation Ni^{2+}. Below 12 K, $Ni_3V_2O_8$ undergoes four distinct magnetic phase transitions. The spin frustration in $Ni_3V_2O_8$ gives rise to high- and low-temperature incommensurate magnetic spin orderings, which is stable below 9.1 and 6.3 K, respectively.

Hexagonal $YMnO_3$ belongs to a unique family of lanthanide manganites RMnO3 that exhibit multiferroic properties in conjunction with two-dimensional frustrated magnetism on triangular lattice (Figure 21). Ferroelectric and antiferromagnetic phase transitions are around 900 and 65 K, respectively [136]. Tilting of MnO_5 bipyramids and buckling of Y layers are responsible for multiferroelectricity in $YMnO_3$. Magnetic moments of Mn^{3+}cations which are

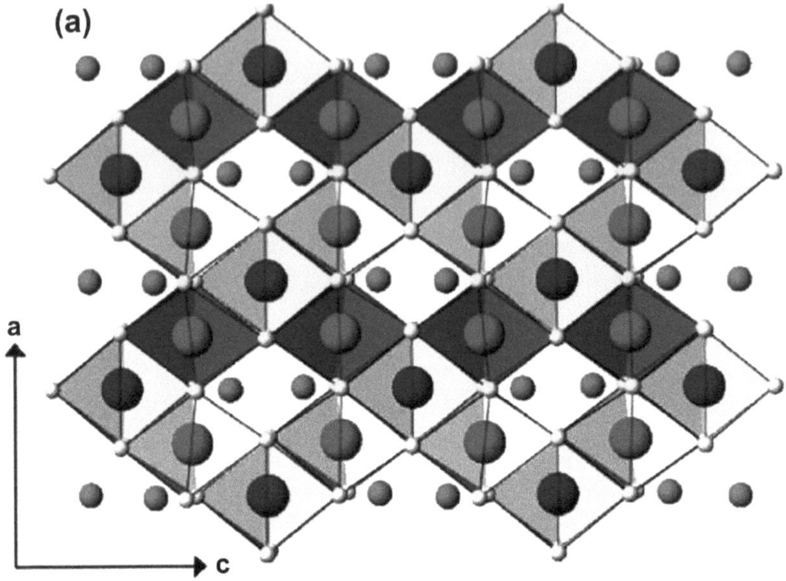

FIGURE 20 $Ni_3V_2O_8$ crystal structure viewed (a) along nearly the **b** axis. The spine Ni^{2+} sites are shown in red, the cross-tie Ni^{2+} sites are shown in blue and the V^{5+} are shown in pink. Also shown are the oxygen octahedra of the spine sites in light and dark blue, and those of the spine sites in yellow. The oxygen tetrahedra around the V^{5+} are shown in pink.

aligned in *ab*-plane, are strongly frustrated below T_N. The microscopic origin of magnetoelectric interactions has remained controversial and an influence of magnetoelastic coupling needs more exact specification [137].

The compound $TbMn_2O_5$ has also shown some interesting multiferroic behavior [138]. This compound is orthorhombic (space group Pbam) at room temperature and contains both Mn^{4+} and Mn^{3+} ions (Figure 22). The Mn^{4+} ions are octahedrally coordinated by oxygen, whereas the Mn^{3+} ions are located at the base center of a square pyramid. Long range antiferromagnetic ordering of the Mn cations occurs at 40K with a spin reorientation at ~25 K. The ordering of the Tb^{3+} ions is suspected to occur around 10 K. The onset of ferroelectricity was seen at 38 K with a dominant polarization along the *b*-axis. The magnetic and ferroelectric ordering temperatures are quite similar and the long range ordering of the Mn^{4+}/Mn^{3+} ions induces the ferroelectric transition. This is believed to occur by inducing an additional distortion of the Jahn–Teller-distorted Mn^{3+} neighbors [139].

The multiferroic behavior of mineral hubnerite $MnWO_4$, a magnetic oxide with monoclinic crystal structure and spiral long-range magnetic order was reported in [140]. Crystal structure composed is of the $Mn^{2+}O_6$ and WO_6 octahedra, which share oxygen anions and form zigzag-like chains along

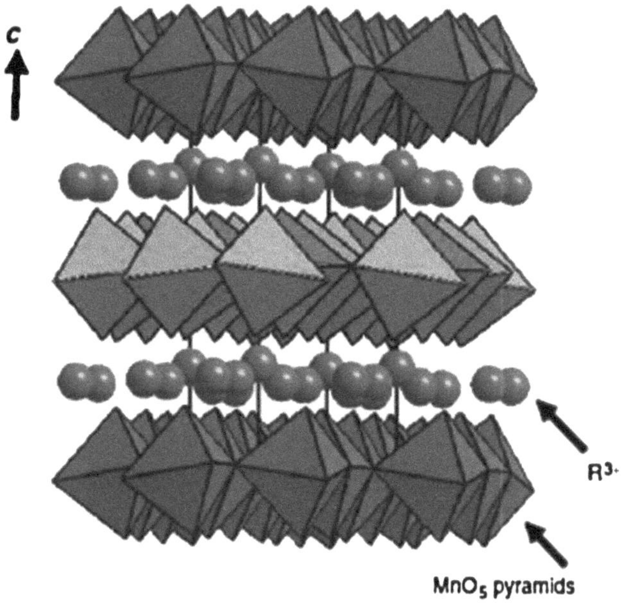

FIGURE 21 Crystal structure of a hexagonal manganite $RMnO_3$, with R a rare-earth element, with ferroelectric distortion.

c axis (see Figure 23). Elliptical spiral magnetic structure exists between 8 and 12 K , while below 8 K collinear antiferromagnetic spin ordering is evident from neutron diffraction data [141]. Based upon recent theoretical predictions, $MnWO_4$ should exhibit ferroelectric polarization coexisting with the spiral magnetic structure. Multiferroic state below 12 K in this compound was confirmed by observing a finite electrical polarization in the magnetically ordered state via pyroelectric current measurements.

Delafossite $CuFeO_2$ has the rhombohedral ($R3m$) structure with a triangular arrangement of Fe^{3+} magnetic cations [142] and complex multiple intralayer and interlayer superexchange pathways which are a source of strong spin frustrations (Figure 24). Multiferroic state can be induced in this crystal by applying a magnetic field or by doping with nonmagnetic Al or Ga impurities. The resulting multiferroic ground state is noncollinear and characterized by the complicated incommensurate propagation vector [143]. Collinear spin ordering along [110] direction which was established below 11 K is transformed to a spiral spin structure in applied magnetic field along the *c* axis.

Several compounds with a spinel structure were also described as multiferroics. So, $CoCr_2O_4$ has the transition from the ferromagnetic to conical spin structure at 25 K which accompanied by the appearance of spontaneous polarization [144]. Recently CdV_2O_4 with a spinel structure also proved to be multiferroic [145].

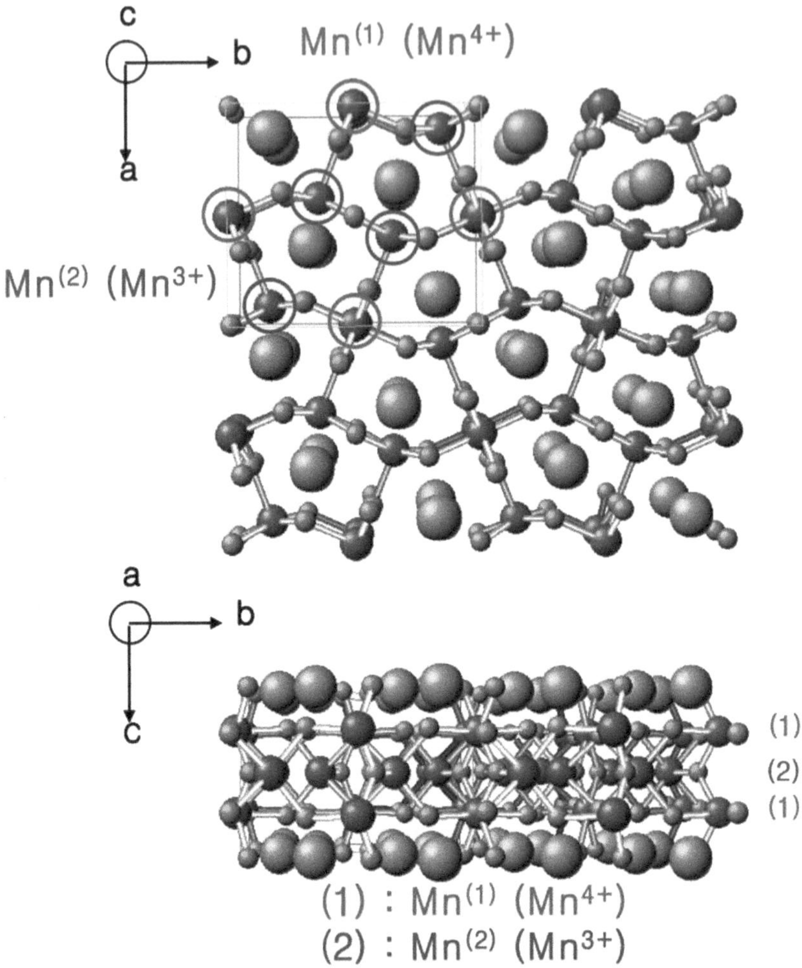

FIGURE 22 Crystal structure of $TbMn_2O_5$. View along the c axis (upper panel) and along the a axis (lower panel). Orange balls indicate Tb ions.

Pyroxenes with the general formula $AMSi_2O_6$ (A = mono- or divalent metal, M = di- or trivalent metal) are shown to be a new class of multiferroic materials [146]. These crystals are composed of layers of SiO4 tetrahedra and MO6 octahedra, which alternate in such way that the zigzag-like chains of octahedral with the magnetic cations inside run along c axis [147] (see Figure 25). Magnetic structure with a triangular arrangement of the spins arises in these chains serving as a possible source of frustration [148]. It was found that $NaFeSi_2O_6$ becomes ferroelectric in a magnetically ordered state below ~6 K. Similarly, magnetically driven ferroelectricity is also detected in $LiFeSi_2O_6$($T_C \sim 18$ K)

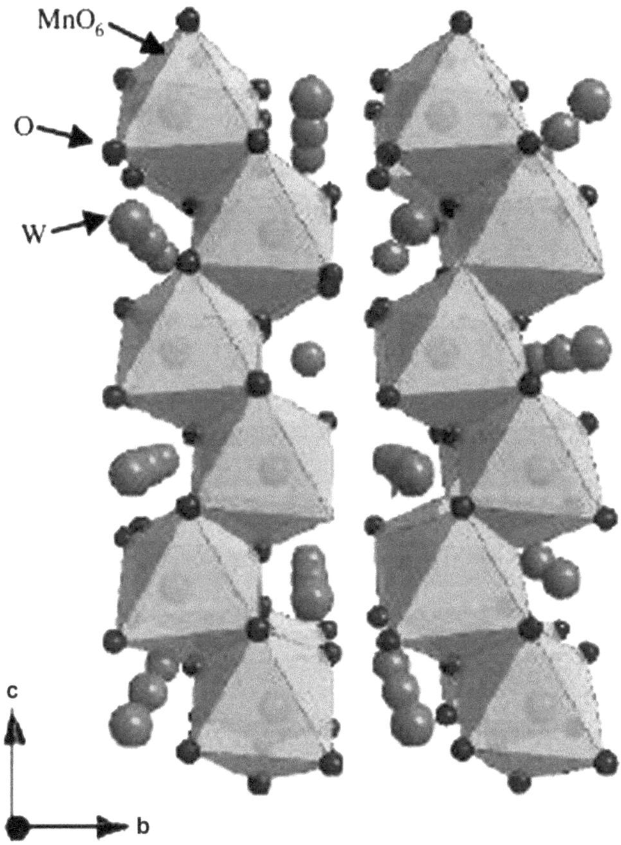

FIGURE 23 Monoclinic structure of $MnWO_4$.

and $LiCrSi_2O_6$(T_C ~ 11 K). In all these monoclinic systems the electric polarization can be strongly modified by magnetic fields. The possibility of a spiral magnetic structure caused by frustration to be the reason for the origin of the multiferroic behavior is proposed. It was concluded that other pyroxenes may also be multiferroic, and that the versatility of this family offers an exceptional opportunity to study general conditions for and mechanisms of magnetically driven ferroelectricity.

The materials discussed above have very limited practical application due to the fact that the frustration-induced ferroelectricity originated only at low-temperature range. It turns out that it is possible to achieve multiferroic properties at room temperature in crystals having a frustrated antiferromagnetic spin structure at this technologically-attractive temperature [149]. Orthorhombic $Bi_4Fe_2O_9$ is such an example of material with non collinear spin frustrated antiferromagnetic structure below 265 K. It is

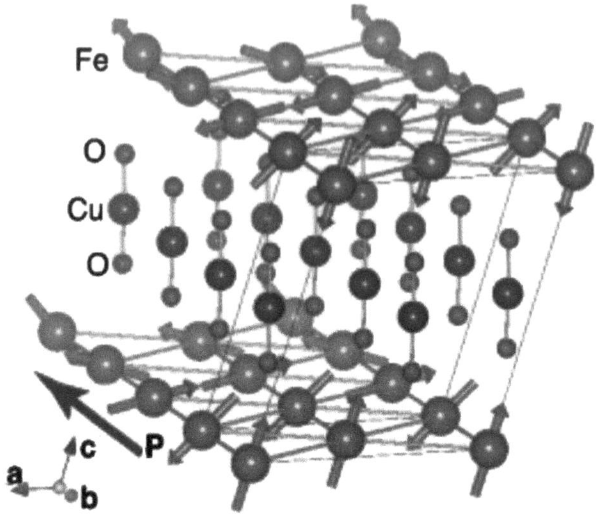

FIGURE 24 Atomic configuration and magnetic structure of $CuFeO_2$.

interesting to note that this compound is a common byproduct in the non-optimized solid state preparation of $BiFeO_3$. The crystal structure of this complex oxide belongs to the orthorhombic mullite-type space group *Pbam* ($z = 2$) and unit cell contains two different kinds of Fe cations. So, there are four Fe cations which are tetrahedrally coordinated and additional four Fe cations located in octahedral (see Figure 26). Spins of first type of Fe cations interact with each other and with the spins of second type of Fe cations antiferromagnetically, whereas the interaction of second types of Fe spins is ferromagnetic [150].

These competing exchange interactions clearly generate spin frustration in this system which leads to a noncollinear antiferromagnetic order in the *ab* plane. Detail investigation of dielectric and magnetic properties of $Bi_4Fe_2O_9$ provides clear evidence for strong coupling of electric and magnetic orders which are caused by spin frustration amongst Fe cation.

C. Tetragonal tungsten bronze structural family

The tetragonal tungsten bronze (TTB) compounds with a general formula $A1_2A2_4C_4B1_2B2_8O_{30}$ have been extensively investigated due to their interesting ferroelectric properties [151]. Incorporation of magnetic cations in these known ferroelectrics is possible in practical way to search for new MF materials. TTB has been known to have high-structure stability and the three different interstitial sites (the 9-, 12-, and 15-coordinated positions) (see

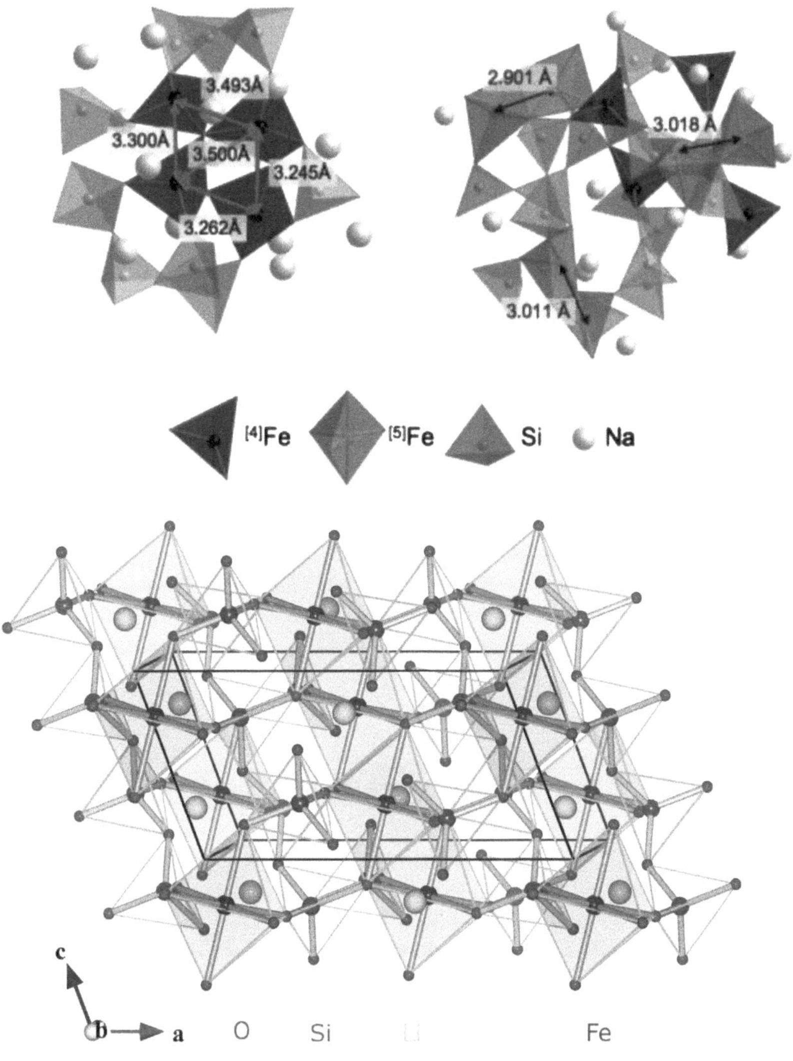

FIGURE 25 (a) Structure $Na_2FeSi_2O_6$. (b) Structure of $LiFeSi_2O_6$ showing the Fe^{3+} chains running along the c-axis linked by SiO_4 tetrahedra.

Figure 27) which can be offered different opportunities to induce electrical and magnetic polarizations using doping of magnetic cations and allow a wide variety of interactions. Moreover, the incommensurate polarization of TTB compounds may favor a coupling between magnetic and ferroelectric order. $Ba_2LnFeNb_4O_{15}$ (Ln=rare earth), and $Ba_4Bi_2Fe_2Nb_8O_{30}$ have been reported to be room temperature multiferroics [152,153]. Preliminary results on $Ba_6FeNb_9O_{30}$ [151] indicated that this ferroelectric has some ordered magnetic

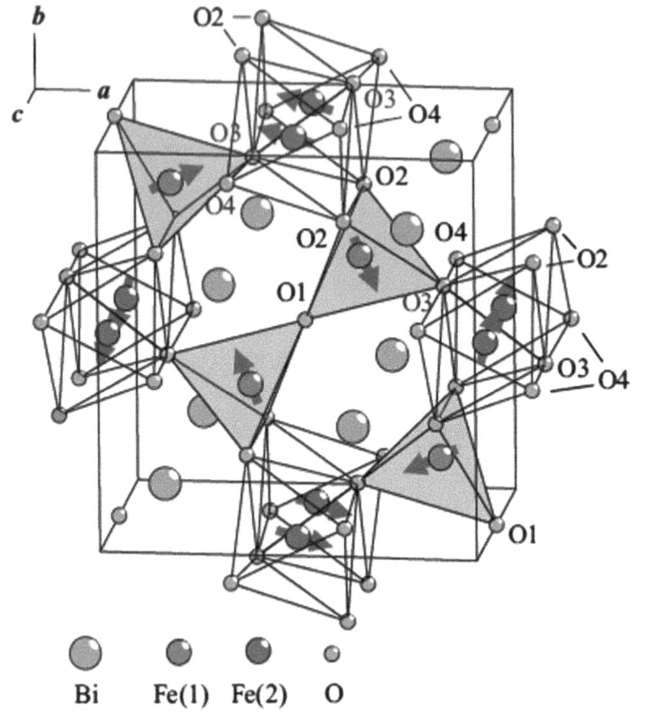

FIGURE 26 The unit cell of $Bi_2Fe_4O_9$ which can be described by columns of wdge-sharing Fe octahedra parallel to the c-axis, linked by corner-sharing Fe tetrahedra. The direction of magnetic moment in the antiferromagnetic state are also shown.

structure at low temperature but the nature of the ordering has yet to be understood. However, due to the complex compositions of TTB compounds, prepared powder samples often contain some amount of impurity phases, some of which may be magnetic, causing confusions about the correct nature of observed properties [154].

D. Langasites

The Fe-based langasites are the first reported compounds presenting a magnetic ordering in this rich family, besides being well known for piezoelectric properties and optical activity. The structural, magnetic, and dielectric properties of the Fe langasite compounds, with various substitution of non-magnetic cations, have been studied in details in [155]. $Ba_3NbFe_3Si_2O_{14}$ (trigonal, space group $P32_1$) is a spin-frustrated antiferromagnet with T_N = 26 K. It displays a helical magnetic order with signatures of frustration below T_N where an anomaly of the dielectric permittivity is observed. In this compound

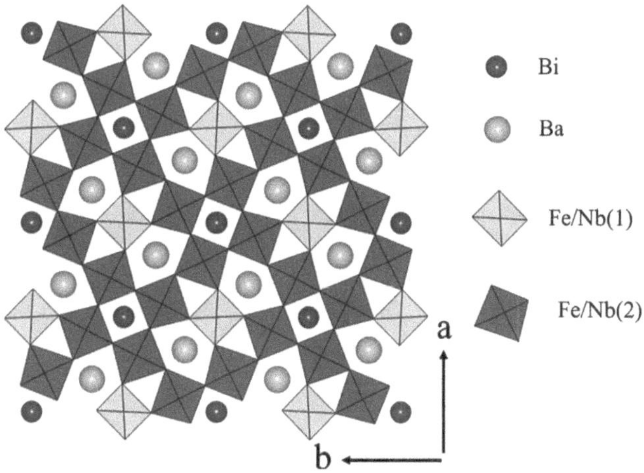

FIGURE 27 The ab projection of the crystal structure of $Ba_4Bi_2Fe_2Nb_8O_{30}$.

each NbO_6 octahedron shares its corners with six FeO_4 tetrahedra to form Fe trigonal prism (see Figure 28). The Fe^{3+} cations form sheets of isolated Fe triangles parallel to the *ab* plane where magnetic moments adopt a triangular configuration. The spin exchanges are frustrated in each *ab* sheet and between adjacent sheets of Fe^{3+} cations. Helical spin rotation by 51° along the c axis occurs to minimize the intersheet spin frustration as well as the intersheet magnetic dipole–dipole interaction.

The general formula is $A_3BC_3D_2O_{14}$, thus containing four different cationic sites. The decahedral A-site and the octahedral B-site form a layer at $z = 0$, while the two tetrahedral sites C and D are located on the plane $z = 1/2$ (see Figure 28). Due to its atomic arrangement, this structure type is able to accommodate a large number of different cations with various sizes and valences, leading to a wide variety of isostructural compounds. Among these, some contain magnetic cation sublattices. A special attention was focused on the effect on the magnetic properties of different non-magnetic cations on sites A, B, and D. The most striking result is that all the magnetic characteristics are roughly identical for all compounds even when the substitution produces a large variation of the cell parameters, except for those containing Sb (in place of Nb or Ta). The presence of this cation systematically increases the Neel temperature, decreases the Curie–Weiss one, increases the value of the ordered magnetic moment [156].

E. Orthophosphates

Several orthophosphates $LiAPO_4$ (A = Mn, Co, Fe, Ni) have the olivine-type structure where AO_6 octahedra are linked into layers and share oxygen atoms with the PO_4 tetrahedra, which also assemble layers (see Figure 29) [157].

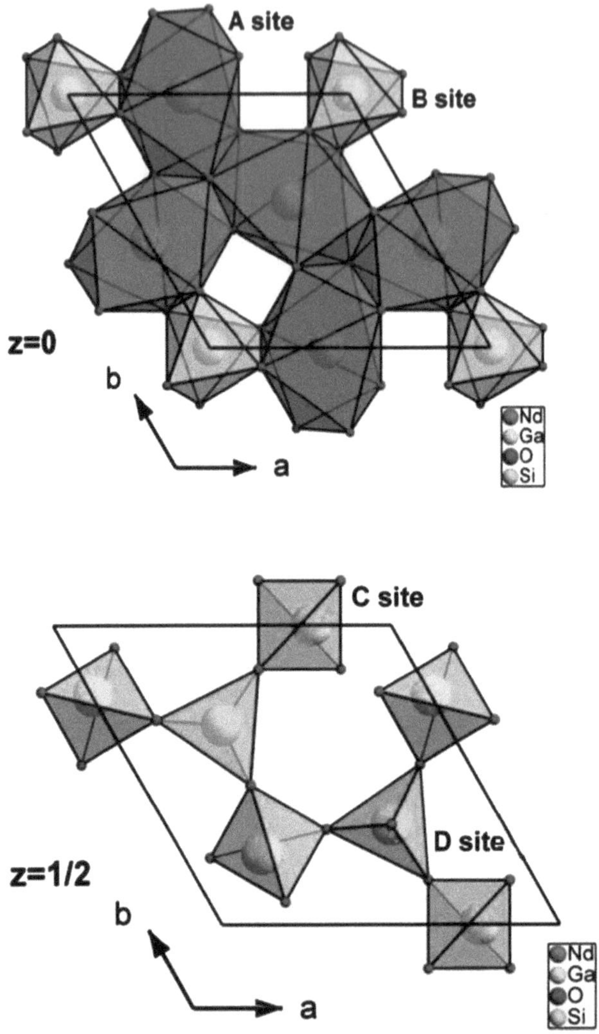

FIGURE 28 Polyhedral representation of the langasite structure with its four different cation sites on two layers. The main features of the crystal structure are chains of edge-sharing [FeO_6] octahedral and chains of corner-sharing [SiO_4] tetrahedral running along the *c*-direction.

Li cations are located in between these two types of layers and its content is responsible for a charge compensation related with the mixed valence state of A cations. These compounds are antiferromagnetics with T_N between 20 and 40 K. Two types of magnetic interactions are responsible for antiferromagnetic ordering in this type of structure: the strong superexchange interaction in the octahedral layers and the weak exchange along the A–O–P–O–A bonds. In

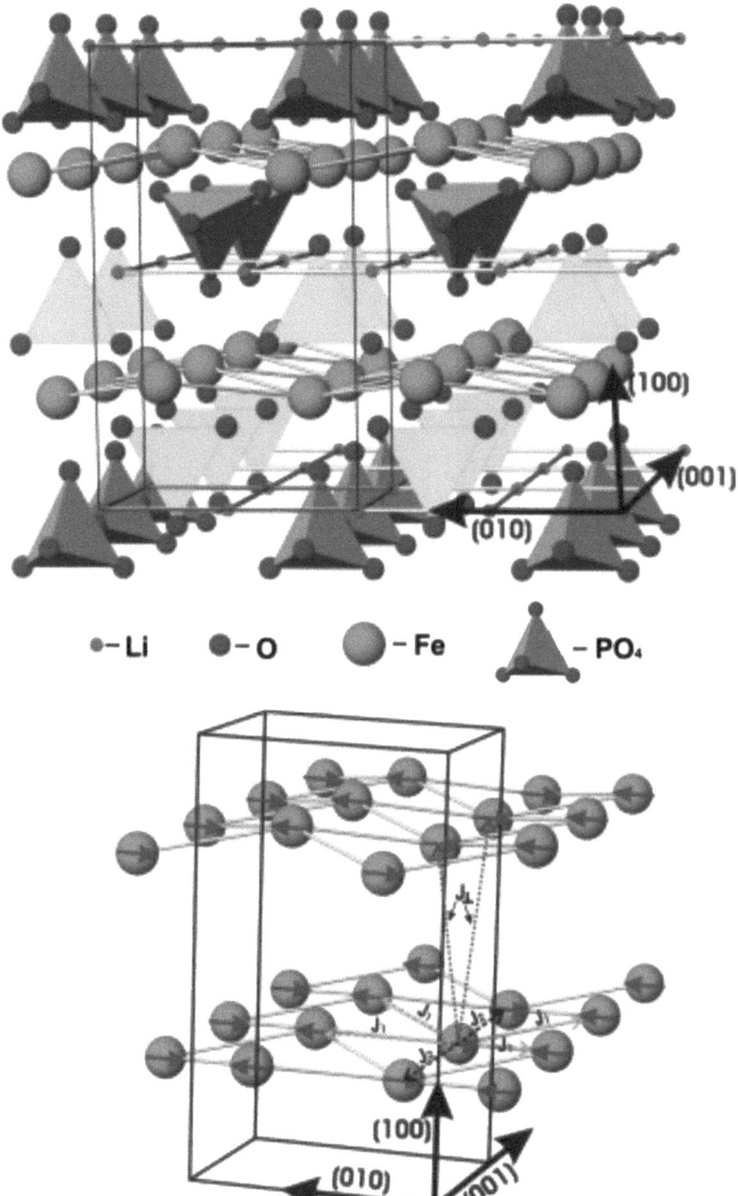

FIGURE 29 Atomic structure of $LiFePO_4$. The Fe^{2+} ions form buckled layers stacked perpendicular to the [99] crystallographic direction. The ground state of $LiFePO_4$ is collinear antiferromagnetic with the average moment along b direction.Spin arrangement of the two Fe^{2+} layers, the in-plane nearest and next-nearest-neighbor interactions J1 and J2 and interplane nearestneighbor interaction J_{inter} are labeled.

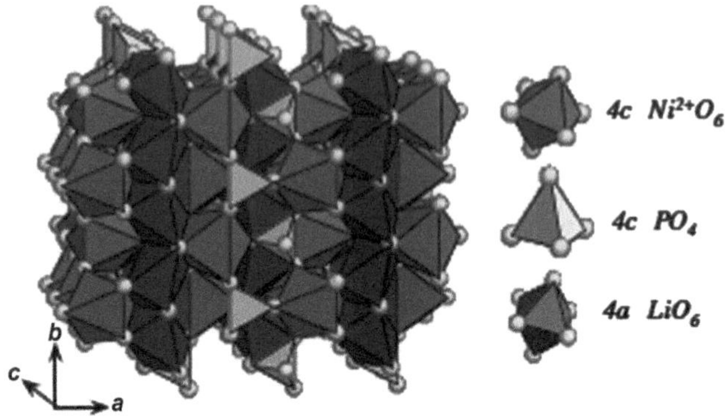

FIGURE 30 The schematic drawing of the crystal structure of $LiNiPO_4$ in the *ab*-plane.The olivine has a hexagonal closed-packed (hcp) oxygen array from NiO_6 octahedra and tetragonal sites occupied by P^{5+}.

$LiNiPO_4$ (Figure 30) was found that the application of magnetic field at the beginning induces the ferroelectric ordering and magnetic structure remains as commensurate. At an additional increase of magnetic field the magnetic structure becomes incommensurate and the spontaneous polarization disappears [158].

F. New potential multiferroics

We will restrict our consideration of these compounds by brief information on the existing potential multiferroics among complex metal oxides.

The simultaneous occurrence of ferroelectricity and ferromagnetism at room temperature in the perovskite-based $Pb_2Fe_2O_5$ ceramics with crystallographic shear planes has been reported in [159].

Heavy rare earth chromites of the formula $LnCrO_3$ with Ln = Ho, Er, Yb, Lu, and Y are shown to be multiferroic, exhibiting canted antiferromagnetism at low temperatures ($T_N \sim 113–140$ K) and a ferroelectric transition in the 472–516 K range [160].

$EuTiO_3$ also exhibits some potentially interesting multiferroic properties [161]. The compound has a simple cubic perovskite structure and is defined as a quantum paraelectric. It consists of Eu^{2+} ions with a spin of 7/2 and it has an antiferromagnetic ordering temperature of $\sim$ 5.5 K. Measurements reveal that the dielectric constant decreases at the magnetic ordering temperature and the dielectric constant changes by about 7% under the application of a 1.5 Tesla field. Additional studies are needed. However, the preliminary results indicate a strong coupling between the Eu^{2+} spins and the dielectric constant. The high-pressure (6 GPa) compound Bi_2NiMnO_6 is a heavily distorted double perovskite with the Ni^{2+} and Mn^{4+} ions ordered in a rock salt configuration.

The compound is reported to be ferromagnetic with an ordering temperature of 140 K and ferroelectric with a T_C of 485 K [162]. The ferroelectricity is proposed to arise from the presence of $6s^2$ lone pairs of Bi^{3+} ions and the erromagnetism is a result of magnetism from the Ni^{2+} and Mn^{4+} cations. In contrast, it is possible to explore the multiferroic candidates in an opposite direction and look for ferroelectricity in the traditional ferromagnetic oxides, which holds perovskite-like lattice units. It opens up a new direction for potential multiferroic candidate of such magnetic materials as $PbFe_{12}O_{19}$ [163], which has a hexagonal structure. Large spontaneous polarization was observed in $PbFe_{12}O_{19}$ ceramic at room temperature, exhibiting a clear ferroelectric hysteresis loop. The remnant polarization of $PbFe_{12}O_{19}$ ceramic is estimated to be $\sim 21\mu C/cm^2$. $PbFe_{12}O_{19}$ ceramics demonstrate strong ferromagnetism at room temperature. Simultaneous occurrence of large ferroelectricity and strong ferromagnetism in $PbFe_{12}O_{19}$ ceramics holds promise for its application in new generation of electronic devices as a practical multiferroic candidate in single phase.

The spatial separation of centers supporting the magnetization and the polarization is an attractive method of overcoming this problem and requires a complex crystal structure with distinct sites for the magnetically and electrically active ions. More recently, corundum-related complex metal oxides A_3TeO_6 (A = Mn, Ni, Co, Cu) (see Figures 31–34) have begun to garner renewed interest in the research community due to their dielectric and magnetic properties [164–167]. These materials are known to exhibit diverse properties as a result of compositional flexibility, however, an understanding of the manipulation of this structure type is still poor compared to perovskites.

The structure of these phases is related to the perovskite structure, however, the presence of new crystallographically non-equivalent sites for magnetic cations provide extra degrees of freedom for manipulation of the structure offering an additional compositional flexibility. Coexistence of long-range ordering of electrical dipoles and magnetic moments makes these phases as the potential multiferroic candidates.

Among known multiferroics there is a series of copper-based complex metal oxides with a frustrated spin structure which is responsible for ferroelectric properties. One of the detailed studies is $LiCu_2O_2$ with orthorhombic structure (Figure 35) in which linear double $Cu^{2+}O_2$ chains separated by Li and Cu^+ cations [168]. This compound is a typical representative of quasi-one dimensional systems. Appearance of spontaneous polarization near $T_C = 23$ K accompanied by magnetic ordering with a spiral spin structure [169].

To introduce a ferromagnetic property into a ferroelectric material originally with no ferromagnetism by itself, one of the common approaches is to dope magnetic impurities into the ferroelectric host material. These materials hold significant potential for spintronic devices if ferromagnetism can be understood and controlled. But ferromagnetic behavior in these systems is controversial. Numerous reports have claimed intrinsic ferromagnetic behavior above room

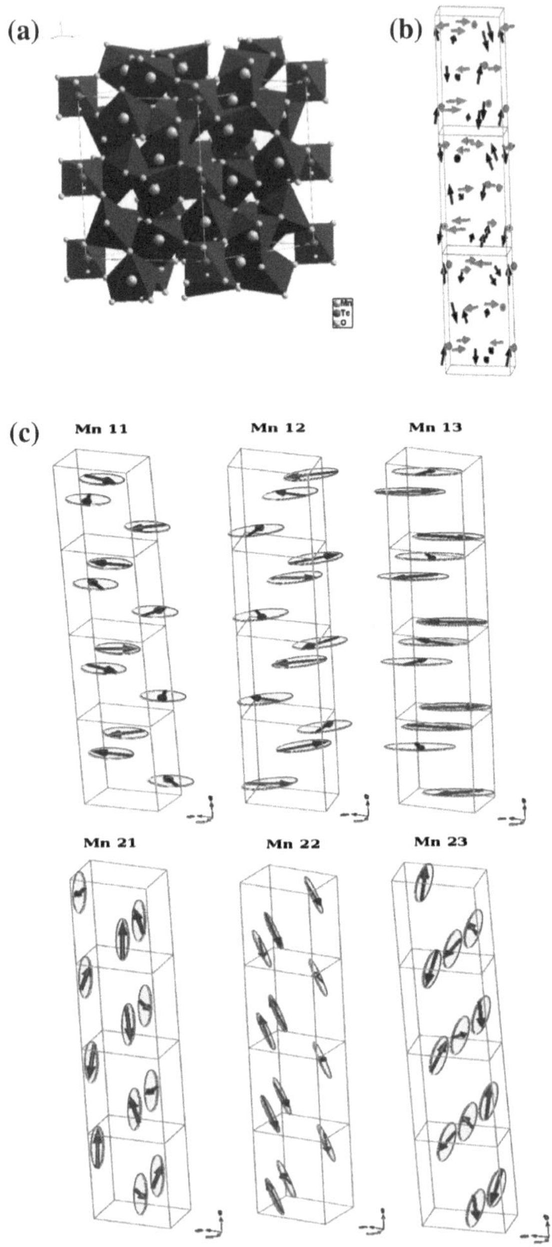

FIGURE 31 a) Polyhedral representation of crystal structure of Mn_3TeO_6. b) Schematic representation of the incommensurate magnetic structure for Mn_3TeO_6 for 3 cells along the direction of the propagation vector.c) A variation of the moment amplitude of all Mn sublattices ($\mathbf{a} \times \mathbf{b} \times 4\mathbf{c}$) in the direction of the propagation vector.

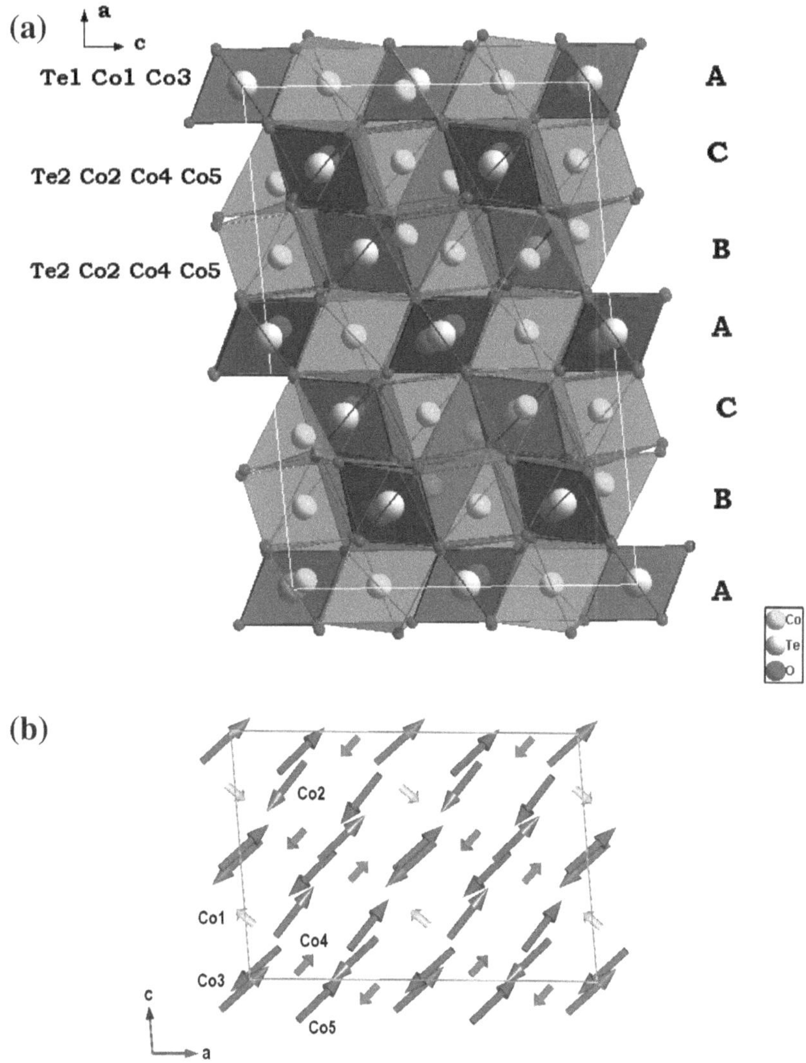

FIGURE 32 a)Polyhedral representation of the crystal structure of Co_3TeO_6 b)Schematic representation of the commensurate magnetic structure for Co_3TeO_6.

temperature, while a large number of others attribute these room temperature effects to impurities.Typical representative of dilute magnetic oxides is Fe-doped $BaTiO_3$. This is ferroelectric with perovskite structure. There was a lot of reports that doping with a few per cent of Fe^{3+} ions leads to observation of room temperature ferromagnetic behavior in this material [137–139]. It was found that $BaTi1\text{-}_xFe_xO_3$ ceramics ($x = 0.07$, 0.3 and 0.7) have hexagonal

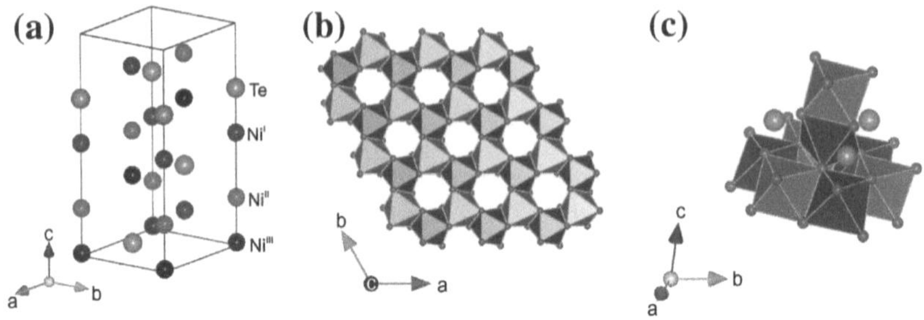

FIGURE 33 Different views of the structure of Ni_3TeO_6: a) the unit cell (oxygen ions are removed for clarity, Ni_1 placed in the origin), b) view along the c axis on the ab plane with its hexagonal structure, c) the link between the Ni_1 –Ni_2 hexagon (blue and red) and the Ni_3 octahedron (black) along the c axis. Ni_2 ion (red octahedron) on top of Ni_3 belongs to the adjacent plane.

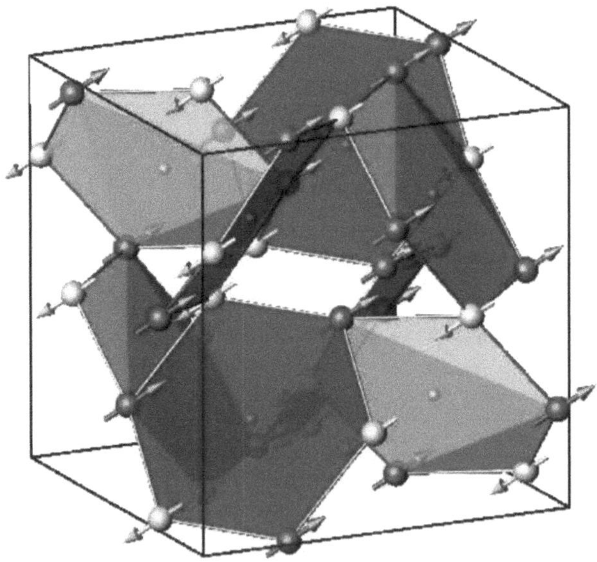

FIGURE 34 Three-dimensional spin structure of Cu_3TeO_6, as determined by a network of copper hexagons and collinear spin arrangement along one (out of four allowed by symmetry) body diagonal. Hexagons are formed around TeO_6 octahedra (only tellurium ions in the middle of the hexagons are shown). There are four pairs of non-equivalent hexagons in the unit cell, shown.

structure and room temperature ferromagnetism was observed for all samples. In spite of numerous reports about room temperature ferromagnetism in Fe-doped $BaTiO_3$ ceramics, the nature of this ferromagnetic behavior (intrinsic or extrinsic) still is not clear. Extrinsic ferromagnetism behaves from clusters

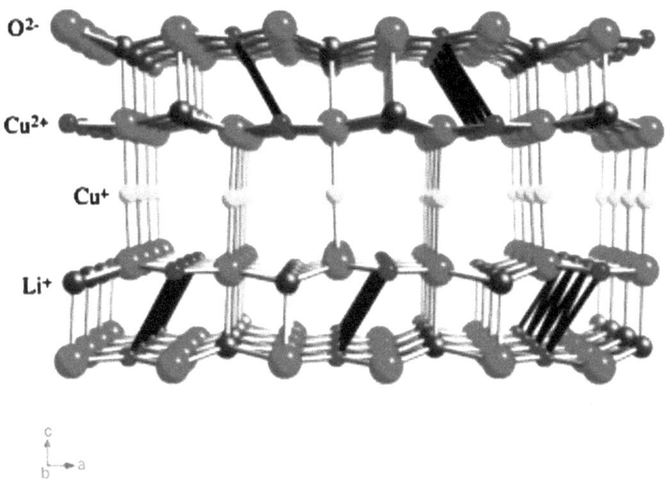

FIGURE 35 Illustration of the crystal structure of $LiCu_2O_2$. The black chemical bonds highlight the zigzag spin chains.

(Fe_2O_3) or secondary phases in these systems. Small clusters easily can show $\mu = 5 * 10^{-6}$ emu magnetic moment. Apart from the selected and above-described multiferroics established among complex metal oxides, additional single-phase compounds with potential magnetoelectric behavior are known. For most of these compounds the presence of multiferroic properties has to be confirmed but due to the limited space of this chapter the characteristic features of these compounds are not described because there is still no unambiguous answer to the question why polarization arises in the magnetically ordered structure.

XVIII. PREDICTIONS FOR NEW MULTIFERROICS

So far, more than 240 single-phase compounds (including the solid solutions) are disclosed to be multiferroic. However, the vast majority of them have low magnetic ordering temperatures or exhibit character of antiferromagnetism or weak ferromagnetism, or are not strong enough insulators to sustain a ferroelectric polarization at room temperature, and the magnitude of the observed magnetoelectric effect is too small for any practical applications. According to the phenomenological theory, the magnetoelectric response α_{ij} of a single-phased material is limited by dielectric constant ε_{ii} and magnetic permeability μ_{jj} through the relationship [12–15]

$$\alpha_{ij}^2 < \varepsilon ii \mu_{jj}.$$

Therefore the multiferroic materials with large and robust ferromagnetic and ferroelectric ordering will have large magnetoelectric coupling effect, to obtain such multiferroic single-phased compound is still a challenge, in particular with both high ordering temperatures.

The current understanding on the mechanism of origin of ferroelectricity and their controlling factors is still on the qualitative (not quantitative) level and is being developed. In proper ferroelectrics, usually ferroelectric active cations without electrons filling on the *d* orbitals such as Ti^{4+}, Zr^{4+}, Nb^{5+}, and Ta^{5+} are present. It was revealed in [25] that the hybridization between B-site ion d^0 and O $2p^6$ plays a dominant role for the instability of ferroelectricity, and that the spontaneous ferroelastic distortion and the hybridization between A-site cation and O could further enhance the ferroelectric polarization and ordering temperature. On the other hand, the ferroelectric- paraelectric phase transition has been often discussed using two limiting cases: the soft-mode or displacive theory, in which a transverse optical mode gradually softens when approaching the phase transition, and an order-disorder model, in which some atoms lying on symmetry- equivalent sites occupy on some preferential ones.

According to Venevtsev [59,63], the temperature of ferroelectric phase transition in perovskite-type oxides can be separated into several main contributions using the following relationship:

$$T_C \sim 1/V(f, \alpha, q)_{A,B}, N_{LEP}, N_{JT}$$

where V- unit cell volume in paraelectric phase, F- a "rattling" of oxygen polyhedra for ferroelectric active A or B cations , α-electronic polarizability, q-charge of for ferroelectric active A or B cations, N_{LEP} – number of cations with a lone-electron pair ($Bi^{3+}, Pb^{2+}, Sb^{3+}, Tl^{1+}$ and Te^{4+}) in perovskite composition, N_{JT} – number of Jahn–Teller cations ($Cu^{2+}, Cr^{2+}, Mn^{3+}$) in perovskite composition. Based on this separation, the following simple guidelines for estimating Tc value were obtained. Spontaneous polarization (or T_C) is most likely increased for perovskites with a smaller volume of perovskite unit cell and a remarkable amounts of Jahn–Teller cations and the cations with a lone-electron pair, strongly correlated with a charge, polarizability and distortions of oxygen polyhedra of ferroelectric active cations.

Another way for estimation of a temperature of ferroelectric phase transition was proposed in [170]. Atomic coordinate analysis allows any well-determined phase to be investigated in order to reveal the atomic displacements required for a ferroelectric phase transition to another phase of higher but related symmetry. The Curie temperature (in K) may be derived from the maximum cation displacement by

$$T_C \sim 2.00(9)x10^4 (\Delta Z_i)^2$$

where ΔZ_i is the largest i-cationic displacement at Tc between its measured location and that along the polar axis at which the spontaneous polarization would be zero. In the 1960s, Goodenough [141] investigated those predominant factors for Néel temperature in the perovskite structures and pointed out that a simple, localized-electron antiferromagnet having one electron per localized orbital interacting via 180° superexchange with similar orbitals on $z$near-neighbor cations via an overlap integral Δ has a Néel temperature:

$$kT_N \approx \left[zqtbt^2 + qebe^2\right](S+1)/S,$$

where *qt*, *qe* are inversely proportional to electrostatic energies associated with electron transfers; $b \sim \varepsilon_0 \Delta t$ and $be \sim \varepsilon_0{\prime}\Delta e$ are the one-electron transfer integral of states having *t2g* and *eg* symmetry, respectively. Accordingly, $dT_N/\mathrm{d}p > 0$, where *p* is the hydrostatic pressure and thus $\mathrm{d}T_N/\mathrm{d}a_0 > 0$, where a_0 is the cation-anion-cation separation, remains in the localized *d* electron systems including $A^{3+}FeO_3$ and $A^{3+}CrO_3$. This fact is attributed to changes in A-O covalent bonding that cause $\Delta$to increase with the more basic A cation for a given lattice parameter. For $BiFeO_3$ and $BiCrO_3$, the antiferromagnetic coupling between the *eg* electrons on neighboring

Fe sites due to the dpσ bonding is very much stronger than that between the t_{2g} electrons on neighboring Cr sites, and thus results in BiFeO3 having a higher magnetic ordering temperature than $BiCrO_3$. On the basis of the simulations for $BiCrO_3, Bi_2FeCrO_6$, and BiFeO3, it was found that high magnetic ordering temperature can be achieved in the antiferromagnets or ferrimagnets by exploiting strong antiferromagnetic superexchange between e_g electrons. Therefore, it is reasonably expected to combine d^5 and d^8 ions for multiferroic compounds to obtain the highest magnetic ordering temperature in the perovskite structure [142].

Although several mechanisms for coexistence of magnetic and ferroelectric ordering have been proposed and extensively studied, it is not so clear yet how to get a single-phased ferromagnetic-ferroelectric. From the results of theoretical calculations of electronic structures it is clear that the electron exchange interaction and correlation are responsible for the stabilization of electric polarization and the formation of magnetic moment, it has been revealed also that the orbital hybridization plays a central role in controlling ferroelectric and magnetic ordering. Therefore, it is possible to balance the coexistence of ferromagnetic and ferroelectric ordering in single-phased MF compounds, and to obtain both high ordering temperatures above room temperature by controlling hybridization degree through controlling geometrical arrangement of cations [142].

XIX. OTHER APPROACHES

Before ending this chapter, it is important to make some remarks on other strategies for integrating of spin and dipole orderings into one single phase compound. These comments may shed light on design and synthesis of novel multiferroics.

Today there are roughly four major classes of multiferroic materials: (1) materials with the perovskite structure, (2) materials with hexagonal structure, (3) boracites, and (4) $BaMF_4$ compounds. The search for new candidate multiferroic compositions and detail study of order parameter coupling are very important for the design of these systems with exotic functionalities and understanding and control the basic chemical nature of these materials. The multiferroicity from spiral spin structure illustrates the fact that the spiral magnetic order often arises from the competing magnetic interactions. These competing interactions usually reduce the ordering temperature of conventional spin ordered phase. Hence, it is hardly possible for the the spiral spin order (phase)-induced ferroelectricity to appear above a temperatures of ~40 K, far below room temperature required for service of most devices. One of the possible ways to overcome this barrier is to search for those magnetic materials with very strong competing magnetic interactions. This effort has been marked with some progress recently and it was found that the magnitude and sign of the main super-exchange interaction J in low-dimensional cuprates depend remarkably on the Cu–O–Cu bond angle ϕ [143]. The most intriguing recent discovery is arguably the comparatively high ferroelectric T_C of 230 K in the simple, binary compound CuO. This is the highest known T_C among the spin spiral induced ferroelectrics and a major advance towards room-temperature magnetoelectric multiferroic materials of this class. In cuprates with $\phi \sim 180°$, J has an order of magnitude of $\sim 10^2$meV, thus favoring ferromagnetic order. Upon decreasing ϕ, J is monotonically suppressed and eventually becomes negative (favoring ferromagnetic order) at $\phi \sim 95°$. Therefore, for those cuprates with ϕ deviating away from 180°, the ferromagnetic interaction (J) competes with the higher-order superexchange interactions, often leading to the spiral magnetic order with relatively high ordering temperature.

Besides CuO, hexaferrite $Ba_{0.5}Sr_{1.5}Zn_2Fe_{12}O_{22}$ is another multiferroic system offering the spiral magnetic and ferroelectric orders at a relatively high temperature [144]. Similarly, hexaferrite $Ba_2Mg_2Fe_{12}O_{22}$ was also found to exhibit magnetic field induced ferroelectricity at relatively high temperature, although it does not show ferroelectricity under zero magneticfield.

The situation would be entirely different if we place the magnetic ions on the A-sites which are ordered similarly so that the magnetic criterion is still satisfied. This corresponds to the A-site magnetism. Placing a ferroelectrically active ion such as Ti^{4+} on the B-site would then satisfy the structural-chemical criterion. The high pressure metastable phases of $FeTiO_3$ and $MnTiO_3$ [144–148] with space group R3c meet these criterions and provide the possibility of realizing the electric field switching of magnetization. Family of compounds $MTiO_3$ (M = Mn, Fe, Ni,Co), in a certain polymorphic structure (acentric R3c), are promising candidates where a polar lattice distortion can induce ferrimagnetism.

Recently, the magnetic field induced ferroelectricity was found in orthorhombic $DyFeO_3$ [149]. The mechanism for ferroelectricity in $DyFeO_3$ depends on both the magnetic structures of Dy and Fe cations. It is postulated that the exchange striction between those adjacent Fe^{3+} layer and Dy^{3+} layer with the interlayer antiferromagnetic interaction results in the multiferroic phase.

XX. FERROELECTRICITY INDUCED BY A-SITE DISORDER

The ferroelectrics with lone pair mechanism have $t < 1$ and the ferroelectricity is from off-centering of the A-site ions. Without Pb or Bi, perovskite structures of $t < 1$ generally have tilted BO_6 octahedra instead of the A-site off-centering. Unfortunately, the magnetic perovskite materials often have $t < 1$ and tilted BO_6 octahedra because those ions with *d* electrons are generally larger than *d*ions, and then not ferroelectrically active. However, for some complex perovskites with the average tolerance factor significantly smaller than one, the octahedral tilting might be prevented because the cations with strongly different sizes distribute randomly in the lattice, which is coined as A-site disorder. The ferroelectricity appears to originate from the large off-centering of A-cations by chemical pressure. It was predicted that $(La, Lu)MnNiO_6$ with $t < 1$ exhibits polar-type lattice distortion [150]. This polar behavior arises from the frustration of the octahedral tilting instabilites due to the mixture of A-site cations of different sizes and the fact that the coherence length for the A-site off-centering is shorter than that for the tilting instabilities [150]. On the other hand, Mn^{3+} and Ni^{3+} ions can occupy the B-sites in an order form, resulting in the double perovskite structure. The superexhange interaction between Mn^{3+} and Ni^{3+} is ferromagnetic [151]. Due to these mechanisms, $(La, Lu)MnNiO_6$ may exhibit large ferroelectric polarization and ferromagnetism simultaneously. However, again it is difficult to synthesize $(La, Lu)MnNiO_6$ because of the phase separation and competing phases, which often occur for perovskite oxide materials with mismatching A-site species.

The preparation and investigation of new MF solid solutions based on $BiFeO_3$, $PbFe_{2/3}W_{1/3}O_3$, Ba_2FeRe0_6, and others, would be timely because the temperatures of ferroelectric and magnetic phase transitions of these compounds lie above room temperature. This allows to hope that other compositions with attractive MF properties may be found namely in solid solutions close to morphotropic phase boundaries [13,25].

Extending the search for multiferroics from simple ferromagnets to systems with incommensurate magnetic order is an important additional direction in finding new materials which have a room temperature transition into a phase with simultaneously appearing magnetic and ferroelectric order . Insulating ferromagnets tend to have very low transition temperatures, but very often incommensurate magnets have ordering temperatures well above 300 K and may be attractive candidates in the search for strongly coupled multiferroics at room temperature.

Moreover, detail knowledge of more suitable combinations of electric and magnetic ordering types in existing MF compounds are also desirable for design of new compositions with a coexisting of spin and dipole orders. Therefore experimental investigations on material science aspects of MF should be continued more intensively. In the same time a theoretical construction of microscopic theory for explanation of magnetoelectric exchange interactions and calculation of the exchange constants is urgent.

XXI. CONCLUDING REMARKS

Search and preparation of a room-temperature multiferroic among complex metal oxides is a very difficult scientific and technological task and, in a first approximation, equivalent to the task of finding a needle in a haystack. Nevertheless, the importance of this chapter is in that it indicates in which "haystack" to search. This review adds information to the earlier reviews containing a variety of useful information on MF complex oxides. It is apparent that these compounds have been studied 50 years ago, but have only recently been rediscovered. Unfortunately, in spite of substantial advancements in preparation and characterization together with the new theoretical approaches (magnetic frustration and breaking of the inversion center due to an AFM ordering), known MF compounds have been studied with different degrees of thoroughness and very often only data on their temperatures of ferroelectric and magnetic phase transitions are available. Their properties have been studied insufficiently and there are many potential MF complex metal oxides which call for additional investigations. Even in the great perovskite family, present knowledge on MF properties is nearly exclusively based on data from ceramic or powder samples, which are often not reliable. The validity of some results is not fully accepted because they cannot be reproduced by others. The main problem with most of these results is that multiferroic properties are observed in samples containing several different phases and it is very difficult to understand which fraction is responsible for its occurrence. In a few cases, however, the phase is known but multiferroism was registered exclusively on the surface. For complex metal oxides, the surface conditions differ from those inside the bulk, and the degree of this difference depends on many parameters, and some of them are extrinsic.

In the case of distorted perovskites with the structure tolerance factor less than 0.96, there are serious problems with a preparation of pure perovskite phase under ambient pressure. Like in the cases of $BiFeO_3$, $BiMnO_3$, and $BiFe_{1/2}Cr_{1/2}O_3$, it is difficult to obtain perfect pure single phase even with high-pressure synthesis. Magnetic cations tend to be fairly easy to oxidize or reduce, and are usually associated with multiple valence states accompanied by anion non-stoichiometry and an appearance of metastable compounds. This fact seriously increases the difficulties in sample preparation of desired pure MF perovskite phases.

Starting with the sunset of multiferroicity, there are two concurrent directions with different final targets. First branch (proper multiferroics) was connected with the ability to use both electrical and magnetical degrees of freedom as two different and non-interacting media. Used as a memory device, on the same chip one could write two sets of memories. The principal requirement then would be to minimize any possible interaction between the electrical and magnetic degrees of freedom. Second branch (improper muliferroics) was related with an electric control of magnetization having some materials with strongest interaction between the electrical and magnetic subsystems. Now it is evident that at present time we have the proper multiferroics with quite weak magnetoelectric effect (small enough for applications) effect and the improper multiferroics with quite large (large enough for application) magnetoelectric effect. So, optimal conditions for multiferroic behavior are not found yet and search and design of new attractive compositions should be harder.

The mechanism of MF coupling for some MF complex oxides remains unclear and existing proposed microscopic models are not fully understood even in a qualitative sense.

Although quite a number of multiferroics have been synthesized and studied, almost all of them exhibit a small spontaneous magnetization or electric polarization. The magnetic order state in MF complex metal oxides usually is antiferromagnetic-type. Moreover, the temperature for the coexistence of ferroelectricity and magnetism remains very low and FE and magnetic phase transitions are separated by large temperatures. Basically, the MF coupling for most MF compounds remains weak. Therefore experimental investigations on material science aspects of MF's and the exploration of new useful physical effects should be continued more intensively.

Although several mechanisms for coexistence of magnetic and ferroelectric ordering have been proposed and extensively studied, it is still unclear how to get a room temperature single-phased ferromagnetic-ferroelectric compound.

In most cases MF were searched for among compounds of those structural types where known ferroelectrics contain a considerable amount of paramagnetic cations. It should however be noted that many magnetic materials have not yet been investigated for FE properties. The significance of these studies cannot be underestimated and it is expected that systematic dielectric studies of such compounds will yield new MF compounds. Each new MF compound is a challenge in itself having its own specific structure–property relationships, which one has to study and explore.

In the same time, recent research on multiferroic systems gives an optimistic scenario wherein alternative mechanisms bring about magnetic ferroelectricity in a variety of materials. Some of the important mechanisms that have emerged relate to tilting of metal–oxygen polyhedra, spiral magnetic ordering, and stereochemical activity of the Bi and Pb lone pairs. In many of the multiferroic oxides, magnetic frustration appears to give rise to novel magnetic properties which then induce ferroelectricity. The role of local non-centrosymmetry and charge-ordering of cations with mixed valence states are several new ideas that have come to the fore. It appears that many ferroelectric materials formally possessing centrosymmetric structure may indeed have local non-centrosymmetry. In principle, this deviation from a centrosymmetry can also be achieved by making use of strain, defects, and other factors.

Owing to the advanced techniques the single phase MF compounds have been extended to new artificial structures, such as multilayers, nanostructures, superlattices, etc. Novel multiferroics connected with new mechanisms for the magnetoelectric coupling (usually a strain as the coupling channel) can be created using this kind of artificial design. These new artificial structures can be very promising in future investigations.

FURTHER READING

Abrahams SC, Kurtz SK, Jamieson PB. Phys Rev 1968;172:551.
Ederer C, Spaldin NA. Phys Rev B 2005;71:060401.
Goodenough JB, Stickler JJ. Phys Rev 1964;164:768.
Harris AB, Aharony A, Wohlman OE. J Phys: Condens Matter 2008;20:434202.
Jian Y, Chu J. Chin Sci Bull 2008;53:2097.
Kimira T, Lawes G, Ramirez AP. Phys Rev Lett 2005;94:137201.
Kimura T, Sekio Y, Nakamura H, Siegrist T, Ramirez AP. Nature Mater 2008;7:291.
Ko J, Prewitt CT. Phys Chem Miner 1988;15:355.
Kumar P. Integrated Ferroelectrics 2011;131:25.
Lin F, Jiang D, Ma X, Shi W. J Magn Magn Mater 2007;320:691.
Mashkina E, McCammon C, Seifert F. J Solid State Chem 2003;177:262.
Ming LC, Kim YH, Uchida T, Wang Y, Rivers M. Am Mineral 2006;91:120.
Ramamoorthy R, Martin L. Multiferroics: synthesis, characterization and applications. Wiley-VCH Verlag; 2012.
Ray S, Mahadevan P, Mandal S, Krishnakumar SR, Kuroda CS, Sasaki T, et al. Phys Rev B 2008;77:104416.
Singh DJ, Park CH. Phys Rev Lett 2008;100:087601.
Tokunaga Y, Iguchi S, Arima T, Tokura Y. Phys Rev Lett 2008;101:097205.
Ueda K, Tabata H, Kawai T. Science 1998;280:1064.

REFERENCES

[1] Aizu KJ. Phys Soc Jp 1969;27:387.
[2] Aizu K. Phys Rev B 1970;29:754.
[3] Newnham RE. Cross LE. Mater Res Bull 1974;9:927.
[4] Newnham RE,Cross LE. Mater Res Bull 1974;9:1021.
[5] Smolenskii GA, Ioffe VA, Communications de Colloque International de Magnetism de Grenoble(France), 2–6 Jullet, Communication No.1; 1958.
[6] Smolenskii GA, Agranovskaya AI, Popov SN, Isupov VA. Zh Tekhnich fiz 1958;28:2152.
[7] Smolenskii GA, Isupov VA, Krainik NN, Agranovskaya AI, Izv Akad Nauk, Ser Fiz 1961;25:1333.

[8] Bokov VA, Myl'nikova I, Smolenskii GA. Sov Phys JETP 1962;15:447.
[9] Bokov VA, Kizhayev SA. Fiz Tverd Tela 1964;6:3038.
[10] Isupov VA, Agranovskaya AI,Khuchua NP, Izv AN SSSR, Ser Fiz 1960;24:1270.
[11] Smolenskii GA, Agranovskaya AI, Isupov VA. Fiz Tverd Tela 1959;1:990.
[12] Smolenskii GA, Chupis IE. Sov Phys Usp 1982;25:475–493.
[13] Venevtsev YuN, Gagulin VV, Zhitomirsky ID. Ferroelectrics, 1987;73:221.
[14] Schmid H. In: Magnetoelectric interaction phenomena in crystals. London, New York, Paris; Springer 1975. p. 121.
[15] YuVenevtsev N, Gagulin VV. Ferroelectrics 1994;162:23.
[16] Ascher E, Rieder H, Schmid H, Stössel H. J Appl Phys 1966;37:1404.
[17] Schmid H. Phys Status Solidi 1970;37:209.
[18] Eibschutz M, Guggenheim HJ. Solid State Commun 1968;6:737.
[19] Scott JF. Ferroelectrics 1980;24:127.
[20] Kato K, Iida S, Yanai K, Mizushima K. J Magn Magn Mater 1983; 31-34:783.
[21] Ismailzade IG, Kizhaev SA. Sov Phys Solid State 1965;7:236.
[22] Teague JR, Gerson R, James WJ. Solid State Commun 1970;8:1073.
[23] Wang KF, Liu JM, Ren ZF. Adv Phys 2009;58:321.
[24] Scott JF. Ferroelectric memories. Berlin: Springer-Verlag; 2000.
[25] Lines ME, Glass A. Principle and application of ferroelectric and related materials. Oxford: Oxford University Press; 2001.
[26] Schmid H. Ferroelectrics 1994;162:317.
[27] Schmid H. J Phys: Condens Matter 2008;20:434201.
[28] Chupis IE. Low Temp Phys 2010;36:597.
[29] Prellier W, Singh MP, Murugavel P. J Phys: Condens Matter 2005;17:R803.
[30] Brink J, Khomskii DI. J Phys: Condens Matter 2008;20:434217.
[31] Kreisel J, Kenzelman M. Europhys News 2009;40:17.
[32] Eerenstein W, Mathur ND, Scott JF. Nature 2006;442:759; Ziese M, Thornton MJ, editors.
[33] Spin electronics. Berlin: Springer-Verlag; 2001.
[34] Blasse GJ. Inorg Nucl Chem 1965;27:993.
[35] Hill NA. J Phys Chem B 12000;04:6694.
[36] Hill NA, Filippetti N. J Magn Magn Mater 2002; 242–245:976.
[37] Filippetti N, Hill NA. Phys Rev B 2002;65:95120.
[38] Goodenough JB, Longo JM. In Landolt-Bornstein, editor. Numerical data and functional relationships in science and technology, New series, vol. III.4. Berlin: Springer; 1970. p. 126.
[39] Mitsui T, Nakamura E, Shiozaki Y. In: Landolt-Bornstein, editor. Numerical data and functional relationships in science and technology, New series, vol.16(1). Berlin: Springer; 1981. p. 126.
[40] Martin LW. Dalton Trans 2010;39:10813.
[41] Kimura T, Coto T, Shinhani H, Ishizaka HK, Arima T, Tokura Y. Nature 2003;426:426.
[42] Hur N, Park S, Sharma PA, Ahn JS, Guha G, Cheong SW. Nature 2004;429:392.
[43] Mostovoy SW. M Phys Rev Lett 2006;96:067601.
[44] Khomskii DI. J Magn Magn Mater 2006;306:1; Khomskii DI. Physics 2009;2:20.
[45] Kimura T. Ann Rev Condens Matter Phys 2012;3:93.
[46] Khomskii DI. Physics 2009;2:20.
[47] Choi YJ, Yi HT, Lee S, Huamg Q, Kiryukhin V, Cheong SW. Phys Rev Lett 2008;100:047601.
[48] Ye Z-G. Key Eng Mater 1998; 155/156:81.
[49] Anderson MT, Greenwood KB, Taylor GA, Poeppelmeier KR. Prog Solid State Chem 1993;22:197.
[50] Amador U, Hetherington CJD, Moran E, Alario-Franco MA. J Solid State Chem 1992;96:132.
[51] Galasso FS. Structure properties and preparation of perovskite-type compounds. Oxford, UK: Pergamon Press; 1970.
[52] Fuentes AF, Ibarra OH, Mendoz-Suarez G, Escalante-Garcıa JI, Amador U. J Solid State Chem 2003;173:319.
[53] Vegas A, Jansen M. Acta Cryst B 2002;58:38.
[54] Shpanchenko RV, Chernaya VV, Tsirlin AA, Chizhov PS, Sklovsky DE, Antipov EV. Chem Mater 2004;16:3267.

[55] Bertaut EF, Pauthenet R, Mercier M. Phys Lett 1963;7:110.
[56] Woodward PM. Acta Cryst B 1997;53:32.
[57] Glazer AM. Acta Cryst B 1972;28:3384.
[58] Howard CJ, Stokes HT. Acta Cryst B 1998;54:782.
[59] Rietveld HM. J Appl Cryst 1969;2:65.
[60] Philipp JB, Majewski P, Reisinger D, Geprags S, Opel M, Erb A, et al. Acta Phy Polonica 2004;105:7.
[61] Venevtsev YuN, Politova ED, Ivanov SA. Ferro- and antiferroelectrics of barium titanate family, Chemistry, Moscow, 1985.
[62] Chen IW. J Phys Chem Solids 2000;61:197.
[63] Stenger CGF, Burggraaf AJ. Phys Status Solidi A 1980;61:653.
[64] Smolenskii GA, Bokov VA, Isupov VA, Krainik NN, Pasynkov RE, Sokolov AI, et al. The physics of ferroelectric phenomena. Leningrad: Nauka; 1985.
[65] Venevtsev YuN, Lyubimov VN. Ferroelectromagnets. Moscow: Nauka; 1990.
[66] Bokov AA, Raevsky IP. Ferroelectrics 1989;90:125.
[67] Bokov AA, Raevsky IP. Ferroelectrics 1993;144(1–4):147.
[68] Brunskill IH, Schmid H, Tissot P. Ferroelectrics 1981;37:547.
[69] Geddo Lehmann A, Sciau Ph. J Phys: Condens Matter 1999;11:1235.
[70] Lampis N, Sciau Ph, Geddo Lehmann A. J Phys: Condens Matter 2000;12:2367.
[71] Watanabe S, Koyama Y. Phys Rev B 2002;65:064108.
[72] Raevski IP, Kubrin SP, Raevskaya SI, Titov VV, Sarychev DA, Malitskaya Zakharchenko MA, et al. Phys Rev. B 2009;80:024108.
[73] Mitoseriu L, Marré D, Siri A, Nanni P. Appl Phys Lett 2003;83:5509.
[74] Randall CA, Bhalla AS. Jpn J Appl Phys, Part 1 1990;29:327.
[75] Kumar A, Murari NM, Katiyar RS. Appl Phys Lett 2007;90:162903.
[76] Ivanov SA, Eriksson S-G, Tellgren R, Rundlof H. Mater Res Bull 2004;39:2317.
[77] Ivanov SA, Eriksson SG, Tellgren R, Rundlöf H. J Solid State Chem 2005;178:3605.
[78] Eng HW, Barnes PW, Auer BM, Woodward PM. J Solid State Chem 2003;175:94.
[79] Negas T, Roth RS. J Solid State Chem 1970;1:409.
[80] Töpfer J, Pippardt V, Voigt I, Kriegel R. Solid State Sci 2004;6:647.
[81] Ivanov SA, Eriksson SG, Tellgren R, Rundlöf H, Nordblad P, Eriksen J. J Solid State Chem 2006;179:2645.
[82] Ramesha K, Prakash AS. Mater Sci Eng B 2009;164:124.
[83] Gatalan G, Scott JF. Adv Mater 2009;21:2463.
[84] Michel C, Moreau JM, Achenbach G, Gerson R, James WJ. Solid State Commun 1969;7:701.
[85] Kubel F, Schmid H, Acta Cryst B 1990;46:698.
[86] Lebeugle D, Colson D, Forget A, Viret M, Bonville P, Marucco JF, et al. Phys Rev B 2007;76:024116.
[87] Kumar MM, Palkar VR, Srinivas K, Suryanarayana SV. Appl Phys Lett 2000;76:2764.
[88] Moreau JM, Michel C, Gerson R, James WJ. J Phys Chem Solids 1971;32:1315.
[89] Sosnowska I, Peterlin-Neumaier T, Steichele E. J Phys C: Solid State Phys 1982;15:4835.
[90] Selbach SM, Tybell T, Einarsrud MA, Grande T. Adv Mater 2008;20:3692.
[91] Akbashev AR, Kaul AR. Russ Chem Rev 2011;80:1159.
[92] Yuan GL, Or SW, Liu JM, Liu ZC. Appl Phys Lett 2006;89:052905.
[93] Yuan GL, Or SW. J Appl Phys 2006;100:024109.
[94] Palkar VR, Kundaliya DC, Malik SK, Bhattacharya S. Phys Rev B 2004;69:212102.
[95] Zhang ST, Zhang Y, Lu MH, Du CL, Chen YF, Liu ZC, et al. Appl Phys Lett 2006;88:162901.
[96] Wang DH, Goh WC, Ning M, Ong CK. Appl Phys Lett 2006;88:212907.
[97] Khomchenko VA, Kiselev DA, Vieira JM, Kholkin AL, Pogorelov YG. Appl Phys Lett 2007;90:242901.
[98] Khomchenko VA, Kiselev DA, Vieira JM, Li J, Kholkin AL, Lopes AML, et al. J Appl Phys 2008;103:024105.
[99] Goodenough JB. Magnetism and the chemical bonds. New York: Wiley Publishers; 1963.
[100] Ueda KH, Tabata H, Kawai T. Science 1998;280:1064.
[101] Baettig P, Spaldin NA. Appl Phys Lett 2005;86:012505.
[102] Baettig P, Ederer C, Spaldin NA. Phys Rev B 2005;72:214105.
[103] Belik AA, Iikubo S, Kodama K, Igawa N, Shamoto S, Takayama-Muromachi E. Chem Mater 2008;20:3765.

[104] Santos AFM, Cheetham AK, Tian W, Pan XQ, Jia YF, Murphy NJ, et al. Appl Phys Lett 2004;84:91.
[105] Belik AA, Iikubo S, Kodama K, Igawa N, Shamoto S, Niitaka S, et al. Chem Mater 2006;18:798.
[106] Azuma M, Carlsson S, Rodgers J, Tucker MG, Tsujimoto M, Ishiwata S, et al. J Am Chem Soc 2007;129:14433.
[107] Suchomel MR, Thomas CI, Allix M, Rosseinsky MJ, Fogg AM, Thomas MF. Appl Phys Lett 2007;90:112909.
[108] Azuma M, Takata K, Saito T, Ishiwata S, Shimakawa Y, Takano M. J Am Chem Soc 2005;127:8889.
[109] Hughes H, Allix MB, Bridges CA, Claridge JB, Kuang XJ, Niu HJ, et al. J Am Chem Soc 2005;127:13790.
[110] Belik AA. J Solid State Chem 2012.
[111] Valant M, Axelsson AK, Alford N. Chem Mater 2007;19:5431.
[112] Kanai T, Ohkoshi S, Nakajima A, Watanabe T, Hashimoto K. Adv Mater 2001;13:487.
[113] Ueda K, Tabata H, Kawai T. Phys Lett 1999;75:555–557.
[114] Gotardo RAM, Santos IA, Cotica LF, Botero ER, Garcia D, Eiras JA. Scripta Mater 2009;61:508.
[115] Comyn TP, McBride SP, Bell AJ. Mater Lett 2004;58:3844.
[116] Zhu WM, Guo HY, Ye ZG. Phys Rev B 2008;78:014401.
[117] Ivanov SA, Nordblad P, Tellgren R, Ericsson T, Korchagina SK, Rybakova LF, et al. Solid State Sci 2008;10:1875.
[118] Ivanov SA, Nordblad P, Tellgren R, Ritter C. Solid State Sci 2010;12:115.
[119] Bochenek D, Płonska M, Dudek J, Surowiak Z. J Therm Anal Calorim 2010;101:623.
[120] Silva J, Reyes A, Esparza H, Camacho H, Fuentes L. Integrated Ferroelectrics 2011;126:47.
[121] Bell A. J Eur Ceramic Soc 2008;28:1307.
[122] Lazarevic Z, Stojanovic BD, Varela JA. Sci Sintering 2005;37:199.
[123] Reaney IM, Taylor DV, Brooks KG. J Sol–Gel Sci Tech 1998;13:813.
[124] Harrison RJ. Rev Miner Geochem 2006;63:113.
[125] Radaelli PG, Chapon LC. Phys Rev B 2007;76:054428.
[126] Harris AB. Phys Rev B 2007;76:054447.
[127] Volkova LM, Polyshchuk SA, J Supercond 2005;18:583.
[128] Volkova LM, Marinin DV. J Phys: Condens Matter 2009;21:015903.
[129] Ikeda N, Kohn K, Myonga N, Takahasi E, Kitoh H, Takekawa S. J Phys Soc Jpn 2000;69:1526.
[130] Subramanian MA, He T, Chen J, Rogado NS, Galvarese TG. Sleight Adv Mater 2006;18:1737.
[131] Brink J, Khomskii DI. J Phys: Condens Matter 2008;20:434217.
[132] Efremov DV, Brink J, Khomskii DI. Nat Mater 2004;3:853.
[133] Lopes AML, Araujo JP, Amaral VS, Correia JG, Tomioka Y, Tokura Y. Phys Rev Lett 2008;100:155702.
[134] Rogado N, Lawes G, Huse DA, Ramirez AP, Cava RJ. Solid State Commun 2002;124:229.
[135] Lawes G, Harris AB, Kimura T, Rogado N, Cava RJ, Arahony A, et al. Phys Rev Lett 2005;95:087205.
[136] Van Aken BB, Meetsma A, Palstra TT. Acta Cryst E 2001;57:101.
[137] Van Aken BB, Palstra TT, Filippetti A, Spaldin NA. Nat Mater 2004;3:164.
[138] Cheong SW, Mostovoy M. Nat Mater 2007;6:13.
[139] Koo J, Li S, Jang TH, Jeong YH, Lee KBJ. Korean Phys Soc 2007;51:562.
[140] Weitzel H, Kristal Z, 1976;144:238.
[141] Lautenschlager G, Weitzel H, Vogt T, Hock R, Bohm A, Bonnet M, et al. Phys Rev B 1993;48:6087.
[142] Sheets WC, Mugnier E, Barnabe A, Marks TJ, Poeppelmeier KR. Chem Mater 2006;18:7.
[143] Marquardt MA, Ashmore NA, Cann DP. Thin Solid Films 2006;496:146.
[144] Choi YJ, Okamoto J, Huang DJ, Chao KS, Lin HJ, Chen CT, et al. Phys Rev Lett 2009;102:067601.
[145] Giovannetti G, Stroppa A, Picozzi S, Baldomir D, Pardo V, Blanco-Canosa S, et al. J Phys Rev B 2011;83:060402.

[146] Jodlauk S, Becker P, Mydosh JA, Khomskii DI, Lorenz T, Streltsov SV, et al. J Phys: Condens Matter 2007;19:432201.
[147] Cameron M, Papike JJ. Am Mineral 1981;66:1.
[148] Redhammer GJ, Roth G, Treutmann W, Hoelzel M, Paulus W, Andre G, et al. Solid State Chem 2009;182:2374.
[149] Singh AK, Patnaik S. Aconf Pro IPc. 2011;1349:29.
[150] Singh AK. Appl Phys Lett 2008;92:132919.
[151] Qi X, Chang JY. Ferroelectrics 2011;421:60.
[152] Josse M, Bidault O, Roulland F, Castel E, Simon A, Michau D, et al. Solid State Sci 2009;11:1118.
[153] Wu YJ, Hong ZJ, Lin YQ, Gu SP, Liu XQ, Chen XM. J. Appl Phys 2010;108:014111.
[154] Ismailzade IH, Huseinov HG, Saltanov GJ, Hajiev EM. Ferroelectrics 1976;13:389.
[155] Heyer O, Hollmann N, Klassen I, Jodlauk S, Bohaty L, Becker P, et al. J Phys: Condens Matter 2006;18:L471.
[156] Lee G, Kan E, Xiang H, Whangbo MH. Chem Mater 2010;22:5290.
[157] Li J, Andersen NH, Lefmann K, Schefer J, Zimmermann M, Zaretsky JL, et al. Phys Rev B 2006;73:024410.
[158] Jensen TB, Christensen NB, Kenzelmann M, Ronnow HM, Niedermayer C, Andersen NH, et al. Phys Rev B 2009;79:092412.
[159] Wang M, Tan G. Mater Res Bull 2011;46:438.
[160] Ramesha K, Llobet A, Proffen T, Serro CR, Rao CNR. J Phys: Condens Matter 2007;19:102202.
[161] Kamba S, Nuzhnyy D, Vanek P, Savinov M, Knizek K, Shen Z, et al. Europhys Lett 2007;80:27002.
[162] Azuma M, Tanaka K, Saito T, Ishiwata S, Shimakawa Y, Takano M. J Am Chem Soc 2005;127:8889.
[163] Tan GL, Wang MJ. Electroceram 2011;26:170.
[164] Golubko NV, Proidakova YuV, Kaleva GM, Ivanov SA, Mosunov AV, Stefanovich YuS, et al. Bull Russ Acad Sci: Phys 2010;74:724.
[165] Zupan J, Kolar D, Urbanc V. Mater Res Bull 1971;6:1353.
[166] Zivkovic I, Prsa K, Zaharko O, Berger H. J Phys: Condens Matter 2010;22:056002.
[167] Choi KY, Lemmens P, Choi ES, Berger H. J Phys: Condens Matter 2008;20:505214.
[168] Berger RJ. Less-Common Met 1991;169:33.
[169] Park S, Choi YJ, Zhang CL, Cheong SW. Phys Rev Lett 2007;98:057601.
[170] Goodenough JB. Phys Rev 1967;164:785.

Index